OCEAN ACIDIFICATION AND MARINE WILDLIFE

OCEAN ACIDIFICATION AND MARINE WILDLIFE
Physiological and Behavioral Impacts

GUANGXU LIU
College of Animal Sciences, Zhejiang University, Hangzhou, P.R. China

ACADEMIC PRESS
An imprint of Elsevier

Academic Press is an imprint of Elsevier
125 London Wall, London EC2Y 5AS, United Kingdom
525 B Street, Suite 1650, San Diego, CA 92101, United States
50 Hampshire Street, 5th Floor, Cambridge, MA 02139, United States
The Boulevard, Langford Lane, Kidlington, Oxford OX5 1GB, United Kingdom

Notices
Knowledge and best practice in this field are constantly changing. As new research and experience broaden our understanding, changes in research methods, professional practices, or medical treatment may become necessary.

Practitioners and researchers must always rely on their own experience and knowledge in evaluating and using any information, methods, compounds, or experiments described herein. In using such information or methods they should be mindful of their own safety and the safety of others, including parties for whom they have a professional responsibility.

To the fullest extent of the law, neither the Publisher nor the authors, contributors, or editors, assume any liability for any injury and/or damage to persons or property as a matter of products liability, negligence or otherwise, or from any use or operation of any methods, products, instructions, or ideas contained in the material herein.

British Library Cataloguing-in-Publication Data
A catalogue record for this book is available from the British Library

Library of Congress Cataloging-in-Publication Data
A catalog record for this book is available from the Library of Congress

ISBN: 978-0-12-822330-7

For Information on all Academic Press publications
visit our website at https://www.elsevier.com/books-and-journals

Publisher: Charlotte Cockle
Acquisitions Editor: Anna Valutkevich
Editorial Project Manager: Megan Ashdown
Production Project Manager: Maria Bernard
Cover Designer: Matthew Limbert

Typeset by MPS Limited, Chennai, India

Working together
to grow libraries in
developing countries

www.elsevier.com • www.bookaid.org

Contents

List of contributors

Shiguo Li
Research Center for Eco-Environmental Sciences, Chinese Academy of Sciences, Beijing, P.R. China; University of Chinese Academy of Sciences, Chinese Academy of Sciences, Beijing, P.R. China

Guangxu Liu
College of Animal Sciences, Zhejiang University, Hangzhou, P.R. China

Yi Qu
Muping Coastal Environmental Research Station, Yantai Institute of Coastal Zone Research, Chinese Academy of Sciences, Yantai, Shandong, P.R. China; Key Laboratory of Coastal Biology and Biological Resources Utilization, Yantai Institute of Coastal Zone Research, Chinese Academy of Sciences, Yantai, Shandong, P.R. China; University of Chinese Academy of Sciences, Beijing, P.R. China

Wei Shi
College of Animal Sciences, Zhejiang University, Hangzhou, P.R. China

Ting Wang
International Research Center for Marine Biosciences at Shanghai Ocean University, Ministry of Science and Technology, Shanghai, P.R. China

Xin Wang
Muping Coastal Environmental Research Station, Yantai Institute of Coastal Zone Research, Chinese Academy of Sciences, Yantai, Shandong, P.R. China; Key Laboratory of Coastal Biology and Biological Resources Utilization, Yantai Institute of Coastal Zone Research, Chinese Academy of Sciences, Yantai, Shandong, P.R. China; University of Chinese Academy of Sciences, Beijing, P.R. China

Youji Wang
International Research Center for Marine Biosciences at Shanghai Ocean University, Ministry of Science and Technology, Shanghai, P.R. China

Qianqian Zhang
Muping Coastal Environmental Research Station, Yantai Institute of Coastal Zone Research, Chinese Academy of Sciences, Yantai, Shandong, P.R. China; Key Laboratory of Coastal Biology and Biological Resources Utilization, Yantai Institute of Coastal Zone Research, Chinese Academy of Sciences, Yantai, Shandong, P.R. China

Tianyu Zhang
Muping Coastal Environmental Research Station, Yantai Institute of Coastal Zone Research, Chinese Academy of Sciences, Yantai, Shandong, P.R. China; Key Laboratory of Coastal Biology and Biological Resources Utilization, Yantai Institute of Coastal Zone Research, Chinese Academy of Sciences, Yantai, Shandong, P.R. China; University of Chinese Academy of Sciences, Beijing, P.R. China

Jianmin Zhao
Muping Coastal Environmental Research Station, Yantai Institute of Coastal Zone
Research, Chinese Academy of Sciences, Yantai, Shandong, P.R. China; Key Laboratory
of Coastal Biology and Biological Resources Utilization, Yantai Institute of Coastal Zone
Research, Chinese Academy of Sciences, Yantai, Shandong, P.R. China; Center for
Ocean Mega-Science, Chinese Academy of Sciences, Qingdao, Shandong, P.R. China

CHAPTER ONE

Physiological impacts of ocean acidification on marine invertebrates

Guangxu Liu and Wei Shi
College of Animal Sciences, Zhejiang University, Hangzhou, P.R. China

Introduction

Due to anthropogenic activities such as deforestation, fossil fuel utilization, cement production, and biomass burning since the Industrial Revolution in the mid-eighteenth century, the concentration of atmospheric carbon dioxide (CO_2) has increased approximately from 280 to 387 parts per million (ppm), which is higher now than it has been for more than 800,000 years (Booth et al., 2012; Caldeira & Wickett, 2003; Feely et al., 2004; Orr et al., 2005). Being the earth's largest carbon sink, the ocean plays an extremely important role in the global carbon cycle (Doney et al., 2009; Le Quéré et al., 2009). Approximately 30%−50% of the CO_2 released into the atmosphere has been absorbed by the earth's ocean, which thus resulted in reductions in seawater pH, a process termed "ocean acidification" (OA) (Caldeira & Wickett, 2003; Sabine et al., 2004). Over the past two centuries, the global average surface seawater pH has already decreased by more than 0.1 units, from approximately pH 8.21 to pH 8.10, which is equivalent to a 30% increase in the hydrogen ion (H^+) concentration in the seawater (Ellis et al., 2017; Sabine et al., 2004). According to the prediction made by the Intergovernmental Panel on Climate Change, if fossil fuel emissions and carbon-sequestration efforts continue at the present rate, the surface seawater pH will drop another 0.3−0.4 units by the end of the 21st century and by 0.7 units around the year 2300 (Pachauri et al., 2014). Besides, oceanic uptake of atmospheric CO_2 also lowers the carbonate concentration and reduces the saturation state of calcium carbonate in seawater, especially aragonite and

Ocean Acidification and Marine Wildlife.
DOI: https://doi.org/10.1016/B978-0-12-822330-7.00003-4

calcite, which are critical for many marine invertebrates in creating their skeletal structures or shells (Caldeira & Wickett, 2003; Fitzer et al., 2016; Thomsen et al., 2013; Zhao et al., 2017). Therefore theoretically, OA will affect a diversity of marine invertebrate species by altering seawater chemistry (Andersson & Gledhill, 2013; Gibson et al., 2011; Mollica et al., 2018).

Invertebrates, which make up about 95% of all animal species, are the largest group of animals on earth. In the ocean, marine invertebrates are not only functionally important in the marine ecosystem but also have significant commercial value worldwide (Marinelli & Williams, 2003). Since living in an acidified environment would constitute stress to marine inhabitants, OA could have profound ramifications on the physiological performance of marine invertebrates (Gallo et al., 2019; Gazeau et al., 2010; Kurihara, 2008; Shi, Han, et al., 2017; Shi et al., 2019). To date, OA is projected to impact marine invertebrates such as mollusks, crustaceans, and echinoderms present in various areas, from the open sea to estuaries and coastal areas (Bechmann et al., 2011; Holcomb et al., 2014). The present chapter focuses on the physiological impacts of OA on marine invertebrates, including gametic traits, fertilization success, embryonic development, biomineralization, metabolism, growth, and immune responses.

Impacts of ocean acidification on gametes and fertilization success of invertebrates

Fertilization, in its simplest form, is the fusion of two specialized gametes to form a single viable cell, which is known as the zygote. The release of gametes into the natural seawater column for external fertilization is an ancestral mating strategy commonly employed by various marine invertebrates (Lotterhos & Levitan, 2010). Once discharged, these gametes are in direct contact with the surrounding seawater. In this regard, the gametes and the subsequent fertilization of these marine broadcast spawners may be particularly vulnerable to OA (Table 1.1).

Sperm velocity is theoretically related to the probability of collision of gametes, and studies have shown that sperm with high velocity would be more effective in fertilizing the egg (Kupriyanova & Havenhand, 2005; Levitan, 2000). For example, as compared to faster sperm of the sea urchin *Lytechinus variegatus*, sperm with 0.01 mm/s decrease in velocity

Table 1.1 Effects of ocean acidification on the gametic traits and fertilization success of marine invertebrates.

Taxon	Species	pH/pCO₂	Objectives	Effects	References
Coelenterata	*Acropora digitifera*	pH 8.03−6.55 (400−21,100 ppm)	Sperm	↓Sperm flagellar motility	Morita et al. (2010)
		pH 7.74 (1000 ppm)	Sperm	↓Sperm motility	Nakamura and Morita (2012)
		pH 7.99−7.60 (438−1111 ppm)	Sperm	Unaffected	Iguchi et al. (2015)
	A. palmata	pH 7.7 (998 ppm)	Sperm	↓Fertilization success	Albright et al. (2010)
Mollusca	*Crassostrea gigas*	pH 8.12−7.85	Sperm and egg	Unaffected	Havenhand and Schlegel (2009)
	Saccostrea glomerata	pH 8.09−7.73 (580−3573 ppm)	Sperm and egg	↓Sperm motility; ↓Fertilization rate	Barros et al. (2013)
		600, 750, and 1000 ppm	Sperm and egg	↓Fertilization success	Parker et al. (2009)
	Mytilus galloprovincialis	pH 7.6 (1000 ppm)	Sperm	↓Sperm swimming speed; ↓Percentage of motile sperm	Vihtakari et al. (2013)
	Tegillarca granosa	pH 8.1−7.4 (589−3582 ppm)	Sperm and egg	↓Sperm swimming speed; ↓Fertilization success; ↓Gamete fusion probability;	Shi et al. (2017)
		pH 8.1−7.4	Sperm and egg	↑Polyspermy risk	Han et al. (2021)
Arthropoda	*Acartia tonsa*	pH 8.23−7.15 (385−6000 ppm)	egg	↓Fertilization success	Cripps et al. (2014)
Urochordata	*Ciona robusta*	pH 8.1−7.8	Sperm	↓Sperm motility; ↓Sperm viability.	Gallo et al. (2019)

(Continued)

Table 1.1 (Continued)

Taxon	Species	pH/pCO₂	Objectives	Effects	References
Echinodermata	*Hemicentrotus pulcherrimus*	pH 7.35 and 6.83 (2000 and 10,000 ppm)	Sperm and egg	↓Fertilization success	Kurihara and Shirayama (2004)
	Heliocidaris erythrogramma	pH 8.06−7.55	Sperm and egg	Unaffected	Zhan et al. (2016)
		pH 7.7	Sperm	↓Sperm swimming speed; ↓Percent sperm motility; ↓Fertilization success	Havenhand et al. (2008)
	Strongylocentrotus franciscanus	pH 7.81 and 7.55 (800 and 1800 ppm)	Sperm and egg	↓Fertilization success; ↑Polyspermy risk	Reuter et al. (2011)
	S. purpuratus	pH 8.03−7.61	Sperm and egg	Unaffected	Kapsenberg et al. (2017)
	Centrostephanus rodgersii	pH 8.1−7.6 (435−1558 ppm)	Sperm	↓Sperm mitochondrial membrane potential; ↓Sperm swimming speed	Schlegel et al. (2015)
	Acanthaster planci	pH 8.11−7.61 (520−1658 ppm)	Sperm	↓Sperm swimming speed; ↓Sperm motility; ↓Fertilization success	Uthicke et al. (2013)
	Holothuria spp.	pH 8.03−6.55 (400−21,100 ppm)	Sperm	↓Sperm flagellar motility	Morita et al. (2010)
	Glyptocidaris crenularis	pH 7.98−7.48 (453−1674 ppm)	Sperm and egg	↓Fertilization success; ↓percentage of abnormal fertilized eggs	Zhan et al.(2016)
	Lytechinus variegatus	pH 7.8	Sperm and egg	Unaffected	Lenz et al. (2019)

require an order of magnitude higher concentration to achieve 50% fertilization (Levitan, 2000). According to previous studies (Gallo et al., 2019; Schlegel et al., 2015; Shi, Han, et al., 2017), the percentage of sperm motility and swimming speed of many marine invertebrates could be negatively affected by OA. Statistically significant reductions in swimming velocity and motility rate upon OA exposure have been observed in echinoderm such as the sea urchins *Heliocidaris erythrogramma* and *Centrostephanus rodgersii*; the sea cucumber *Holothuria* spp.; and the sea star *Acanthaster planci* (Havenhand et al., 2008; Morita et al., 2010; Schlegel et al., 2015; Uthicke et al., 2013). For example, exposure to acidified seawater at pH 7.7 (1000 ppm pCO_2) resulted in 11.7% and 16.3% reductions in the sperm swimming speed (mm/s) and sperm motility, respectively, in the sea urchin *H. erythrogramma*, which would decrease the fertilization success by approximately 25% (Havenhand et al., 2008). The percentage of motile sperm cells and sperm swimming speed of the sea star *A. planci* were both decreased after 30 minutes of exposure under OA conditions, which would lead to 29% and 75% reductions in their fertilization success at pH 7.9 and 7.6, respectively (Uthicke et al., 2013). Similarly, more than 70% of the sperm cells of the sea cucumber *Holothuria* spp. were motile at pH 8.0 and 7.8, while less than 30% of the sperm cells were motile at pH levels ranging from 7.7 to 6.6 (Morita et al., 2010). The OA-induced negative effects on sperm swimming performance were also detected in Coelenterata, Mollusca, Urochordata, and Arthropoda (Gallo et al., 2019; Lenz et al., 2019; Morita et al., 2010; Shi, Han, et al., 2017, Shi, Zhao, et al., 2017). It was demonstrated that the velocity average path, curvilinear velocity, and velocity straight line of the sperm of the blood clam *Tegillarca granosa* were decreased to 69.0%, 73.7%, and 60.9% of the control at pH 8.1, respectively, upon 1 hour of OA exposure at pH 7.4 (Shi, Han, et al., 2017).

Morita et al. (2010) found that a relatively slight decrease in the seawater pH (pH 7.7) would hamper the sperm flagellar motility of the coral *Acropora digitifera* dramatically (69% of the sperm cells were motile at pH 8.0; only 46% remained motile at pH 7.8; and less than 20% of the sperm cells were motile at pH 7.7). It is reported that the external fertilization of sessile marine organisms would result in a rapid dilution of sperm concentration (Levitan, 2000; Styan, 1998). Therefore a slight decrease in sperm motility after exposure to OA conditions could seriously threaten the life cycle of various marine organisms, due to inefficiency with respect to fertilization (Fig. 1.1).

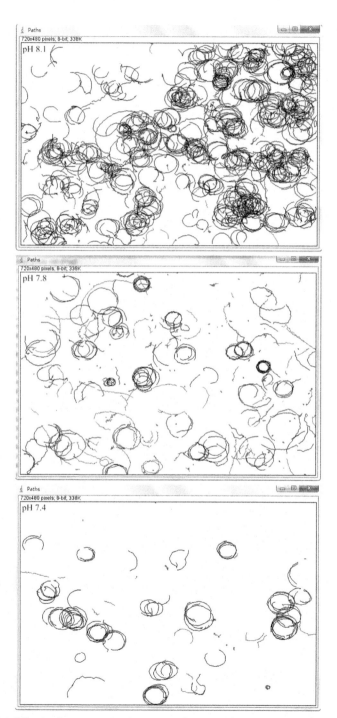

Figure 1.1 Effects of seawater pH (8.1, 7.8 and 7.4) on the sperm swimming behavior of blood clam *Tegillarca granosa*.

Since the sperm swimming behavior is an energy-consuming process, the inhibitory effects of acidified seawater on sperm performance may partially result from the declined energy available for motility (Kasai et al., 2002; Schlegel et al., 2015). For example, a previous study conducted on the sea urchin *C. rodgersii* has shown that the sperm energy metabolism, reflected by the mitochondrial membrane potential (MMP), was significantly reduced by 35% and 48% after exposure to acidified seawater at pH 7.8 and 7.6, respectively (Schlegel et al., 2015), which may reduce the proportion of motile sperm and hamper sperm flagellar motility. Similarly, a significant decrease in the MMP value, to 63.9% of that of the control at pH 8.1 (Gallo et al., 2019), was also observed in the sperm of the ascidian *Ciona robusta* after OA exposure (pH 7.8). The mitochondrial dysfunction of sperm under OA may result from the OA-induced mitochondrial ultrastructural damage (Gallo et al., 2019). It was shown that exposure to acidified seawater at pH 7.8 led to the detachment and sliding of the mitochondrion from its typical position up to abnormal morphology with fragmentation, rounding, and sliding of the mitochondrion in the sperm of ascidian *C. robusta* (Gallo et al., 2019). In addition, for many marine invertebrates like sea urchins, the sperm are stored immotile inside the testes in an acidified environment that inhibits respiration and motility; an elevation of intracellular pH in the sperm is crucial for sperm activation (Johnson & Epel, 1981; Tosti, 1994). On one hand, this activation of sperm is dependent on the influx of external Na^+ and subsequently the release of H^+ ions within the cell. This increase in the internal pH of the sperm cells activates dynein ATPase and the subsequent sperm motility (Tosti & Ménézo, 2016). Thus excessive CO_2 inside the cell under future high $p$$CO_2$ conditions may disturb the intracellular acid—base balance and thereby the activation of sperm (Kurihara, 2008). On the other hand, the enzymes regulating this activation process have restricted optimal pH values, the activity of which could be reduced by the intracellular pH alteration under OA conditions, thus leading to disruption of sperm motility (Gallo et al., 2019).

According to the fertilization kinetics model and several laboratory experiments, the determinants of the quality of eggs, such as size, quantity, and membrane integrity, are crucial for the fertilization success of broadcast spawning invertebrates as well (Kurihara & Shirayama, 2004; Vogel et al., 1982). A previous experiment conducted in the blood clam *T. granosa* demonstrated that the fertilization success of OA-treated eggs compared with untreated sperm was approximately only 40.9% (pH 7.8) and

25.4% (pH 7.4) of the control at pH 8.1, respectively (Shi, Zhao, et al., 2017), indicating that OA would also inhibit the fertility of eggs. However, relatively few studies have investigated the impacts of OA on the egg quality of marine organisms. Only a few case studies found that OA might decrease the egg production rates of adult female individuals such as the copepods *Acartia steueri* and *A. erythraea* (Kurihara, 2008).

To date, transcriptomics and proteomic analysis carried out in marine invertebrates suggest that the expression of genes and proteins can be severely affected by OA (Evans & Watson-Wynn, 2014; O'Donnell et al., 2010; Timmins-Schiffman et al., 2014). Notably, it has been suggested that exposure to OA could induce significant changes in the structure of peptides and the electrostatic properties of receptor proteins in marine invertebrates (Roggatz et al., 2016; Timmins-Schiffman et al., 2014). As a result, there is a possibility that OA may also affect the gametic recognition and binding-related proteins located on the cell membrane of the eggs, thus hampering the successful recognition and binding of gametes (Shi, Han, et al., 2017). This inference was supported by the decreased probability of gamete fusion per collision in the blood clam *T. granosa* after exposure of gametes to acidified seawater (Shi, Han, et al., 2017; Shi, Zhao, et al., 2017). However, apart from indirect evidence (the estimation of gamete fusion probability), more direct evidence such as structural alterations in these fertilization-related proteins is still needed to further verify this inference.

Since the gametic quality lays the basis for successful fertilization, the reduction in gametic quality caused due to OA exposure would undoubtedly lead to a decrease in fertilization success (Gallo et al., 2019; Han et al., 2019; Scanes et al., 2014; Shi, Zhao, et al., 2017; Vogel et al., 1982). It has been reported that exposure to near-future OA scenarios led to a significant reduction in the fertilization success of a variety of marine invertebrates such as the sea urchins *Echinometra mathaei* and *Hemicentrotus pulcherrimus* (Kurihara & Shirayama, 2004); the scallop *Mimachlamys asperrima* (Scanes et al., 2014); the blood clam *T. granosa* (Shi, Han, et al., 2017); and the oyster *Saccostrea glomerata* (Parker et al., 2009). Havenhand et al. (2008) found that the fertilization success of *H. erythrogramma* dropped to approximately 76% of the control group upon OA (pH 7.7) exposure. Similarly, the fertilization success of the oyster *S. glomerata* was significantly inhibited as the seawater pCO_2 increased (600 and 750 ppm) (Parker et al., 2009). Additionally, Shi, Zhao, et al. (2017) suggested that exposure to OA (pH 7.4) could reduce the fertilization success of *T. granosa* by approximately 28.4% by decreasing the sperm–egg collision probability, lowering gamete fusion probability, and disrupting intracellular Ca^{2+} oscillations during the process of fertilization.

Apart from hampering the fertilization potency of gametes, OA exposure may also elevate the polyspermy risk in marine invertebrates (Desrosiers et al., 1996; Reuter et al., 2011; Sewell et al., 2014). For example, Reuter et al. (2011) reported that the estimated time for the egg of the sea urchin *Strongylocentrotus franciscanus* to effectively block the entry of superfluous sperm significantly increased after OA treatment (pH 7.55; 1800 ppm), which could lead to a significant increase in the rate of polyspermy. Similarly, it was also found that the time for the egg to build up a complete block to polyspermy significantly increased (70%–100%) for the sea urchin *Sterechinus neumayeri* at elevated pCO_2 levels (480 and 660 ppm) compared to the control (380 ppm) (Sewell et al., 2014). More direct evidence via estimating and comparing the polyspermy rate demonstrated that exposure to OA at pH 7.4 (pCO_2 at 2900 ppm) led to a significantly greater risk of polyspermy, approximately 2.38 times of that of the control (pH 8.1; 496 ppm) (Han et al., unpublished data), in the blood clam *T. granosa*. The OA-induced increases in polyspermy risk in marine invertebrates may result from the hampered polyspermy-blocking mechanisms (Han et al., unpublished data). Two polyspermy-blocking mechanisms have successfully evolved in many broadcast spawning species to ensure monospermy, namely, a fast but transient electrical block created by the depolarization of the oocyte membrane and a slow but permanent physical block created by the formation of the fertilization membrane through the cortical reaction (Cheeseman et al., 2016). However, Han et al. (unpublished data) found that the exposure of oocytes to future OA scenarios (pH 7.4) can lead to significant reductions in both the amplitude and duration of the membrane potential change during depolarization and the dramatic delay of cortical granule exocytosis, which may facilitate the entry of superfluous sperm into the oocyte and result in increased polyspermy (Fig. 1.2).

Interestingly, the gametes and fertilization of some marine invertebrate species are shown to be robust against OA, suggesting that the sensitiveness of the fertilization process to OA may be species-specific (Havenhand & Schlegel, 2009; Kurihara, 2008). For example, it was shown that the fertilization rate of the Mediterranean mussel *Mytilus galloprovincialis* was not influenced by seawater acidified with 2000 ppm pCO_2 (pH 7.4)

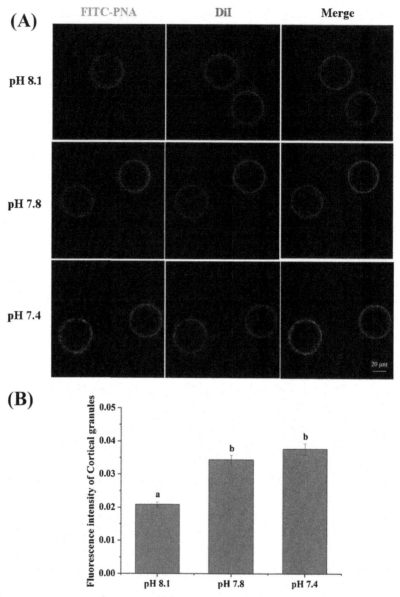

Figure 1.2 Impacts of ocean acidification on cortical granule (CG) exocytosis in *Tegillarca granosa*. (A) Examples showing CG-specific fluorescent staining of oocytes 3 min postgamete mixing for the control (pH 8.1) and the two experimental groups (pH 7.8 and pH 7.4). (B) The relative CG-specific fluorescent intensity in the oocytes 3 min postgamete mixing for the control (pH 8.1) and the two experimental groups (pH 7.8 and pH 7.4). $N = 6$ and data are presented as the mean $\pm$ SEM for (B) Mean values not sharing the same superscript letter are significantly different [Tukey's (honestly significant difference), $P < .05$].

(Kurihara et al., 2007). In addition, some contrasting results have been reported in species from allopatric and even sympatric populations (Byrne et al., 2010; Havenhand et al., 2008). For instance, Byrne et al. (2010) found no significant effects of OA on the fertilization success of the sea urchin *H. erythrogramma*, whereas a dramatic reduction in the fertilization success upon OA exposure was detected for the same species from sympatric populations (Havenhand et al., 2008).

These contrasting results could have resulted from the differences in the experimental methodology (Styan, 1998). According to Vogel's fertilization kinetics model (Vogel et al., 1982), the fertilization success of marine invertebrates can be estimated using Eq. (1.1), given as follows:

$$p = 1 - \exp\left(\frac{\beta S_0}{\beta_0 E_0} \times \left(1 - e^{-\beta_0 E_0 t_c}\right)\right), \tag{1.1}$$

where p is the fertilization success rate; S_0 and E_0 are the initial concentrations of the sperm and the egg, respectively; t_c is the gamete contact time; β is the fertilization rate constant; and β_0 is the sperm–egg collision rate constant. Therefore a slight change in these factors during the investigation could yield significantly different results (Byrne et al., 2010; Dong et al., 2012). Although previous fertilization studies generally employed similar experimental seawater pH levels, the methodology that was adopted varied significantly with regard to the gamete concentrations, the sperm–egg ratio, and the gamete contact time (Byrne et al., 2010), which might have led to contrasting results among investigations. Therefore to facilitate comparison among different studies investigating the impacts of OA on fertilization success, a standardized experiment method is strongly recommended.

Due to the impact of dilution and the sheer force of seawater, the fertilization success of broadcasting spawners is suggested to be relatively low for wild populations (Levitan, 1991). It has been predicted that only 3% of gamete collisions can result in successful fertilization for some species (Farley & Levitan, 2001); the addition of pCO_2 into the ocean may further reduce the probability of successful fertilization, thus posing far-reaching consequences for the population dynamics of marine invertebrates. Although the adaptability of marine organisms to future OA remains a matter of debate (Baird et al., 2008), relatively long generation times and a higher sensitivity of gametes and the subsequent fertilization to OA may restrict their potential for adaptation.

Impacts of ocean acidification on embryonic development of invertebrates

The embryonic and/or larval development of marine invertebrates are suggested to be highly susceptible to OA as well, especially for invertebrate calcifiers that start to form calcareous skeleton during their early life (Kurihara, 2008; Orr et al., 2005; Thomsen et al., 2010; Wang et al., 2018). Since embryonic and/or larval development can affect future adult population densities greatly, many studies have been conducted to explore the potential effects of OA on this crucial life event (Bechmann et al., 2011; Dupont et al., 2008; Kurihara, 2008; Orr et al., 2005; Parker et al., 2009; Wang et al., 2018). To date, OA-induced negative impacts on early-stage development have been reported in a variety of marine invertebrates such as the mussels *M. galloprovincialis*, *Mytilus californianus*, and *Mytilus edulis*; the oysters *S. glomerata* and *Pinctada martensii*; the brittle star *Ophiothrix fragilis*; and the crabs *Paralithodes camtschaticus* and *Chionoecetes bairdi* (Frieder et al., 2014; Guo et al., 2015; Kurihara, 2008; Kurihara & Shirayama, 2004) (Table 1.2).

According to previous studies, the embryonic and/or larval developmental time of marine invertebrates is crucial for their survival, as it is highly associated with their susceptibility to predation (Allen, 2008). However, the embryonic and/or larval development of marine invertebrates could be slowed down by exposure to near-future OA scenarios (Frieder et al., 2014; Gazeau et al., 2010; Guo et al., 2015; Kurihara & Shirayama, 2004). For instance, during the embryonic stage of *H. pulcherrimus* and *E. mathaei*, the percentage of embryos at later stages tended to decrease with an increase in pCO_2 concentrations (pH 8.2–pH 6.79). Furthermore, at 210 minutes after insemination, some of these fertilized eggs kept at pCO_2 concentrations higher than 5000 ppm did not cleave at all (Kurihara & Shirayama, 2004). The exposure of embryos to OA also resulted in a delay of transition from the trochophore to the veliger stage in various marine invertebrates, such as the mussels *M. californianus* and *M. galloprovincialis*; the abalones *Haliotis diversicolor* and *Haliotis discus hannai*; the clam *Mercenaria mercenaria*; and the oyster *Crassostrea angulata* (Frieder et al., 2014; Guo et al., 2015; Talmage & Gobler, 2010). For instance, Guo et al. (2015) found that more than 90% of the abalone *H. discus hannai* larvae that reared in ambient seawater (pH 8.15; 447 ppm) developed into veligers, while significantly fewer larvae, approximately only 77.4%

Table 1.2 Effects of ocean acidification on embryonic development of marine invertebrates.

Taxon	Species	pH/pCO_2	Effects	References
Coelenterata	*Acropora digitifera*	pH 8.05–7.33 (331–3100 ppm)	↓Metamorphosis rate	Nakamura et al. (2011)
	A. gemmifera	pH 8.1–7.5	↓Settlement rate	Yuan et al. (2018)
	A. millepora	pH 8.04–7.60 (401–1299 ppm)	↓Settlement rate	Doropoulos et al. (2012)
Mollusca	*Saccostrea glomerata*	600, 750, and 1000 ppm	↓D-veligers percentage and size; ↑Abnormal D-veligers	Parker et al. (2009)
	Haliotis discus hannai	pH 8.15–7.43 (447–2780 ppm)	↓Trochophore development; ↓Veliger survival and metamorphosis	Guo et al. (2015)
	Mercenaria mercenaria	pH 8.02–7.49 (354–1437 ppm)	↓Survivorship; ↓Metamorphosis	Talmage and Gobler (2010)
	Crassostrea virginica	pH 8.07–7.50 (354–1477 ppm)	↓Metamorphosis; ↓Growth	Talmage and Gobler (2010)
	Argopecten irradians	pH 8.08–7.48 (354–1605 ppm)	↓Survivorship; ↓Metamorphosis	Talmage and Gobler (2010)
	M. galloprovincialis	pH 8.13 and 7.42 (380 and 2000 ppm)	↑Morphological abnormalities	Kurihara et al. (2008)
	M. edulis	pH 8.12–7.15 (334–3712 ppm)	↑Mortality rate; ↑Abnormally developing larvae percentage	Ventura et al. (2016)

(Continued)

Table 1.2 (Continued)

Taxon	Species	pH/pCO$_2$	Effects	References
	Doryteuthis pealeii	pH 7.87–7.30 (390–2200 ppm)	↓D-veligers percentage	Kaplan et al. (2013)
Arthropoda	*Acartia steueri*	pH 8.14–6.86 (365–10365 ppm)	↓D-veligers percentage; ↑Nauplius mortality	Kurihara et al. (2004)
	A. erythraea	pH 8.14–6.86 (365–10,365 ppm)	↓D-veligers percentage; ↑Nauplius mortality	Kurihara et al. (2004)
	Calanus finmarchicus	pH 8.10–6.95 (380–8000 ppm)	↓D-veligers percentage	Mayor et al. (2007)
	Stenopus hispidus	400–850 ppm	↓Ability to recognize settlement cues	Lecchini et al. (2017)
	Balanus amphitrite	pH 8.2 and 7.6	No influence on settlement	Campanati et al. (2016)
Echinodermata	*Hemicentrotus pulcherrimus*	pH 8.01–6.83 (365–10,365 ppm)	↓Cleavage rate; ↓Developmental speed; ↑Abnormal morphology	Kurihara and Shirayama (2004)
	Echinometra mathaei	pH 8.01–6.83 (365–10,365 ppm)	↓Cleavage rate; ↓Developmental speed; ↑Abnormal morphology	Kurihara and Shirayama (2004)
	Ophiothrix fragilis	pH 8.1–7.7	↑Larval mortality; ↓Larval size; ↑Abnormal development	Dupont et al. (2008)
	Acanthaster planci	pH 8.2–7.9 (341–783 ppm)	↓Larval development; ↓Settlement rate	Uthicke et al. (2013)

and 75.8%, reared in seawater acidified to pH 7.71 (1500 ppm) and pH 7.61 (2000 ppm), respectively, completed this developmental process in groups exposed to OA. After rearing in areas with elevated pCO_2 levels, the percentage of the embryos of the oyster *S. glomerata* to reach D-veliger was significantly decreased, and meanwhile, the percentage of individuals with abnormal D-veliger was significantly increased (Parker et al., 2009). For the blue mussel *M. edulis*, the percentage of embryos reaching D-veliger under OA conditions (pH 7.6; 1900 ppm) was approximately 25% lower than that of the control (pH 8.1; 540 ppm) (Gazeau et al., 2010). Similarly, all individuals of the group of the Mediterranean mussels *M. galloprovincialis* developed into D-shaped larvae in the ambient seawater at 54 hours postfertilization, in contrast to approximately only 20% larvae doing so in groups exposed to elevated pCO_2 (pH 7.4), indicating significant retardation in embryonic development. It was also shown that almost all (> 98%) larvae of the coral *A. digitifera* metamorphosized normally in ambient seawater, whereas, in the case of larvae exposed to OA scenarios, nearly 20% of larvae did not metamorphosize completely (pH 7.6 and 7.3) (Nakamura et al., 2011). In addition, it was found in a study that 51% of the clam *M. mercenaria* larvae had fully metamorphosized in ambient seawater (pH 8.1) after 14 days of development, while less than 7% had successfully metamorphosized when exposed to OA scenarios (pH 7.8 and 7.5) (Talmage & Gobler, 2010). The study carried out by Dupont et al. (2008) also demonstrated that 50% of the brittle star *O. fragilis* larvae cultured under ambient condition (pH 8.1) were four-armed after 1.83 days, whereas it took 2.07 and 2.25 days to reach this developmental stage for those larvae reared in acidified seawater at pH 7.9 and 7.7, respectively. As a result, the arrested embryonic and/or larval development induced by elevated pCO_2 would increase the chance of loss by predation due to their incapability to evade attacking predators and ultimately lead to higher larval mortality of these marine invertebrates (Lundvall et al., 1999; Rumrill, 1990; Uthicke et al., 2013).

Along with the delay in developmental time, exposure of embryos to OA often leads to growth inhibition as well (Kurihara, 2008; Kurihara & Shirayama, 2004). It has been demonstrated that both the shell length and the height of larvae of the Mediterranean mussel *M. galloprovincialis* reared in acidified seawater (pH 7.42) were significantly smaller (pH 8.13), approximately 74% and 80%, respectively, when compared to the control group (Kurihara, 2008). The D-veliger shells of the mussel *M. edulis* were approximately 12.7% smaller at pH 7.6 than at control pH during the first

2 days of development (from eggs to D-shaped larvae) (Gazeau et al., 2010). For the American lobster *Homarus americanus* reared in CO_2-acidified seawater (1200 ppm), the cumulative number of days for their larvae to reach successive molts was always higher than that of the individuals in ambient seawater (400 ppm) (Keppel et al., 2012). Exposure to high amounts of CO_2 also caused a significant reduction in the shell length of the bay scallops *Argopecten irradians*, which were 84.1%, 92.5%, and 88.5% of the mean lengths of the shells that underwent ambient CO_2 treatment on days 1, 3, and 7, respectively (White et al., 2013). Similarly, the embryos of *H. pulcherrimus* and *E. mathaei* reared in elevated pCO_2 developed into larvae with smaller larval sizes, including a reduction in overall length, postoral arm length, and body length compared to the control (Kurihara & Shirayama, 2004). Since the encounter and clearance rates of food particles are positively correlated to the larval body size of marine organisms, the smaller larvae upon exposure to OA scenarios are more susceptible to starvation, which in turn constrains the energy available for growth and eventually poses a threat to the larval survival and population recruitment (Anger, 1987; Kurihara et al., 2007).

Apart from the prolonged developmental time and inhibited growth, the exposure of embryos of invertebrates to OA scenarios also often leads to morphological abnormalities (Dupont et al., 2008; Kurihara, Asai, et al., 2008; McDonald et al., 2009; Wang et al., 2018). It was reported that although all embryos of the Mediterranean mussel *M. galloprovincialis* raised in acidified seawater developed into D-shaped larvae after 120 hours post-fertilization, almost all ($> 99\%$) larvae had an abnormal morphology, such as indentation of the shell margin, protrusion of the mantle from the shell, and a convex formation of the hinge (Kurihara, Asai, et al., 2008). Similarly, approximately 60% of the blue mussel *M. edulis* larvae developed into abnormally D-shaped larvae with elevated pCO_2 (pH 7.35), while this occurred in only 14% of the control (Ventura et al., 2016). The proportions of the degrees of asymmetry in the asymmetric sea urchin *Strongylocentrotus intermedius* larvae increased with a decrease in seawater pH as compared to those of the control (Zhan et al., 2016) It has been shown that in the brittle star *O. fragilis*, a large percentage ($> 50\%$) of the larvae were unable to develop into normal pluteus larvae or asymmetric larvae at 5 days post fertilization under low pH levels (pH 7.9 and 7.7), while abnormalities were completely absent in the larvae in ambient seawater (Dupont et al., 2008). In addition, it has been shown that a reduction of 0.7 units in seawater pH also resulted in major morphological abnormalities in the larvae of the oyster *Crassostrea gigas* (only 4%–5% developed normally). Since the calcite skeleton of larval marine organisms, including larval brittle stars and sea urchins, has been proposed to lead to several adaptive developmental benefits, including the maintenance of body shape, support in morphogenesis, improved feeding, and defense against predators, abnormal development of the skeleton of these marine organisms due to exposure to OA conditions would constrain these essential physiological aspects of their larvae (Pennington & Strathmann, 1990) (Fig. 1.3).

With all these negative impacts on embryonic development, it is not surprising that increased larval mortality caused by OA exposure was observed in a wide variety of marine invertebrates (Dupont et al., 2008;

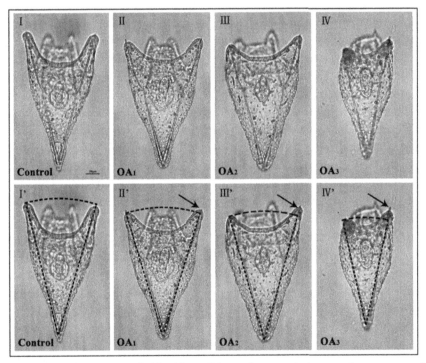

Figure 1.3 The impact of CO_2-driven OA on the morphology and symmetry of *Strongylocentrotus intermedius* larvae 70 h postfertilization. (I, I'): Control group (pH = 8.00 ± 0.03); (II, II'): OA_1 group ($\triangle pH = -0.3$ units); (III, III'): OA_2 group ($\triangle pH = -0.4$ units); (IV, IV'): OA_3 group ($\triangle pH = -0.5$ units). Scale bar: 50 μm. *From Zhan, Y., Hu, W., Zhang, W., Liu, M., Duan, L., Huang, X., Chang, Y., & Li, C. (2016). Driven ocean acidification on early development and calcification in the sea urchin Strongylocentrotus intermedius. Marine Pollution Bulletin, 112(1–2), 291–302. https://doi.org/10.1016/j.marpolbul.2016.08.003 295.*

Kurihara, Asai, et al., 2008; Talmage & Gobler, 2010; Ventura et al., 2016). A representative case is when the arrival of low-pH seawater along the West Coast of the United States caused the mass mortality of oyster larvae in hatcheries located in the area between 2005 and 2009, which shocked the oyster industry with a loss of 110 million United States dollars (Ekstrom et al., 2015; Mabardy et al., 2015). In addition, OA-induced larvae mortality was also revealed by a series of laboratory studies (Talmage & Gobler, 2010; Ventura et al., 2016). It was shown that the larvae of the hard clam *M. mercenaria* and the bay scallop *A. irradians* exhibited dramatic declines in survivorship, dropping to approximately 50% of that of the larvae reared in seawater with ambient pH (pH 8.02), with increased $p\mathrm{CO_2}$ levels (pH 7.84; 650 ppm) (Talmage & Gobler, 2010). Similarly, the mortality rates of the mussel *M. edulis* larvae were found to have significantly increased by the elevation in $p\mathrm{CO_2}$ levels in seawater, and all larvae reared in acidified seawater at pH 7.1 failed to survive beyond 29 days (Ventura et al., 2016). In addition, Dupont et al. (2008) reported that a decrease in the pH value by 0.2 units (pH 7.9) led to the mortality of all larvae of the brittle star *O. fragilis* within 8 days, while more than 70% of the counterparts reared in ambient seawater were still alive at the same time. More seriously, since larvae with extended metamorphosis times and smaller sizes would be susceptible to greater rates of predation and natural mortality (Rumrill, 1990; Uthicke et al., 2013), the larvae of marine invertebrates would be expected to experience even greater mortality rates under future OA conditions than the rates observed from laboratory experiments.

Accumulating data indicate that the exposure of larvae to OA scenarios may also exert negative effects on the hatching and settlement of marine invertebrates (Albright & Langdon, 2011; Gazeau et al., 2010; Kaplan et al., 2013; Kurihara et al., 2004; Mayor et al., 2007). Kurihara et al. (2004) found the hatching rates of two copepods, *A. steueri* and *A. erythraea*, tended to decrease after 24 hours with an increase in seawater $p\mathrm{CO_2}$ levels (10,000 ppm). Only 0.7% of the eggs of the squid *Doryteuthis pealeii* hatched in seawater with elevated $p\mathrm{CO_2}$ (pH 7.3; 2200 ppm) after 14 days of exposure, whereas more than 60% of the eggs hatched in ambient seawater (Kaplan et al., 2013). This dramatic reduction in the success rate of hatching (37%) as a result of exposure to seawater with elevated $p\mathrm{CO_2}$ (8000 ppm) was also reported in the keystone copepod *Calanus finmarchicus*, in which the success rate of hatching was approximately only half of those reared in ambient seawater (34% at OA *vs* 68% at ambient seawater) (Mayor et al., 2007).

Successful recruitment is critical to maintaining the population of marine invertebrates, during which successful larval settlement becomes a crucial prerequisite (Lecchini et al., 2017). However, this critical event that occurs in various marine invertebrate species has been shown to be susceptible to OA (Gazeau et al., 2010; Uthicke et al., 2013). The larvae of the blue mussel *M. edulis* exhibited a significant reduction in the success of settlement when there is a 0.25- to 0.34-unit decrease in the pH value of the seawater (Gazeau et al., 2010). Similarly, it was shown that the settlement of the sea star *A. planci* on crustose coralline algae under high pCO_2 conditions (pH 7.9) declined by about 50% as compared to the control groups (pH 8.10) (Uthicke et al., 2013). In the coral *Porites astreoides*, the rates of successful settlement significantly declined by 11% and 28% for larvae reared in acidified seawater with elevated pCO_2 at 560 and 800 ppm, respectively, as compared to those of the control group (380 ppm) (Albright & Langdon, 2011). Moreover, it has been shown that the ability of the crustacean *Stenopus hispidus* larvae to recognize the settlement cues was significantly hampered by the exposure to OA scenarios (700 and 1000 ppm) (Lecchini et al., 2017). Apart from affecting the ability of larvae to settle directly, it is worth noting that OA can also exert impacts on the settlement process of marine invertebrates indirectly by altering the composition of settlement inducers (Webster et al., 2013). For example, Webster et al. (2013) observed a significant shift in the microbial communities of the biofilms on crustose coralline algae under OA scenarios (pH 7.7; 1187 ppm), which could make it less favorable for the corals *Acropora millepora* and *Alternaria tenuis* to settle on.

Although determining the underlying affecting mechanism of OA's impacts on marine invertebrates can be extremely complicated and still awaits further investigation, the currently available results suggest that OA may exert negative impacts on the embryonic and/or larval development of marine invertebrates through the following pathways. Firstly, under high pCO_2 conditions, the larvae may allocate more energy to crucial processes such as osmotic regulation, which would inevitably put a constraint on the energy available for development (Stumpp et al., 2012, 2013). For example, Stumpp et al. (2012) found that the larvae of the sea urchin *S. purpuratus* spent an average of 39%—45% of the available energy for somatic growth under high pCO_2 conditions, while more than 78% of the total energy can be allocated to somatic growth in ambient seawater. Secondly, the exposure of larvae to OA scenarios could influence the pH in the digestive tract of the larvae and thus may reduce the digestive efficiency (Zhao et al., 2017). Since the optimum pH value for the digestive enzymes (i.e., proteases and phosphatases) in the midgut of several species is approximately 11 (Sharma et al., 1984), OA-induced intracellular pH decline may reduce the activities of these enzymes and therefore constrain energy intake. Stumpp et al. (2013) reported that the larvae of the sea urchin *Strongylocentrotus droebachiensis* exposed to acidified seawater at pH 7.4 suffered a 0.5-unit drop in gastric pH as compared to those (approximately pH 9.5) reared in seawater at pH 8.0, which may have significant impacts on the digestive function. Similarly, significantly inhibited activity of citrate synthase, a rate-limiting enzyme in aerobic metabolism, was observed in larvae of the coral *Pocillopora damicornis* in seawater with elevated pCO_2 (950 ppm), indicating a significant decrease in the metabolic rate upon OA exposure (Rivest & Hofmann, 2014). Thirdly, the expressions of genes that play crucial roles in the regulation of embryonic and/or larval development may be disrupted due to the exposure of embryos to OA scenarios (Hammond & Hofmann, 2012; Stumpp, Dupont, et al., 2011). For instance, it was found that the expression level of the gene *Wnt8*, which is important for endomesodermal specification and early embryo patterning, was significantly increased after the exposure of the sea urchin *S. purpuratus* to acidified seawater (pH 7.6; 1350 ppm) (Hammond & Hofmann, 2012). This alteration in the expression of key genes may result in abnormalities such as the presence of multiple invagination sites around the embryo and therefore offers a possible explanation for the observed negative impacts on embryonic development induced by OA (Hammond & Hofmann, 2012; Wikramanayake et al., 2000).

It has been suggested that relatively small perturbations in larval populations of marine invertebrates may be significantly enlarged for those of the adult. Therefore near-future OA scenarios, which have been shown to hamper the embryonic and/or larval development of many marine invertebrates, may have far-reaching effects on the wild populations of these species (Hettinger et al., 2012; Kurihara, 2008). For example, it was observed that once the larvae of the oyster *Ostrea lurida* experienced OA during larval development, their subsequent growth, even under normal conditions, would continue to lag behind when compared to that of the control oysters that never experienced OA (Hettinger et al., 2012). Therefore to evaluate the real impact of OA on embryonic development as a part of the life cycle of a species, data on longer term exposure, that is, from the egg to the end of the life cycle, should be considered in future experiments.

Impacts of ocean acidification on biomineralization of invertebrates

In order to defend against predators, the calcifiers, such as gastropods, bivalves, and sea urchins, in the ocean must build mechanically strong and structurally integrated shells as soon as possible using $CaCO_3$ crystals through the biomineralization process (Sadler et al., 2018; Zhao et al., 2017, 2020). According to previous studies, the biomineralization process of marine organisms can be largely affected by the saturation state of calcium carbonate (Ω) (Fabry et al., 2008; Orr et al., 2005; Waldbusser et al., 2015), which, with respect to calcareous mineral phases, can be calculated using Eq. (1.2), given as follows:

$$\Omega_i = \frac{[CO_3^{2-}] \times [Ca^{2+}]}{K_{sp}}, \tag{1.2}$$

where i is the mineralogy (e.g., calcite, aragonite); and K_{sp} is the apparent solubility product of aragonite or calcite, which depends on temperature, salinity, pressure, and the particular mineral phase (Mucci, 1983).

Since calcium ion (Ca^{2+}) concentration is relatively constant in seawater (approximately 10 mmol/L), Ω is largely determined by variations in carbonate ion (CO_3^{2-}) concentrations (Feely et al., 2004). The formation of the shell and the skeleton in calcifying organisms is favored when $\Omega > 1$, and dissolution occurs when $\Omega < 1$ (Waldbusser et al., 2015). Given that increased CO_2 concentrations in the ocean will decrease the CO_3^{2-} concentration, Ω for aragonite (Ω_{ara}) and calcite (Ω_{cal}) will be altered thereupon (Waldbusser et al., 2015). According to predictions, the CO_3^{2-} concentration has dropped by 10% since the Industrial Revolution and will drop another 35% by 2100 (Orr et al., 2005). Consequently, OA may pose a great threat to the calcification process of many calcifying organisms. A series of previous studies have demonstrated that some marine invertebrate species, including gastropods, bivalves, sea urchins, and corals, exposed to elevated pCO$_2$ levels would experience significant decreases in net calcification and mechanical properties of the shell (Harvey et al., 2018; Kurihara, Asai, et al., 2008; Kurihara et al., 2007; Melzner et al., 2011; Zhao et al., 2020) (Table 1.3).

To date, the OA-induced inhibition on shell synthesis has been reported in various marine invertebrates (Albright & Langdon, 2011; Kurihara, 2008). For example, when reared in acidified seawater, both the

Table 1.3 Effects of ocean acidification on biomineralization of marine invertebrates.

Taxon	Species	pH/pCO$_2$	Effects	References
Coelenterata	Oculina patagonica	pH 8.3–7.3	↑Skeleton dissolution	Fine and Tchernov (2007)
	Stylophora pistillata	pH 8.2–7.3 (387–3898 ppm)	↑Systematic crystallographic changes	Coronado et al. (2019)
Mollusca	Crassostrea gigas	pH 8.21–7.42 (348–2268 ppm)	↓Shell mineralization; ↓Shell length and height	Kurihara et al. (2007)
	Haliotis discus hannai	450–1200 ppm	↑Shell malformation rate;	Onitsuka et al. (2018)
	Mytilus edulis	380–1000 ppm	↓Shell length ↓constraint in crystallographic orientation	Fitzer et al. (2014)
		380–1000 ppm	↓Crystallographic control of shell formation; ↑Amorphous calcium carbonate formation	Fitzer et al. (2016)
		pH 8.03 and 7.78 (642–1213 ppm)	↓Shell thickness; ↓Shell growth	Gazeau et al. (2010)
	M. galloprovincialis	pH 8.1 and 7.8	↓Shell length; ↑Shell damage; ↓Shell thickness	Bressan et al. (2014)
	Mercenaria mercenaria	pH 8.02–7.49 (354–1437 ppm)	↑Shell malformation rate; ↓Shell thickness; ↑Shell dissolution	Talmage and Gobler (2010)

	Species	pH (ppm)	Effects	Reference
	Argopecten irradians	pH 8.02−7.49 (354−1437 ppm)	↓Shell malformation rate; ↓Shell thickness; ↑Shell dissolution	Talmage and Gobler (2010)
	Tegillarca granosa	pH 8.13−7.39 (554−3120 ppm)	↓Inner shell surface integrity; ↓Calcification rate	Zhao et al. (2017)
Arthropoda	*Hydroides elegans*	pH 8.08−7.34 (514−3438 ppm)	↑Shell calcite/aragonite ratio; ↓Integrity of shell	Chan et al. (2012)
	Callinectes sapidus	409−2856 ppm	↓Net calcification	Ries et al. (2009)
	Penaeus plebejus	409−2856 ppm	↓Net calcification	(Ries, Cohen, & McCorkle, 2009)
	Lysmata californica	pH 7.99 and 7.53 (462 and 1297 ppm)	↑Exoskeleton mineralization	Taylor et al. (2015)
Echinodermata	*Amphiura filiformis*	pH 8.0−6.8	↑Calcification rate	Wood et al. (2008)

(Gallo et al., 2019; Schlegel et al., 2015; Shi et al., 2017) shell length and height of the D-shaped larvae of the oyster *C. gigas* were significantly smaller than those of the control oysters (Kurihara et al., 2007). Furthermore, 18% of the larvae exhibited birefringence over the entire surface at 24 hours, and 48% of the larvae exhibited no birefringence after OA treatment (pH 7.4; 2268 ppm), while the corresponding values for the same phenomena in the control group (pH 8.1; 348 ppm) were 66% and 24%, respectively (Kurihara et al., 2007). Significantly smaller shells were also detected in the larvae of the abalone *H. discus hannai* under a high pCO$_2$ condition (pH 7.6; 1200 ppm) as compared to that of the control (pH 8.0; 450 ppm) (Onitsuka et al., 2018). It was also shown that the shell growth of the blue mussel *M. edulis* significantly reduced in acidified seawater at pCO$_2$ levels of 550 and 750 ppm as compared with the control group with a pCO$_2$ level of 380 ppm (Fitzer et al., 2014). Similarly, compared to the control group (pH 8.03), the final shell size and thickness of the blue mussel *M. edulis* after 15 days (from D-shape to pediveliger larvae) of exposure to OA scenarios (pH 7.8) were 6% smaller and 12% thinner than those reared in seawater at pH 8.03, respectively (Gazeau et al., 2010). In addition, it was demonstrated that the shell thickness of the clam *M. mercenaria* was approximately 17 μm when reared in ambient seawater (250 ppm), whereas it was only 6.7 and 3.8 μm for those exposed to acidified seawater at 750 and 1500 ppm of pCO$_2$, respectively (Talmage & Gobler, 2010). The thicknesses of the shells of the scallop *A. irradians* were decreased significantly to approximately 12, 11, and 6 μm after 17 days of development, with the increasing pCO$_2$ levels being 390, 750, and 1500 ppm, respectively; meanwhile, the shell thickness was around 20 μm under 250-ppm pCO$_2$ concentration (Talmage & Gobler, 2010). Significant decreases in shell length, as well as thickness, were also observed in many of the juveniles of the Mediterranean mussel *M. galloprovincialis* and the clam *Chamelea gallina* after 6 months of OA exposure (Bressan et al., 2014; Talmage & Gobler, 2010).

Moreover, OA can dissolve and damage the existing shells or skeleton of marine calcifiers (Bressan et al., 2014; Duquette et al., 2017; Garilli et al., 2015; Zhao et al., 2017). For instance, evident shell damage upon OA (pH 7.4; 2700 ppm) exposure was observed in the Mediterranean mussel *M. galloprovincialis* and the clam *C. gallina*, which had approximately 35% and 11% of the outer shell area damaged compared to the intact shell of the control (pH 8.1), respectively (Bressan et al., 2014). It was also shown that the inner shell surfaces of the blood clam *T. granosa* were partly corroded and dissolved after OA exposure for 40 days (pH 7.4; 3120 ppm), while the inner shell surfaces were unaffected in those from the treatment group at pH 8.1 (Zhao et al., 2017) (Fig. 1.4). Similarly, the inner shell surfaces of bivalves such as the blue mussel *M. edulis* and the striped venus clam *C. gallina* were also found to be corroded and dissolved due to OA (Chan et al., 2012; McDonald et al., 2009). In addition, the gastropod *Charonia lampas* had clearly undergone progressive shell dissolution under high pCO_2 conditions (pH 7.8). Individuals

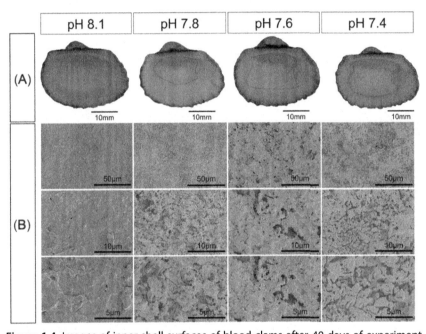

Figure 1.4 Images of inner shell surfaces of blood clams after 40 days of experiment. (A) stereomicroscopic images of inner shell surfaces. The corroded parts (irregular white plaques) of the inner shell surfaces occurred in the blood clams at the three lowered pH levels (7.8, 7.6, and 7.4), and were most apparent and clear at pH 7.4. Blood clams at pH 8.1 showed normal inner shell surfaces. The black dashed frames indicate the corroded parts of the inner shell surfaces. (B) Scanning electron microscope (SEM) images of inner shell surfaces at different magnifications. For blood clams at the three lowered pH levels (7.8, 7.6, and 7.4), the SEM images were taken from the corroded parts. *From Zhao, X., Shi, W., Han, Y., Liu, S., Guo, C., Fu, W., Chai, X., & Liu, G. (2017). Ocean acidification adversely influences metabolism, extracellular pH and calcification of an economically important marine bivalve, Tegillarca granosa. Marine Environmental Research, 125, 82–89. https://doi.org/10.1016/j.marenvres.2017.01.007 87.*

collected from high-pCO_2 seawater sites had no epiphytes revealing a predominantly white shell coloration that was highly conspicuous against the substratum. In addition, high-pCO_2 seawater exposure also resulted in damage and/or truncation to the apex and older shell regions of *C. lampas*, which even made the soft tissue of some individuals under exposure to the surrounding seawater through holes in the shell apex (Harvey et al., 2018). Interestingly, Fine and Tchernov (2007) observed that the corals *Oculina patagonica* and *Madracis pharencis* lost their exoskeleton and lived as soft-bodied polyps when exposed to acidified seawater at pH 7.4. It is worth noting that the OA-induced dissolution in calcifying reef species may alter the ocean community composition, which could result in the loss of biodiversity in the marine environment (Garilli et al., 2015; Mollica et al., 2018).

The mineralogy and ultrastructure of the calcareous products of marine calcifiers may also be influenced by OA, which may thereby affect the mechanical properties of the calcareous skeleton (McDonald et al., 2009; Zhao et al., 2017). Research conducted by McDonald et al. (2009) demonstrated that the central shell wall plates of the barnacle *Amphibalanus amphitrite* exposed to an OA scenario at pH 7.4 were significantly weaker than those raised in ambient seawater at pH 8.2, which would render the organisms more vulnerable to predation. Increases in calcite content as well as alternations in the ultrastructure of the calcareous tubes were observed in the juvenile tubeworms *Hydroides elegans* under elevated seawater pCO_2 conditions. For example, as compared to those under ambient pH (pH 8.1), the tubes produced by *H. elegans* reared under reduced pH levels (pH 7.6 and 7.4) had higher porosity and layer irregularity with more signs of pitting, and less structured crystallites were detected in the irregularly oriented prismatic structure layer of the tubes after exposure to OA (pH 7.9, 7.6 and 7.4), which led to a holistic deterioration of the hardness and elasticity of the tube (Chan et al., 2012). For the mussel *Mytilus coruscus* reared under seawater with different pH levels for 40 days, the internal shell surface was intact, with a typical glossy appearance under ambient seawater (pH 8.1), while the internal shell surface had become dull and white in the individuals under acidified seawater (pH 7.8 and 7.4). In addition, the aragonite tablets of the nacreous layer and the calcite crystals of the prismatic layer on the normal internal shell surface of the mussels had uniform structural orientations; in contrast, mussels under acidified conditions appeared to be disorientated or dissolved, which subsequently weakened the defense capacity of the shell

(Zhao et al., 2020). Similarly, Fitzer et al. (2016) confirmed that seawater at 1000 -ppm pCO_2 levels can induce significant changes in the hydrated and dehydrated forms of amorphous calcium carbonate in the aragonite and calcite layers of the shell of the blue mussel *M. edulis*, indicating a suppressed ability to produce the protective crystalline shell in this species. Systematic crystallographic changes such as better constrained crystal orientation and anisotropic distortions of bio-aragonite lattice parameters were also detected in the coral *Stylophora pistillata* under high pCO_2 conditions (pH 7.6 and 7.3) (Coronado et al., 2019).

OA may affect the biomineralization of marine invertebrates through several possible mechanisms. Firstly, hampered shell synthesis can result from the OA-induced acid–base imbalance in marine calcifiers (Zhao et al., 2017). As reported in previous studies, OA can lead to extracellular acidosis in many marine invertebrates, such as the blue mussel *M. edulis*, the sea urchin *S. droebachiensis*, and the blood clam *T. granosa* (Michaelidis et al., 2005; Stumpp et al., 2012; Thomsen et al., 2013; Zhao et al., 2017). For example, the hemolymph pH values of the blood clam *T. granosa* declined significantly from about 7.51−7.19 when seawater pH decreased from 8.1 to 7.4 (Zhao et al., 2017). The hemolymph pH level of the thick mussel *M. coruscus* was also markedly decreased due to OA, which was reduced from 7.52 in the control group to nearly 7.38 and 7.24 in the groups with pH 7.8 and 7.4, respectively (Zhao et al., 2020). Furthermore, the pH of the coelomic fluid of the sea star *Asterias rubens* was also significantly reduced to approximately 7.35 after 1 week of OA exposure (pH 7.7) as compared to the pH of 7.5 of the control (pH 8.1) (Hernroth et al., 2011). Consequently, extracellular acidosis may reduce the saturation state of $CaCO_3$ at the calcification sites in the body of these organisms. For instance, the hemolymph Ca^{2+} concentration of the mussel *M. coruscus* decreased to about 283 and 278 mg/L after 40 days of exposure to acidified seawater at pH 7.8 and 7.4, respectively, while it maintained at 297 mg/L under ambient seawater (pH 8.1) (Zhao et al., 2020). Thus, these alternations in hemolymph pH levels and Ca^{2+} concentrations would result in the observed reduction in net calcification rates and shell formations. Secondly, constrained energy supply under OA scenarios may also suppress the shell formation of marine invertebrates (Pan et al., 2015; Shang et al., 2018). A study conducted in the blue mussel *M. edulis* showed that more serious corrosion of the inner shell surface would occur in individuals facing starvation when compared to those with enough food supply (Melzner et al., 2011). Since OA may suppress the metabolism of marine invertebrates and the energy allocated for the fabrication of calcareous shells, a reduction in energy supply may inhibit the process of shell formation of these calcifiers (Hüning et al., 2013). Thirdly, the alternations in the expressions of biomineralization-related genes caused by OA would affect the biomineralization ability of marine invertebrates as well (Kurihara et al., 2012; Zhao et al., 2020). For example, the expression level of the gene *msp130*, which has been proposed to transport Ca^{2+} to the skeletogenesis site, was significantly downregulated in the sea urchin *H. pulcherrimus* (Kurihara et al., 2012). Similarly, Stumpp, Dupont, et al. (2011) found that genes such as *SM30B*, *SM50*, and *msp130* and *metalloproteinases* that encode proteins for larval spicule biomineralization in the sea urchin *S. purpuratus* were significantly downregulated by up to 36% upon 4 days of exposure to acidified seawater at pH 7.7 (1318 ppm pCO_2) compared to that of the control (pH 8.1; 399 ppm). Moreover, it was shown that the expression of genes plays a crucial role in ion regulation during biomineralization, with those encoding lysosomal H^+-ATPase, vacuolar H^+-ATPase, sarco/endoplasmic reticulum calcium-transporting ATPase (SERCA), and Na^+/K^+-ATPase also being severely impacted by OA in the larvae of *S. purpuratus* (Stumpp, Dupont, et al., 2011). In addition, the expression of *chitinase*, which is potentially important for the calcification process in the mantle of the blue mussel *M. edulis*, was strongly suppressed due to the exposure to OA scenarios (4000 ppm pCO_2), which would put a constraint on the availability of organic components (chitin) for crystal formation (Hüning et al., 2013).

However, the biomineralization responses of marine calcifiers to OA exposure were shown to vary greatly among taxa and species, which may depend on factors such as the ability to regulate pH at the calcification site, the extent of organic-layer coverage of the external shell, and the biomineral solubility of the skeleton of the calcifiers (Duquette et al., 2017; Parker et al., 2013). For example, the effects of OA on the biomineralization of limpets were found to be significantly different even among the same genus (*Patella caerulea and Patella rustica*) (Duquette et al., 2017). The percentages of aragonite and calcite in the shells differed significantly between the limpet *P. rustica* collected from an area with a high pCO_2 (pH 7.7) concentration and that collected from the reference area (pH 8.1); meanwhile, no significant differences were detected in *P. caerulea* (Duquette et al., 2017). Stemmer et al. (2013) demonstrated that the shell growth and crystal microstructure of the bivalve *Arctica islandica* did not differ between individuals exposed to seawater at pCO_2 levels ranging from 380 to 1120 ppm. Similarly, it was shown that the brittle star *Amphiura filiformis* could maintain its skeletal integrity upon OA exposure by elevating calcification rates (Wood et al., 2008).

In conclusion, the impacts of OA on marine calcification are more complicated than previously thought and vary significantly among species; therefore they need to be explored in a wider range of invertebrate organisms. Moreover, the mechanisms underpinning the inhibition of shell formation under OA scenarios and the possibility of the adaption of marine calcifiers to OA should also be further investigated.

 Impacts of ocean acidification on metabolism and growth of invertebrates

In response to environmental stress like OA, marine invertebrates usually spend extra energy to protect individuals against hypercapnia-induced disturbances (Guppy & Withers, 1999; Pörtner, 2008). Since oxygen and CO_2 levels are often inversely correlated in aquatic habitats, the effects of OA on the oxygen consumption rate and metabolism of marine invertebrates have been extensively studied (Carter et al., 2013; Peng et al., 2017; Wang et al., 2015; Zhao et al., 2017) (Table 1.4).

In recent years, OA-induced metabolic depression has been reported in various marine invertebrates (Wang et al., 2015; Zhao et al., 2017). Nakamura et al. (2011) found that the metabolism of the coral *A. digitifera* larvae declined with reduced seawater pH (8.0−7.3) to enhance their survival rate in the short term; however, this process may diminish the survival of the coral larvae over long periods. Suppressed metabolic rate upon OA exposure was also reported in the Mediterranean mussel *M. galloprovincialis* (Michaelidis et al., 2005). It was shown that the oxygen consumption rate was reduced to 35% and 65% of the control (pH 8.05) for adults and juveniles of *M. galloprovincialis* exposed to acidified seawater at pH 7.3 for 20 hours, respectively (Michaelidis et al., 2005). The respiration rates of the larvae of the oyster *C. gigas* also showed a tendency to decrease with the decreasing pH. While the weight-specific respiration at pH 7.7 was statistically similar to that at the ambient pH 8.1, it was significantly reduced in the extreme treatment with a pH level of 7.4 (Ginger et al., 2013). Similarly, exposure to acidified seawater at pH 7.58 led to a

Table 1.4 Effects of ocean acidification on metabolism and growth of marine invertebrates.

Taxon	Species	pH/pCO_2	Effects	References
Coelenterata	*Acropora digitifera*	pH 8.05–7.33 (331–3100 ppm)	↓Metabolic rate	Nakamura et al. (2011)
Mollusca	*Tegillarca granosa*	pH 8.1–7.4 (554–3120 ppm)	↓Feeding activity; ↓Aerobic metabolism	Zhao et al. (2017)
	Mytilus galloprovincialis	pH 8.05–7.3	↓Oxygen consumption rate; ↓Growth rate	Michaelidis et al. (2005)
	M. edulis	pH 8.1–6.7	↓Growth rate	Berge et al. (2006)
	Ruditapes philippinarum	pH 7.96–7.39 (505–2086 ppm)	↓Oxygen-to-nitrogen ratio; ↓Clearance rate; ↓Scope for growth	Xu et al. (2016)
	Crassostrea virginica	pH 8.3–7.5 (385–3523 ppm)	↓Shell and soft-body growth; ↑Metabolic rate	Beniash et al. (2010)
Arthropoda	*Petrolisthes cinctipes*	pH 7.93–7.58 (574–1361 ppm)	↓Metabolic rate	Carter et al. (2013)
	Paralithodes camtschaticus	pH 8.04–7.5 (437–1637 ppm)	↓Growth rate	Long et al. (2013)
	Hyas araneus	pH 8.11–7.34 (380–3000 ppm)	↓Growth rate	Walther et al. (2010)
	Palaemon pacificus	pH 8.17–7.64 (380–1900 ppm)	↓Antenna/total length ratio	Kurihara et al. (2008)
Echinodermata	*Tripneustes gratilla*	pH 8.15–7.6 (448–1886 ppm)	↓Growth rate	Brennand et al. (2010)
	Lytechinus variegatus	pH 8.10–7.83 (367–765 ppm)	↓Growth rate	Albright et al. (2012)
	Strongylocentrotus purpuratus	pH 8.18–7.70 (375–1264 ppm)	↑Metabolic rates; ↓Growth rate	Stumpp, Dupont, et al. (2011)

significant reduction in the mean metabolic rate (approximately 11%) of the crab *Petrolisthes cinctipes* compared with those reared in the control (pH 7.93) (Carter et al., 2013). The respiration rates of the blood clam *T. granosa* was reduced due to exposure to OA scenarios as well, which dropped to approximately 89% and 84% of that of the control group (pH 8.1) for individuals treated with seawater at pH 7.6 and 7.4, respectively (Zhao et al., 2017).

The reduction in metabolic rates induced by OA exposure may have resulted from the following reasons. Firstly, OA may inhibit the metabolism of marine invertebrates by interfering with food intake and digestive efficiency (Xu et al., 2016; Zhao et al., 2017). The feeding behavior, as indicated by the clearance rates, of the clam *Ruditapes philippinarum* was significantly suppressed by the exposure to acidified seawater at pH 7.7 as compared to those under ambient conditions (pH 8.0) (Xu et al., 2016). Similarly, the clearance rates of the blood clam *T. granosa* significantly declined to approximately 78%, 66%, and 53% of those reared in ambient seawater (pH 8.1) for individuals reared in acidified seawater at pH 7.8, 7.6, and 7.4, respectively (Zhao et al., 2017). Besides, as the activities of digestive enzymes are pH-sensitive, OA may also affect the digestive efficiency of these organisms as mentioned previously in this chapter (Kong et al., 2019; Stumpp et al., 2013). As a result, the OA-induced suppression of energy acquisition may therefore result in energy limitation to fuel the essential processes for metabolism. Secondly, OA may inhibit the metabolism of marine organisms by affecting the expression of metabolism-related genes (Stumpp, Dupont, et al., 2011). For example, the expression of ten metabolism-related genes (e.g., *citrate synthase, pyruvate kinase,* and *arylformamidase*) from glycolysis, fatty acid biosynthesis, tryptophan metabolism, and tricarboxylic acid cycle (TCA cycle) pathways of the razor clam *Sinonovacula constricta* were significantly altered after 7 days of exposure to near-future OA scenarios (pH 7.6 and 7.4), offering an explanation for the observed slowdown in the metabolic activity upon OA treatment (Peng et al., 2017). Similar alternations in the expression of metabolism-related genes induced by OA exposure were also detected in the larvae of the sea urchin *S. purpuratus* and the coral *Pocillopora damicornis* (Todgham & Hofmann, 2009; Vidal-Dupiol et al., 2013). In addition, metabolic depression may be adopted by marine organisms to decrease overall energy demand and to provide sufficient energy for only essential biological processes to enable survival under OA conditions (Zhao et al., 2020). For example, Zhao et al. (2020) found that the gene expression levels of Na^+/K^+-ATPase, K^+ channel, Na^+-related cotransporters, and solute carrier organic anion transporter, which mediate ATP-consuming processes, were markedly upregulated, indicating that the mussels in the study reallocated the additional energy for ion and acid—base regulation.

In contrast, many marine invertebrates suffering from OA for short times can compensate for the elevated energy demand by increasing food intake and enhancing metabolism (Wood et al., 2008). For example, the brittle star *A. filiformis* was shown to increase its metabolism rates, indicated by a higher oxygen uptake rate, to compensate for the negative impacts caused by elevated seawater acidity at the cost of reduction in muscle mass (Beniash et al., 2010; Carter et al., 2013). Similarly, Beniash et al. (2010) demonstrated that the exposure of the oyster *Crassostrea virginica* to OA led to a significantly higher standard metabolic rate, which may meet the extra energy demand to maintain in vivo homeostasis. Furthermore, it was suggested that the substrates (carbohydrates, lipids, or proteins) used in the energy metabolism of marine invertebrates may also be altered under OA scenarios to cover the extra energy budget required to maintain extracellular acid—base balance. For instance, Zhao et al. (2017) found that the largest fraction of energy expenditure of *T. granosa* under OA (pH 7.4) was covered by amino acid catabolism instead of carbohydrate and lipid catabolism at normal. Similarly, a switch from lipids to proteins as the metabolic substrate was also observed in the crab *P. cinctipes* after being reared in acidified seawater at pH 7.58 for 6 days (Carter et al., 2013). In addition, the OA-induced alternation in metabolic pathways was detected in the oyster *C. gigas* as well (Lannig et al., 2010). However, since this activity of short-term upregulation of metabolism may occur at the cost of reduced skeletal integrity and muscle mass, it is unlikely to be sustainable in the long term (Lannig et al., 2010; Wood et al., 2008).

Metabolic depression and/or compensation associated with low pH are often coupled with inhibited growth in marine invertebrates (Stumpp, Dupont, et al., 2011; Stumpp, Wren, et al., 2011). For example, compared to that of the control (pH 8.1), exposure of the blue mussel *M. edulis* to acidified seawater (pH 7.4 and 7.1) for 44 days resulted in significant reductions in the mean increment of shell length (Berge et al., 2006). Similarly, the crab *C. bairdi* reared in ambient seawater (pH 8.04) had a wet mass that was approximately 16% and 118% larger than that of the individuals reared in acidified seawater at pH 7.8 and 7.5 after 1750 degree-days (days °C), respectively (Long et al., 2013). The growth performance of the shrimp *Palaemon pacificus* was also found to be significantly inhibited by an elevated pCO_2 level (1000 ppm) as compared to that reared in ambient seawater (pCO_2 level at 380 ppm), indicated by a significantly smaller ratio of antenna length to total length after 7 weeks of experimental treatment (Kurihara, Matsui, et al., 2008). It was also shown that the gain in biomass of the sea star *A. rubens* was significantly reduced after 6 weeks of exposure to high pCO_2 levels (4000 ppm), which was approximately only one-third in those reared in seawater with an ambient pCO_2 level (380 ppm) (Appelhans et al., 2014). Similar growth inhibition caused by OA exposure was also detected in the sea urchins *H. pulcherrimus* and *E. mathaei*; the blue mussel *M. edulis*; the clam *M. mercenaria*; the scallop *A. irradians*; the crab *P. camtschaticus*; and the lobster *Homarus gammarus* (Arnold et al., 2009; Berge et al., 2006; Long et al., 2013; Shirayama & Thornton, 2005; Talmage & Gobler, 2010). This reduction in growth could be caused by the disturbance of cellular protein turnover induced by OA (Langenbuch et al., 2006). For example, a significant reduction (of 60%) in protein synthesis was observed in the worm *Sipunculus nudus* reared in acidified seawater at pH 6.7, which may pose a threat to the cellular homeostasis of structural as well as functional proteins and eventually lead to an impairment of growth (Langenbuch et al., 2006).

Impacts of ocean acidification on immune responses of invertebrates

Living in a complex environment, marine invertebrate species are often challenged by various pathogenic microorganisms, and therefore possessing a sound immune response is crucial for the survival of these species (Asplund et al., 2014; Liu et al., 2016). Unlike vertebrates, the immune response of marine invertebrates is mainly reliant on the innate defense executed by their hemocytes and humoral factors (Liu et al., 2016; Loker et al., 2004). However, accumulating evidence suggests that near-future OA scenarios may hinder the immune responses of marine invertebrates such as bivalves, echinoderms, and crustaceans, which could render these invertebrate species more susceptible to pathogenic infection (Brothers et al., 2016; Hernroth et al., 2011; Leite Figueiredo et al., 2016; Liu et al., 2016; Migliaccio et al., 2019) (Table 1.5).

Table 1.5 Effects of ocean acidification on immune responses of marine invertebrates.

Taxon	Species	pH/pCO$_2$	Effects	References
Mollusca	*Mytilus edulis*	pH 7.9–6.5 (665–3316 ppm)	↓Phagocytosis capacity	Bibby et al. (2008)
	Tegillarca granosa	pH 8.1–7.4 (549–3064 ppm)	↓Hemocyte count; ↓Hemocyte phagocytosis; Blood cell-type composition alternation	Liu et al. (2016)
	Crassostrea gigas	pH 8.17–7.55 (438–2062 ppm)	↓Antioxidant enzyme activities; ↑Apoptosis in hemocytes	Wang et al. (2016)
Arthropoda	*Nephrops norvegicus*	pH 8.1–7.9 (319–622 ppm)	↓Hemocyte count; ↓Phagocytic capacity	Hernroth et al. (2011)
	Chionoecetes bairdi	pH 8.09–7.50 (391–1597 ppm)	↑Dead, circulating hemocyte count	Meseck et al. (2016)
Echinodermata	*Heliocidaris erythrogramma*	pH 8.1 and 7.6 (389 and 1684 ppm)	↓Coelomocyte concentration; ↓Phagocytic capacity;	Brothers et al. (2016)
	Asterias rubens	pH 8.1 and 7.7 (330 and 921 ppm)	↓Phagocytic capacity; ↓Activation of p38 MAP-kinase	Hernroth et al. (2011)
	Strongylocentrotus droebachiensis	pH 8.2 and 7.7 (351 and 1275 ppm)	↑Coelomocyte concentration; ↓Vibratile cells	Dupont and Thorndyke (2012)

Species	pH conditions	Effects	Reference
Leptasterias polaris	pH 8.2 and 7.7 (351 and 1275 ppm)	↑Coelomocyte concentration	Dupont and Thorndyke (2012)
Lytechinus variegatus	pH 8.0–7.3	↓Phagocytic capacity; ↓Cell spreading area	Figueiredo et al. (2016)
Echinometra lucunter	pH 8.0–7.3	↓Phagocytic capacity; ↓Cell spreading area; Coelomocyte proportion alternation	Figueiredo et al. (2016)
Paracentrotus lividus	pH 8.02 and 7.8	Total antioxidant capacity; Phagosome expression	Migliaccio et al. (2019)

Hemocytes (coelomocytes), which carry out the host defense response, are regarded as the main immune cells in marine invertebrates (Loker et al., 2004). Furthermore, due to the intrinsic difference in their role and efficiency in immune responses, the presence of different types of hemocytes (coelomocytes) in the right ratio is crucial for the immunity of marine invertebrates (Loker et al., 2004). However, it has been shown that exposure to near-future OA scenarios may reduce the total hemocyte count (THC) and alter the cell-type compositions of the hemocytes or coelomocytes of marine invertebrates, such as the sea star *A. rubens*, the crab *C. bairdi*, the sea urchin *H. erythrogramma*, the pearl oyster *Pinctada fucata*, the blue the mussel *M. edulis*, the striped venus clam *C. gallina*, and the blood clam *T. granosa*, indicating a hampered immune response of these species under near-future OA scenarios (Leite Figueiredo et al., 2016; Migliaccio et al., 2019; Wang et al., 2016). For example, the coelomocyte concentrations in the sea urchin *H. erythrogramma* reared in acidified seawater (pH 7.6) for 30 days were significantly reduced to approximately 30% of that of the control (Brothers et al., 2016). A dissimilarity as large as 22.39% in the cell-type composition in the coelomic fluid has been detected between the sea urchin *H. erythrogramma* reared in seawater at pH 8.15 and that at pH 7.6 (Brothers et al., 2016). The hemocyte counts of the lobster *Nephrops norvegicus* were reduced by almost half after 4 months of exposure to OA at pH 7.9 (Hernroth et al., 2011). Similarly, it was shown that the average THC of the blood clam *T. granosa* significantly decreased from 7.61×10^6 cells/mL to 5.33×10^6 cells/mL as the seawater pH reduced from pH 8.1 to pH 7.4 (Liu et al., 2016). Furthermore, significantly fewer red granulocytes but a higher proportion of basophil granulocytes were detected in blood clam individuals after OA treatment, which might be a result of an OA-induced apoptosis and proliferation suppression of hemocytes (Liu et al., 2016). For instance, the exposure of the oyster *C. gigas* to high pCO_2 (2000 ppm) concentrations for 28 days was found to stimulate the apoptosis of hemocytes by inducing the production of reactive oxygen species (ROS) (Cao et al., 2018; Wang et al., 2016). In addition, the expression levels of the genes of the apoptosis pathway such as *caspase-1* and *caspase-3* in the hemocytes of *C. gigas* were also significantly upregulated in response to OA treatment, which may be caused by the overproduction of ROS and may have led to the detected reduction in THC (Wang et al., 2016). However, previous studies have shown that the effects of OA on coelomocyte counts of echinoderms varied significantly among species (Brothers et al., 2016;

Dupont & Thorndyke, 2012; Hernroth et al., 2011). For example, exposure to acidified seawater at pH 7.7 resulted in an increase in the total coelomocyte count in the sea urchin *S. droebachiensis* and the sea star *Leptasterias polaris*, whereas the same caused a twofold decrease in the total coelomocyte count in the sea star *A. rubens* (Dupont & Thorndyke, 2012; Hernroth et al., 2011).

Phagocytosis executed by hemocytes is considered as the primary line of cellular defense against invasive pathogens in marine invertebrates (Loker et al., 2004; Matozzo et al., 2012; Su et al., 2018). A series of previous studies have demonstrated that the phagocytic rate of hemocytes in various marine invertebrates could be suppressed by the exposure of the organisms to near-future OA scenarios (Hernroth et al., 2011; Liu et al., 2016). Bibby et al. (2008) found that the hemocyte phagocytosis of the blue mussel *M. edulis* was significantly hampered after 32 days of exposure to an elevated seawater pCO_2 level (pH < 7.7). The phagocytic capacity of coelomocytes of the sea star *A. rubens* was also inhibited, to approximately only 30% of that of the control group (pH 8.1; 330 ppm), after 6 months of exposure to acidified seawater (pH 7.7; 921 ppm) (Hernroth et al., 2011). Similarly, compared with the control (pH 8.1; 550 ppm), an approximately 25% reduction in phagocytosis was detected in the blood clam *T. granosa* after exposure to an OA scenario at pH 7.4 (3064 ppm pCO_2) (Liu et al., 2016).

The inhibition of hemocyte mediated phagocytosis in marine organisms under OA conditions may be due to the following reasons (Fig. 1.5). Firstly, OA causing suppression on the phagocytic activity of hemocytes can be partly attributed to the lack of energy available for phagocytosis (Lannig et al., 2010; Liu et al., 2016). For example, the energy of marine invertebrates was reported to be allocated to maintain protein integrity via the chaperon Hsp70 (Mayer & Bukau, 2005). However, Hsp70 was

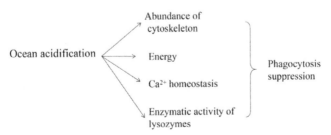

Figure 1.5 Possible underlying affecting mechanisms of ocean acidification–induced phagocytosis suppression.

found to be elevated in the sea star *A. rubens* after 1 week of OA exposure, which suggested a reduced energy budget for immune responses that may subsequently affect their defensive immune responses (Hernroth et al., 2011). Since the phagocytosis of foreign materials by hemocytes is an energy-consuming process, the reduction in the energy budget for immune responses under OA would result in decreased phagocytic activity (Turvey & Broide, 2010). Secondly, OA can induce intracellular Ca^{2+} disturbance and may subsequently hinder the phagocytosis of marine invertebrates (Shi et al., 2018). Zhao et al. (2017) observed that the hemolymph Ca^{2+} concentration of the blood clam *T. granosa* was significantly reduced by OA treatment (pH 7.4; 2072 ppm) compared to the control (pH 8.1; 554 ppm). Similar reductions caused by OA in the hemolymph Ca^{2+} concentration were also reported in mussels, in which the hemolymph Ca^{2+} concentration was reduced to about 283 and 278 mg/L after 40 days of exposure to seawater with pH 7.8 and 7.4, respectively, while it was 297 mg/L in the control group (pH 8.1) (Zhao et al., 2020). According to previous studies, the Ca^{2+} signal is required for almost all steps of phagocytosis including pathogen recognition and adhesion, phagosomal maturation, and phagocytic ingestion in marine invertebrates (Nunes & Demaurex, 2010; Shi et al., 2018). For instance, phagocytosis has been confirmed to be regulated by intracellular Ca^{2+} concentrations (Nunes & Demaurex, 2010). In addition, many immune-related signaling pathways, such as the nuclear factor-κ B (NF-κB)—signaling pathway, were also activated by Ca^{2+} regulation—related genes, such as calmodulin (CaM), CaMK2, and CaMKK2. Therefore theoretically, OA-induced aberrant Ca^{2+} homeostasis will undoubtedly alter the immune responses of these organisms. Thirdly, OA may reduce the phagocytosis of marine invertebrates by constraining the process of engulfment and through degradation of engulfed pathogenic particles (Su et al., 2018). During this process, the engulfment of the pathogenic particles by the hemocytes is facilitated by the cytoskeleton, mainly via the actin-myosin contractile system; alternation in the content of the cytoskeleton may have a significant impact on the process of phagocytosis (Su et al., 2018). Besides, the engulfed target is destroyed within phagosomes mainly by lysozymes after engulfment (oxygen-independent degradation) (Buggé et al., 2007). However, it was shown that the abundance of the cytoskeleton as well as the concentration and enzymatic activity of lysozymes in the hemocytes of marine organisms could be significantly impacted by exposure to acidified seawater (Su et al., 2018). For example, the cytoskeleton component abundance in hemocytes in terms of microfilament was detected to be significantly reduced in the blood clam *T. granosa* after 14 days of OA exposure. Moreover, the expression levels of genes related to the actin cytoskeleton regulation pathway, such as actin-related protein 2 (*ARPC2*), actin-related protein 3 (*ARPC3*), GTPase Kras (*KRAS*), GTPase Mras (*MRAS*), and Ras-related C3 botulinum toxin substrate 1 (*Rac1*) that functions as a complex to assemble the monomers into polymers in the process of producing F-actin from G-actin in *T. granosa*, were shown to be significantly induced after OA exposure (Su et al., 2018). At last, relevant immunogenic pathways of marine invertebrates would be activated upon pathogen challenge and subsequently trigger downstream immune responses. At last, relevant immunogenic pathways of marine invertebrates would be activated upon pathogen challenge, which is essential for the subsequent triggering of (Li & Xiang, 2013; Seth et al., 2006). However, these genes involved in the regulation of immune responses can be influenced by OA exposure (Liu et al., 2016). For instance, Liu et al. (2016) reported that the exposure of the blood clam *T. granosa* to elevated pCO_2 (pH 7.4; 3064 ppm) exerted significant impacts on the expression levels of key genes of the NFκB-signaling pathway and Toll-like receptor (TLR) signaling pathway including myeloid differentiation primary response protein MyD88 (MYD88), retinoic acid-inducible gene 1 (RIG1), TNF receptor-associated factor 2 (TRAF2), mitogen-activated protein kinase 7 (MAPK7), and TLRs.

In addition to cellular immunity, humoral pathways also play an important role in the immune responses of marine invertebrate species in fighting against a pathogen or foreign particle challenges (Holmskov et al., 2003). However, the activities of most immune-related molecules, such as antioxidant enzymes and phosphatases, are dependent on internal pH and thus sensitive to OA (Fan et al., 2006). For example, the activities of superoxide dismutase, catalase, glutathione peroxidase, glutathione, alkaline phosphatases, acid phosphatases, and glutamic pyruvic transaminase in the gills and digestive glands were found to be significantly influenced by the exposure of the thick shell mussel *M. coruscus* to pCO_2-acidified seawater (pH 7.3; 4492 ppm) (Hu et al., 2015). Moreover, the expressions of genes encoding these immune-related molecules could be altered by OA exposure as well, thus reducing these immune enzyme activities (Liu et al., 2016; Wang et al., 2016).

Theoretically, even slight changes in the chemical characteristics of the seawater, such as a drop in pH value, may favor distinct bacterial groups and lead to a compositional alteration of the bacterial community (Allgaier et al., 2008; Lidbury et al., 2012; Zha et al., 2017). Notably, several studies have suggested that OA may have significant impacts on the microbial community in the marine environment. For example, the bacterial community may shift between different pCO_2 concentrations (300, 400, 560, and 1140 ppm), as indicated by the decrease in the relative abundance of *Alphaproteobacteria* and the increase of *Flavobacteriales* (*Bacteroidetes*) with elevated pCO_2 (Witt et al., 2011). Zha et al. (2017) reported that OA has the potential to promote the development and survival of the pathogen *Vibrio*, which may increase the risk of pathogenic infection in marine organisms.

However, these results were mainly obtained from laboratory experiments and await further verification in the natural environment. In this aspect, the effects of OA on the pathogen-host interaction and the subsequent influence on the immunity of marine invertebrates should be considered for future investigation.

Although the physiological impacts of OA on marine invertebrates have been studied for almost two decades already and the negative effects on various aspects such as gametic traits, fertilization success, embryonic development, biomineralization, metabolism, growth, and immune responses have been demonstrated in many species, the underlying mechanism of the OA effects is still largely unclear. Furthermore, the physiological responses of marine invertebrates after OA exposure are varied between species; however, only a few organisms among the great number of marine invertebrate species have been investigated. Therefore more experiments addressing these aspects should be conducted in the future. In addition, the carry-over effects of OA resulting from exposures in the early developmental stage and the combined effects of OA and other pollutants should also be taken into account.

References

Albright, R., & Langdon, C. (2011). Ocean acidification impacts multiple early life history processes of the Caribbean coral *Porites astreoides*. *Global Change Biology*, *17*(7), 2478−2487. Available from https://doi.org/10.1111/j.1365-2486.2011.02404.x.

Albright, R., Mason, B., Miller, M., & Langdon, C. (2010). Ocean acidification compromises recruitment success of the threatened Caribbean coral *Acropora palmata*. *Proceedings of the National Academy of Sciences of the USA*, *107*, 20400−20404.

Albright, R., Bland, C., Gillette, P., Serafy, J. E., Langdon, C., & Capo, T. R. (2012). Juvenile growth of the tropical sea urchin *Lytechinus variegatus* exposed to near-future ocean acidification scenarios. *Journal of Experimental Marine Biology and Ecology*, *426-427*, 12−17.

Allen, J. D. (2008). Size-specific predation on marine invertebrate larvae. *The Biological Bulletin*, *214*(1), 42−49. Available from https://doi.org/10.2307/25066658.

Allgaier, M., Riebesell, U., Vogt, M., Thyrhaug, R., & Grossart, H. P. (2008). Coupling of heterotrophic bacteria to phytoplankton bloom development at different pCO_2 levels: A mesocosm study. *Biogeosciences*, *5*(4), 1007−1022. Available from https://doi.org/10.5194/bg-5-1007-2008.

Andersson, A. J., & Gledhill, D. (2013). Ocean acidification and coral reefs: Effects on breakdown, dissolution, and net ecosystem calcification. *Annual Review of Marine Science*, *5*, 321−348. Available from https://doi.org/10.1146/annurev-marine-121211-172241.

Anger, K. (1987). The D0 threshold: A critical point in the larval development of decapod crustaceans. *Journal of Experimental Marine Biology and Ecology*, *108*(1), 15−30. Available from https://doi.org/10.1016/0022-0981(87)90128-6.

Appelhans, Y. S., Thomsen, J., Opitz, S., Pansch, C., Melzner, F., & Wahl, M. (2014). Juvenile sea stars exposed to acidification decrease feeding and growth with no acclimation potential. *Marine Ecology Progress Series*, *509*, 227−239. Available from https://doi.org/10.3354/meps10884.

Arnold, K. E., Findlay, H. S., Spicer, J. I., Daniels, C. L., & Boothroyd, D. (2009). Effect of CO_2-related acidification on aspects of the larval development of the European lobster, *Homarus gammarus* (L.). *Biogeosciences*, *6*(8), 1747−1754. Available from https://doi.org/10.5194/bg-6-1747-2009.

Asplund, M. E., Baden, S. P., Russ, S., Ellis, R. P., Gong, N., & Hernroth, B. E. (2014). Ocean acidification and host-pathogen interactions: Blue mussels, *Mytilus edulis*, encountering *Vibrio tubiashii*. *Environmental Microbiology*, *16*(4), 1029−1039. Available from https://doi.org/10.1111/1462-2920.12307.

Baird, A., Maynard, J. A., Hoegh-Guldberg, O., Mumby, P. J., Hooten, A. J., Steneck, R. S., Greenfield, P., Gomez, E., Harvell, D. R., Sale, P. F., Edwards, A. J., Caldeira, K., Knowlton, N., Eakin, C. M., Iglesias-Prieto, R., Muthiga, N., Bradbury, R. H., Dubi, A., & Hatziolos, M. E. (2008). Coral adaptation in the face of climate change [with response]. *Science*, *320*(5874), 315−316. Available from http://www.jstor.org/stable/20055014.

Barros, P., Sobral, P., Range, P., Chicharo, L., & Matias, D. (2013). Effects of sea-water acidification on fertilization and larval development of the oyster *Crassostrea gigas*. *Journal of Experimental Marine Biology and Ecology*, *440*, 200−206.

Bechmann, R. K., Taban, I. C., Westerlund, S., Godal, B. F., Arnberg, M., Vingen, S., Ingvarsdottir, A., & Baussant, T. (2011). Effects of ocean acidification on early life stages of shrimp (*Pandalus borealis*) and mussel (*Mytilus edulis*). *Journal of Toxicology and Environmental Health - Part A: Current Issues*, *74*(7−9), 424−438. Available from https://doi.org/10.1080/15287394.2011.550460.

Beniash, E., Ivanina, A., Lieb, N. S., Kurochkin, I., & Sokolova, I. M. (2010). Elevated level of carbon dioxide affects metabolism and shell formation in oysters *Crassostrea virginica*. *Marine Ecology Progress Series*, *419*, 95−108. Available from https://doi.org/10.3354/meps08841.

Berge, J. A., Bjerkeng, B., Pettersen, O., Schaanning, M. T., & Øxnevad, S. (2006). Effects of increased sea water concentrations of CO_2 on growth of the bivalve *Mytilus edulis* L. *Chemosphere*, *62*(4), 681−687. Available from https://doi.org/10.1016/j.chemosphere.2005.04.111.

Bibby, R., Widdicombe, S., Parry, H., Spicer, J., & Pipe, R. (2008). Effects of ocean acidification on the immune response of the blue mussel *Mytilus edulis*. *Aquatic Biology*, *2*(1), 67−74.

Booth, B. B. B., Dunstone, N. J., Halloran, P. R., Andrews, T., & Bellouin, N. (2012). Aerosols implicated as a prime driver of twentieth-century North Atlantic climate variability. *Nature*, *484*(7393), 228−232. Available from https://doi.org/10.1038/nature10946.

Brennand, H. S., Soars, N., Dworjanyn, S. A., Davis, A. R., & Byrne, M. (2010). Impact of ocean warming and ocean acidification on larval development and calcification in the sea urchin *Tripneustes gratilla. PloS one, 5*e11372.

Bressan, M., Chinellato, A., Munari, M., Matozzo, V., Manci, A., Marčeta, T., Finos, L., Moro, I., Pastore, P., & Badocco, D. (2014). Does seawater acidification affect survival, growth and shell integrity in bivalve juveniles? *Marine Environmental Research, 99*, 136−148.

Brothers, C. J., Harianto, J., McClintock, J. B., & Byrne, M. (2016). Sea urchins in a high-CO_2 world: The influence of acclimation on the immune response to ocean warming and acidification. *Proceedings of the Royal Society B: Biological Sciences, 283* (1837). Available from https://doi.org/10.1098/rspb.2016.1501.

Buggé, D. M., Hégaret, H., Wikfors, G. H., & Allam, B. (2007). Oxidative burst in hard clam (*Mercenaria mercenaria*) haemocytes. *Fish and Shellfish Immunology, 23*(1), 188−196. Available from https://doi.org/10.1016/j.fsi.2006.10.006.

Byrne, M., Soars, N., Selvakumaraswamy, P., Dworjanyn, S. A., & Davis, A. R. (2010). Sea urchin fertilization in a warm, acidified and high pCO_2 ocean across a range of sperm densities. *Marine Environmental Research, 69*(4), 234−239. Available from https://doi.org/10.1016/j.marenvres.2009.10.014.

Caldeira, K., & Wickett, M. E. (2003). Anthropogenic carbon and ocean pH. *Nature, 425* (6956), 365.

Campanati, C., Yip, S., Lane, A., & Thiyagarajan, V. (2016). Combined effects of low pH and low oxygen on the early-life stages of the barnacle *Balanus amphitrite. ICES Journal of Marine Science, 73*, 791−802.

Cao, R., Wang, Q., Yang, D., Liu, Y., Ran, W., Qu, Y., Wu, H., Cong, M., Li, F., & Ji, C. (2018). CO_2-induced ocean acidification impairs the immune function of the Pacific oyster against Vibrio splendidus challenge: An integrated study from a cellular and proteomic perspective. *Science of The Total Environment, 625*, 1574−1583.

Carter, H. A., Ceballos-Osuna, L., Miller, N. A., & Stillman, J. H. (2013). Impact of ocean acidification on metabolism and energetics during early life stages of the intertidal porcelain crab petrolisthes cinctipes. *Journal of Experimental Biology, 216*(8), 1412−1422. Available from https://doi.org/10.1242/jeb.078162.

Chan, V. B. S., Li, C., Lane, A. C., Wang, Y., Lu, X., Shih, K., Zhang, T., & Thiyagarajan, V. (2012). CO_2-driven ocean acidification alters and weakens integrity of the calcareous tubes produced by the serpulid Tubeworm, Hydroides elegans. *PLoS One, 7*(8). Available from https://doi.org/10.1371/journal.pone.0042718.

Cheeseman, L. P., Boulanger, J., Bond, L. M., & Schuh, M. (2016). Two pathways regulate cortical granule translocation to prevent polyspermy in mouse oocytes. *Nature Communications, 7*(1), 1−13.

Coronado, I., Fine, M., Bosellini, F. R., & Stolarski, J. (2019). Impact of ocean acidification on crystallographic vital effect of the coral skeleton. *Nature Communications, 10*(1), 1−9.

Cripps, G., Lindeque, P., & Flynn, K. (2014). Parental exposure to elevated pCO_2 influences the reproductive success of copepods. *Journal of Plankton Research, 36*, 1165−1174.

Desrosiers, R. R., Désilets, J., & Dubé, F. (1996). Early developmental events following fertilization in the giant scallop Placopecten magellanicus. *Canadian Journal of Fisheries and Aquatic Sciences, 53*(6), 1382−1392. Available from https://doi.org/10.1139/f96-071.

Doney, S. C., Fabry, V. J., Feely, R. A., & Kleypas, J. A. (2009). Ocean acidification: The other CO_2 problem. *Annual Review of Marine Science, 1*, 169−192. Available from https://doi.org/10.1146/annurev.marine.010908.163834.

Dong, Y., Yao, H., Lin, Z., & Zhu, D. (2012). The effects of sperm-egg ratios on polyspermy in the blood clam, *Tegillarca granosa*. *Aquaculture Research*, *43*(1), 44−52. Available from https://doi.org/10.1111/j.1365-2109.2011.02799.x.

Doropoulos, C., Ward, S., Diaz-Pulido, G., Hoegh-Guldberg, O., & Mumby, P. J. (2012). Ocean acidification reduces coral recruitment by disrupting intimate larval-algal settlement interactions. *Ecology letters*, *15*, 338−346.

Dupont, S., Havenhand, J., Thorndyke, W., Peck, L., & Thorndyke, M. (2008). Near-future level of CO_2-driven ocean acidification radically affects larval survival and development in the brittlestar *Ophiothrix fragilis*. *Marine Ecology Progress Series*, *373*, 285−294. Available from https://doi.org/10.3354/meps07800.

Dupont, S., & Thorndyke, M. (2012). Relationship between CO_2-driven changes in extracellular acid-base balance and cellular immune response in two polar echinoderm species. *Journal of Experimental Marine Biology and Ecology*, *424−425*, 32−37. Available from https://doi.org/10.1016/j.jembe.2012.05.007.

Duquette, A., McClintock, J. B., Amsler, C. D., Pérez-Huerta, A., Milazzo, M., & Hall-Spencer, J. M. (2017). Effects of ocean acidification on the shells of four mediterranean gastropod species near a CO_2 seep. *Marine Pollution Bulletin*, *124*(2), 917−928. Available from https://doi.org/10.1016/j.marpolbul.2017.08.007.

Ekstrom, J. A., Suatoni, L., Cooley, S. R., Pendleton, L. H., Waldbusser, G. G., Cinner, J. E., Ritter, J., Langdon, C., Van Hooidonk, R., & Gledhill, D. (2015). Vulnerability and adaptation of United States shellfisheries to ocean acidification. *Nature Climate Change*, *5*(3), 207−214.

Ellis, R. P., Urbina, M. A., & Wilson, R. W. (2017). Lessons from two high CO_2 worlds−future oceans and intensive aquaculture. *Global Change Biology*, *23*(6), 2141−2148. Available from https://doi.org/10.1111/gcb.13515.

Evans, T. G., & Watson-Wynn, P. (2014). Effects of seawater acidification on gene expression: Resolving broader-scale trends in sea urchins. *Biological Bulletin*, *226*(3), 237−254. Available from https://doi.org/10.1086/BBLv226n3p237.

Fabry, V. J., Seibel, B. A., Feely, R. A., & Orr, J. C. (2008). Impacts of ocean acidification on marine fauna and ecosystem processes. *ICES Journal of Marine Science*, *65*(3), 414−432. Available from https://doi.org/10.1093/icesjms/fsn048.

Fan, H., Kashi, R. S., & Middaugh, C. R. (2006). Conformational lability of two molecular chaperones Hsc70 and gp96: Effects of pH and temperature. *Archives of Biochemistry and Biophysics*, *447*(1), 34−45. Available from https://doi.org/10.1016/j.abb.2006.01.012.

Farley, G. S., & Levitan, D. R. (2001). The role of jelly coats in sperm-egg encounters, fertilization success, and selection on egg size in broadcast spawners. *American Naturalist*, *157*(6), 626−636. Available from https://doi.org/10.1086/320619.

Feely, R. A., Sabine, C. L., Lee, K., Berelson, W., Kleypas, J., Fabry, V. J., & Millero, F. J. (2004). Impact of anthropogenic CO_2 on the $CaCO_3$ system in the oceans. *Science*, *305*(5682), 362−366. Available from https://doi.org/10.1126/science.1097329.

Fine, M., & Tchernov, D. (2007). Scleractinian coral species survive and recover from decalcification. *Science*, *315*(5820), 1811.

Fitzer, S. C., Chung, P., Maccherozzi, F., Dhesi, S. S., Kamenos, N. A., Phoenix, V. R., & Cusack, M. (2016). Biomineral shell formation under ocean acidification: A shift from order to chaos. *Scientific Reports*, *6*. Available from https://doi.org/10.1038/srep21076.

Fitzer, S. C., Phoenix, V. R., Cusack, M., & Kamenos, N. A. (2014). Ocean acidification impacts mussel control on biomineralisation. *Scientific Reports*, *4*. Available from https://doi.org/10.1038/srep06218.

Frieder, C. A., Gonzalez, J. P., Bockmon, E. E., Navarro, M. O., & Levin, L. A. (2014). Can variable pH and low oxygen moderate ocean acidification outcomes for mussel larvae? *Global Change Biology*, *20*(3), 754−764. Available from https://doi.org/10.1111/gcb.12485.

Gallo, A., Boni, R., Buia, M. C., Monfrecola, V., Esposito, M. C., & Tosti, E. (2019). Ocean acidification impact on ascidian Ciona robusta spermatozoa: New evidence for stress resilience. *Science of the Total Environment*, *697*. Available from https://doi.org/10.1016/j.scitotenv.2019.134100.

Garilli, V., Rodolfo-Metalpa, R., Scuderi, D., Brusca, L., Parrinello, D., Rastrick, S. P. S., Foggo, A., Twitchett, R. J., Hall-Spencer, J. M., & Milazzo, M. (2015). Physiological advantages of dwarfing in surviving extinctions in high-CO_2 oceans. *Nature Climate Change*, *5*(7), 678−682. Available from https://doi.org/10.1038/nclimate2616.

Gazeau, F., Gattuso, J. P., Dawber, C., Pronker, A. E., Peene, F., Peene, J., Heip, C. H. R., & Middelburg, J. J. (2010). Effect of ocean acidification on the early life stages of the blue mussel *Mytilus edulis*. *Biogeosciences*, *7*(7), 2051−2060. Available from https://doi.org/10.5194/bg-7-2051-2010.

Gibson, R., Atkinson, R., Gordon, J., Smith, I., & Hughes, D. (2011). Impact of ocean warming and ocean acidification on marine invertebrate life history stages: Vulnerabilities and potential for persistence in a changing ocean. *Oceanography and Marine Biology: An Annual Review*, *49*, 1−42.

Ginger, K. W. K., Vera, C. B. S. R. , D., Dennis, C. K. S., Adela, L. J., Yu, Z., & Thiyagarajan, V. (2013). Larval and post-larval stages of Pacific oyster (*Crassostrea gigas*) are resistant to elevated CO_2. *PloS One*, *8*(5), e64147. Available from https://doi.org/10.1371/journal.pone.0064147.

Guo, X., Huang, M., Pu, F., You, W., & Ke, C. (2015). Effects of ocean acidification caused by rising CO_2 on the early development of three mollusks. *Aquatic Biology*, *23*(2), 147−157. Available from https://doi.org/10.3354/ab00615.

Guppy, M., & Withers, P. (1999). Metabolic depression in animals: Physiological perspectives and biochemical generalizations. *Biological Reviews*, *74*(1), 1−40. Available from https://doi.org/10.1111/j.1469-185X.1999.tb00180.x.

Hammond, L. M., & Hofmann, G. E. (2012). Early developmental gene regulation in *Strongylocentrotus purpuratus* embryos in response to elevated CO_2 seawater conditions. *Journal of Experimental Biology*, *215*(14), 2445−2454.

Han, Y., Shi, W., Rong, J., Zha, S., Guan, X., Sun, H., & Liu, G. (2019). Exposure to Waterborne $nTiO_2$ reduces fertilization success and increases polyspermy in a bivalve mollusc: A threat to population recruitment. *Environmental Science and Technology*, *53*(21), 12754−12763. Available from https://doi.org/10.1021/acs.est.9b03675.

Han, Y., Shi, W., Tang, Y., Zhao, X., Du, X., Sun, S., . . . Liu, G. (2021). Ocean acidification increases polyspermy of a broadcast spawning bivalve species by hampering membrane depolarization and cortical granule exocytosis. *Aquatic Toxicology*, *231*, 105740.

Harvey, B. P., Agostini, S., Wada, S., Inaba, K., & Hall-Spencer, J. M. (2018). Dissolution: The achilles' heel of the triton shell in an acidifying ocean. *Frontiers in Marine Science*, *5*, 371.

Havenhand, J. N., Buttler, F. R., Thorndyke, M. C., & Williamson, J. E. (2008). Near-future levels of ocean acidification reduce fertilization success in a sea urchin. *Current Biology*, *18*(15), R651−R652. Available from https://doi.org/10.1016/j.cub.2008.06.015.

Havenhand, J., & Schlegel, P. (2009). Near-future levels of ocean acidification do not affect sperm motility and fertilization kinetics in the oyster *Crassostrea gigas*. *Biogeosciences*, *6*, 12.

Hernroth, B., Baden, S., Thorndyke, M., & Dupont, S. (2011). Immune suppression of the echinoderm *Asterias rubens* (L.) following long-term ocean acidification. *Aquatic Toxicology*, *103*(3−4), 222−224. Available from https://doi.org/10.1016/j.aquatox.2011.03.001.

Hettinger, A., Sanford, E., Hill, T. M., Russell, A. D., Sato, K. N. S., Hoey, J., Forsch, M., Page, H. N., & Gaylord, B. (2012). Persistent carry-over effects of planktonic exposure to ocean acidification in the Olympia oyster. *Ecology*, *93*(12), 2758−2768. Available from https://doi.org/10.1890/12-0567.1.

Holcomb, M., Venn, A. A., Tambutté, E., Tambutté, S., Allemand, D., Trotter, J., & McCulloch, M. (2014). Coral calcifying fluid pH dictates response to ocean acidification. *Scientific Reports*, *4*. Available from https://doi.org/10.1038/srep05207.

Holmskov, U., Thiel, S., & Jensenius, J. C. (2003). Collectins and ficolins: Humoral lectins of the innate immune defense. *Annual Review of Immunology*, *21*, 547−578. Available from https://doi.org/10.1146/annurev.immunol.21.120601.140954.

Hüning, A. K., Melzner, F., Thomsen, J., Gutowska, M. A., Krämer, L., Frickenhaus, S., Rosenstiel, P., Pörtner, H. O., Philipp, E. E. R., & Lucassen, M. (2013). Impacts of seawater acidification on mantle gene expression patterns of the Baltic Sea blue mussel: Implications for shell formation and energy metabolism. *Marine Biology*, *160*(8), 1845−1861. Available from https://doi.org/10.1007/s00227-012-1930-9.

Hu, M., Li, L., Sui, Y., Li, J., Wang, Y., Lu, W., & Dupont, S. (2015). Effect of pH and temperature on antioxidant responses of the thick shell mussel *Mytilus coruscus*. *Fish and Shellfish Immunology*, *46*(2), 573−583. Available from https://doi.org/10.1016/j.fsi.2015.07.025.

Iguchi, A., Suzuki, A., Sakai, K., & Nojiri, Y. (2015). Comparison of the effects of thermal stress and CO_2-driven acidified seawater on fertilization in coral *Acropora digitifera*. *Zygote*, *23*, 631−634.

Johnson, C. H., & Epel, D. (1981). Intracellular pH of sea urchin eggs measured by the dimethyloxazolidinedione (DMO) method. *Journal of Cell Biology*, *89*(2), 284−291. Available from https://doi.org/10.1083/jcb.89.2.284.

Kaplan, M. B., Mooney, T. A., McCorkle, D. C., & Cohen, A. L. (2013). Adverse effects of ocean acidification on early development of squid (*Doryteuthis pealeii*). *PLoS ONE*, *8*(5). Available from https://doi.org/10.1371/journal.pone.0063714.

Kapsenberg, L., Okamoto, D. K., Dutton, J. M., & Hofmann, G. E. (2017). Sensitivity of sea urchin fertilization to pH varies across a natural pH mosaic. *Evolutionary Ecology*, *7*, 1737−1750.

Kasai, T., Ogawa, K., Mizuno, K., Nagai, S., Uchida, Y., Ohta, S., Fujie, M., Suzuki, K., Hirata, S., & Hoshi, K. (2002). Relationship between sperm mitochondrial membrane potential, sperm motility, and fertility potential. *Asian Journal of Andrology*, *4*(2), 97−103.

Keppel, E., Scrosati, R., & Courtenay, S. (2012). Ocean acidification decreases growth and development in American lobster (*Homarus americanus*) larvae. *Journal of Northwest Atlantic Fishery Science*, *44*, 61−66. Available from https://doi.org/10.2960/J.v44.m683.

Kong, H., Wu, F., Jiang, X., Wang, T., Hu, M., Chen, J., Huang, W., Bao, Y., & Wang, Y. (2019). Nano-TiO_2 impairs digestive enzyme activities of marine mussels under ocean acidification. *Chemosphere*, *237*. Available from https://doi.org/10.1016/j.chemosphere.2019.124561.

Kupriyanova, E. k, & Havenhand, J. n (2005). Effects of temperature on sperm swimming behaviour, respiration and fertilization success in the serpulid polychaete, galeolaria caespitosa (annelida: Serpulidae). *Invertebrate Reproduction and Development*, *48* (1−3), 7−17. Available from https://doi.org/10.1080/07924259.2005.9652166.

Kurihara, H. (2008). Effects of CO_2-driven ocean acidification on the early developmental stages of invertebrates. *Marine Ecology Progress Series*, *373*, 275−284. Available from https://doi.org/10.3354/meps07802.

Kurihara, H., Asai, T., Kato, S., & Ishimatsu, A. (2008). Effects of elevated pCO_2 on early development in the mussel *Mytilus galloprovincialis*. *Aquatic Biology*, *4*(3), 225−233. Available from https://doi.org/10.3354/ab00109.

Kurihara, H., Kato, S., & Ishimatsu, A. (2007). Effects of increased seawater pCO_2 on early development of the oyster *Crassostrea gigas*. *Aquatic Biology, 1*(1), 91−98. Available from https://doi.org/10.3354/ab00009.

Kurihara, H., Matsui, M., Furukawa, H., Hayashi, M., & Ishimatsu, A. (2008). Long-term effects of predicted future seawater CO_2 conditions on the survival and growth of the marine shrimp *Palaemon pacificus*. *Journal of Experimental Marine Biology and Ecology, 367* (1), 41−46. Available from https://doi.org/10.1016/j.jembe.2008.08.016.

Kurihara, H., Shimode, S., & Shirayama, Y. (2004). Sub-lethal effects of elevated concentration of CO_2 on planktonic copepods and sea urchins. *Journal of Oceanography, 60*(4), 743−750. Available from https://doi.org/10.1007/s10872-004-5766-x.

Kurihara, H., & Shirayama, Y. (2004). Effects of increased atmospheric CO_2 on sea urchin early development. *Marine Ecology Progress Series, 274,* 161−169. Available from https://doi.org/10.3354/meps274161.

Kurihara, H., Takano, Y., Kurokawa, D., & Akasaka, K. (2012). Ocean acidification reduces biomineralization-related gene expression in the sea urchin, *Hemicentrotus pulcherrimus*. *Marine Biology, 159*(12), 2819−2826. Available from https://doi.org/ 10.1007/s00227-012-2043-1.

Langenbuch, M., Bock, C., Leibfritz, D., & Pörtner, H. O. (2006). Effects of environmental hypercapnia on animal physiology: A 13C NMR study of protein synthesis rates in the marine invertebrate *Sipunculus nudus*. *Comparative Biochemistry and Physiology - A Molecular and Integrative Physiology, 144*(4), 479−484. Available from https://doi.org/ 10.1016/j.cbpa.2006.04.017.

Lannig, G., Eilers, S., Pörtner, H. O., Sokolova, I. M., & Bock, C. (2010). Impact of ocean acidification on energy metabolism of oyster, *Crassostrea gigas* - Changes in metabolic pathways and thermal response. *Marine Drugs, 8*(8), 2318−2339. Available from https://doi.org/10.3390/md8082318.

Lecchini, D., Dixson, D. L., Lecellier, G., Roux, N., Frédérich, B., Besson, M., Tanaka, Y., Banaigs, B., & Nakamura, Y. (2017). Habitat selection by marine larvae in changing chemical environments. *Marine Pollution Bulletin, 114*(1), 210−217.

Leite Figueiredo, D. A., Branco, P. C., dos Santos, D. A., Emerenciano, A. K., Iunes, R. S., Shimada Borges, J. C., & Machado Cunha da Silva, J. R. (2016). Ocean acidification affects parameters of immune response and extracellular pH in tropical sea urchins *Lytechinus variegatus* and *Echinometra luccunter*. *Aquatic Toxicology, 180,* 84−94. Available from https://doi.org/10.1016/j.aquatox.2016.09.010.

Lenz, B., Fogarty, N. D., & Figueiredo, J. (2019). Effects of ocean warming and acidification on fertilization success and early larval development in the green sea urchin *Lytechinus variegatus*. *Marine Pollution Bulletin, 141,* 70−78. Available from https://doi. org/10.1016/j.marpolbul.2019.02.018.

Levitan, D. R. (1991). Influence of body size and population density on fertilization success and reproductive output in a free-spawning invertebrate. *Biological Bulletin, 181* (2), 261−268. Available from https://doi.org/10.2307/1542097.

Levitan, D. R. (2000). Sperm velocity and longevity trade off each other and influence fertilization in the sea urchin *Lytechinus variegatus*. *Proceedings of the Royal Society B: Biological Sciences, 267*(1443), 531−534. Available from https://doi.org/10.1098/rspb.2000.1032.

Lidbury, I., Johnson, V., Hall-Spencer, J. M., Munn, C. B., & Cunliffe, M. (2012). Community-level response of coastal microbial biofilms to ocean acidification in a natural carbon dioxide vent ecosystem. *Marine Pollution Bulletin, 64*(5), 1063−1066. Available from https://doi.org/10.1016/j.marpolbul.2012.02.011.

Liu, S., Shi, W., Guo, C., Zhao, X., Han, Y., Peng, C., Chai, X., & Liu, G. (2016). Ocean acidification weakens the immune response of blood clam through hampering the NF-kappa β and toll-like receptor pathways. *Fish & Shellfish Immunology, 54,* 322−327.

Li, F., & Xiang, J. (2013). Signaling pathways regulating innate immune responses in shrimp. *Fish & Shellfish Immunology, 34*(4), 973−980.

Loker, E. S., Adema, C. M., Zhang, S. M., & Kepler, T. B. (2004). Invertebrate immune systems—Not homogeneous, not simple, not well understood. *Immunological Reviews, 198*, 10−24. Available from https://doi.org/10.1111/j.0105-2896.2004.0117.x.

Long, W. C., Swiney, K. M., Harris, C., Page, H. N., & Foy, R. J. (2013). Effects of ocean acidification on juvenile red king crab (*Paralithodes camtschaticus*) and tanner crab (*Chionoecetes bairdi*) growth, condition, alcification, and survival. *PLoS ONE, 8*(4). Available from https://doi.org/10.1371/journal.pone.0060959.

Lotterhos, K. E., & Levitan, D. R. (2010). Gamete release and spawning behavior in broadcast spawning marine invertebrates. *The Evolution of Primary Sexual Characters in Animals, 99*, 120.

Lundvall, D., Svanbäck, R., Persson, L., & Byström, P. (1999). Size-dependent predation in piscivores: Interactions between predator foraging and prey avoidance abilities. *Canadian. Journal of Fisheries Aquatic Science, 56*(7), 1285−1292. Available from https://doi.org/10.1139/f99-058.

Mabardy, R. A., Waldbusser, G. G., Conway, F., & Olsen, C. S. (2015). Perception and response of the United States west coast shellfish industry to ocean acidification: The voice of the canaries in the coal mine. *Journal of Shellfish Research, 34*(2), 565−572. Available from https://doi.org/10.2983/035.034.0241.

Marinelli, R. L., & Williams, T. J. (2003). Evidence for density-dependent effects of infauna on sediment biogeochemistry and benthic-pelagic coupling in nearshore systems. *Estuarine, Coastal and Shelf Science, 57*(1−2), 179−192. Available from https://doi.org/10.1016/S0272-7714(02)00342-6.

Matozzo, V., Chinellato, A., Munari, M., Finos, L., Bressan, M., & Marin, M. G. (2012). First evidence of immunomodulation in bivalves under seawater acidification and increased temperature. *PLoS One, 7*.

Mayer, M. P., & Bukau, B. (2005). Hsp70 chaperones: Cellular functions and molecular mechanism. *Cellular and Molecular Life Sciences, 62*(6), 670. Available from https://doi.org/10.1007/s00018-004-4464-6.

Mayor, D. J., Matthews, C., Cook, K., Zuur, A. F., & Hay, S. (2007). CO2-induced acidification affects hatching success in Calanus finmarchicus. *Marine Ecology Progress Series, 350*, 91−97. Available from https://doi.org/10.3354/meps07142.

McDonald, M. R., McClintock, J. B., Amsler, C. D., Rittschof, D., Angus, R. A., Orihuela, B., & Lutostanski, K. (2009). Effects of ocean acidification over the life history of the barnacle *Amphibalanus amphitrite*. *Marine Ecology Progress Series, 385*, 179−187. Available from https://doi.org/10.3354/meps08099.

Melzner, F., Stange, P., Trübenbach, K., Thomsen, J., Casties, I., Panknin, U., Gorb, S. N., & Gutowska, M. A. (2011). Food supply and seawater pCO 2 impact calcification and internal shell dissolution in the blue mussel *Mytilus edulis*. *PLoS ONE, 6*(9). Available from https://doi.org/10.1371/journal.pone.0024223.

Meseck, S. L., Alix, J. H., Swiney, K. M., Long, W. C., Wikfors, G. H., & Foy, R. J. (2016). Ocean acidification affects hemocyte physiology in the Tanner Crab (*Chionoecetes bairdi*). *PloS one*, e0148477.

Michaelidis, B., Ouzounis, C., Paleras, A., & Pörtner, H. O. (2005). Effects of long-term moderate hypercapnia on acid-base balance and growth rate in marine mussels *Mytilus galloprovincialis*. *Marine Ecology Progress Series, 293*, 109−118. Available from https://doi.org/10.3354/meps293109.

Migliaccio, O., Pinsino, A., Maffioli, E., Smith, A. M., Agnisola, C., Matranga, V., Nonnis, S., Tedeschi, G., Byrne, M., & Gambi, M. C. (2019). Living in future ocean acidification, physiological adaptive responses of the immune system of sea urchins resident at a CO2 vent system. *Science of The Total Environment, 672*, 938−950.

Mollica, N. R., Guo, W., Cohen, A. L., Huang, K. F., Foster, G. L., Donald, H. K., & Solow, A. R. (2018). Ocean acidification affects coral growth by reducing skeletal density. *Proceedings of the National Academy of Sciences of the United States of America, 115* (8), 1754−1759. Available from https://doi.org/10.1073/pnas.1712806115.

Morita, M., Suwa, R., Iguchi, A., Nakamura, M., Shimada, K., Sakai, K., & Suzuki, A. (2010). Ocean acidification reduces sperm flagellar motility in broadcast spawning reef invertebrates. *Zygote, 18*(2), 103−107. Available from https://doi.org/10.1017/S0967199409990177.

Mucci, A. (1983). The solubility of calcite and aragonite in seawater at various salinities, temperatures, and one atmosphere total pressure. *American Journal of Science, 283*(7), 780−799. Available from https://doi.org/10.2475/ajs.283.7.780.

Nakamura, M., Ohki, S., Suzuki, A., & Sakai, K. (2011). Coral larvae under ocean acidification: Survival, metabolism, and metamorphosis. *PLoS One, 6*(1). Available from https://doi.org/10.1371/journal.pone.0014521.

Nakamura, M., & Morita, M. (2012). Sperm motility of the scleractinian coral Acropora digitifera under preindustrial, current, and predicted ocean acidification regimes. *Aquatic Biology, 15*, 299−302.

Nunes, P., & Demaurex, N. (2010). The role of calcium signaling in phagocytosis. *Journal of Leukocyte Biology, 88*(1), 57−68. Available from https://doi.org/10.1189/jlb.0110028.

Onitsuka, T., Takami, H., Muraoka, D., Matsumoto, Y., Nakatsubo, A., Kimura, R., Ono, T., & Nojiri, Y. (2018). Effects of ocean acidification with pCO$_2$ diurnal fluctuations on survival and larval shell formation of Ezo abalone, *Haliotis discus hannai*. *Marine Environmental Research, 134*, 28−36. Available from https://doi.org/10.1016/j.marenvres.2017.12.015.

Orr, J. C., Fabry, V. J., Aumont, O., Bopp, L., Doney, S. C., Feely, R. A., Gnanadesikan, A., Gruber, N., Ishida, A., & Joos, F. (2005). Anthropogenic ocean acidification over the twenty-first century and its impact on calcifying organisms. *Nature, 437*(7059), 681−686.

O'Donnell, M. J., Todgham, A. E., Sewell, M. A., Hammond, L. M., Ruggiero, K., Fangue, N. A., Zippay, M. L., & Hofmann, G. E. (2010). Ocean acidification alters skeletogenesis and gene expression in larval sea urchins. *Marine Ecology Progress Series, 398*, 157−171.

Pachauri, R.K., Allen, M.R., Barros, V.R., Broome, J., Cramer, W., Christ, R., Church, J.A., Clarke, L., Dahe, Q., & Dasgupta, P. (2014). *Climate change 2014: Synthesis report. Contribution of working groups I, II and III to the fifth assessment report of the Intergovernmental Panel on Climate Change*. IPCC.

Pan, T. C. F., Applebaum, S. L., & Manahan, D. T. (2015). Experimental ocean acidification alters the allocation of metabolic energy. *Proceedings of the National Academy of Sciences of the United States of America, 112*(15), 4696−4701. Available from https://doi.org/10.1073/pnas.1416967112.

Parker, L. M., Ross, P. M., & O'Connor, W. A. (2009). The effect of ocean acidification and temperature on the fertilization and embryonic development of the Sydney rock oyster *Saccostrea glomerata* (Gould 1850). *Global Change Biology, 15*(9), 2123−2136. Available from https://doi.org/10.1111/j.1365-2486.2009.01895.x.

Parker, L. M., Ross, P. M., O'Connor, W. A., Pörtner, H. O., Scanes, E., & Wright, J. M. (2013). Predicting the response of molluscs to the impact of ocean acidification. *Biology, 2*(2), 651−692. Available from https://doi.org/10.3390/biology2020651.

Peng, C., Zhao, X., Liu, S., Shi, W., Han, Y., Guo, C., Peng, X., Chai, X., & Liu, G. (2017). Ocean acidification alters the burrowing behaviour, Ca^{2+}/Mg^{2+}-ATPase activity, metabolism, and gene expression of a bivalve species, *Sinonovacula constricta*. *Marine Ecology Progress Series, 575*, 107−117. Available from https://doi.org/10.3354/meps12224.

Pennington, J. T., & Strathmann, R. R. (1990). Consequences of the calcite skeletons of planktonic echinoderm larvae for orientation, swimming, and shape. *The Biological Bulletin*, *179*(1), 121−133.

Pörtner, H. O. (2008). Ecosystem effects of ocean acidification in times of ocean warming: A physiologist's view. *Marine Ecology Progress Series*, *373*, 203−217. Available from https://doi.org/10.3354/meps07768.

Le Quéré, C., Raupach, M. R., Canadell, J. G., Marland, G., Bopp, L., Ciais, P., Conway, T. J., Doney, S. C., Feely, R. A., & Foster, P. (2009). Trends in the sources and sinks of carbon dioxide. *Nature Geoscience*, *2*(12), 831−836.

Reuter, K. E., Lotterhos, K. E., Crim, R. N., Thompson, C. A., & Harley, C. D. G. (2011). Elevated pCO_2 increases sperm limitation and risk of polyspermy in the red sea urchin *Strongylocentrotus franciscanus*. *Global Change Biology*, *17*(1), 163−171. Available from https://doi.org/10.1111/j.1365-2486.2010.02216.x.

Ries, J. B., Cohen, A. L., & McCorkle, D. C. (2009). Marine calcifiers exhibit mixed responses to CO_2-induced ocean acidification. *Geology*, *37*, 1131−1134.

Rivest, E. B., & Hofmann, G. E. (2014). Responses of the metabolism of the larvae of *Pocillopora damicornis* to ocean acidification and warming. *PLoS ONE*, *9*(4). Available from https://doi.org/10.1371/journal.pone.0096172.

Roggatz, C. C., Lorch, M., Hardege, J. D., & Benoit, D. M. (2016). Ocean acidification affects marine chemical communication by changing structure and function of peptide signalling molecules. *Global Change Biology*, *22*(12), 3914−3926. Available from https://doi.org/10.1111/gcb.13354.

Rumrill, S. S. (1990). Natural mortality of marine invertebrate larvae. *Ophelia*, *32*(1−2), 163−198. Available from https://doi.org/10.1080/00785236.1990.10422030.

Sabine, C. L., Feely, R. A., Gruber, N., Key, R. M., Lee, K., Bullister, J. L., Wanninkhof, R., Wong, C., Wallace, D. W., & Tilbrook, B. (2004). The oceanic sink for anthropogenic CO_2. *Science*, *305*(5682), 367−371.

Sadler, D. E., Lemasson, A. J., & Knights, A. M. (2018). The effects of elevated CO_2 on shell properties and susceptibility to predation in mussels *Mytilus edulis*. *Marine Environmental Research*, *139*, 162−168. Available from https://doi.org/10.1016/j.marenvres.2018.05.017.

Scanes, E., Parker, L. M., O'Connor, W. A., & Ross, P. M. (2014). Mixed effects of elevated pCO_2 on fertilisation, larval and juvenile development and adult responses in the mobile subtidal scallop *Mimachlamys asperrima* (Lamarck, 1819). *PLoS One*, *9*(4). Available from https://doi.org/10.1371/journal.pone.0093649.

Schlegel, P., Binet, M. T., Havenhand, J. N., Doyle, C. J., & Williamson, J. E. (2015). Ocean acidification impacts on sperm mitochondrial membrane potential bring sperm swimming behaviour near its tipping point. *Journal of Experimental Biology*, *218*(7), 1084−1090. Available from https://doi.org/10.1242/jeb.114900.

Seth, R. B., Sun, L., & Chen, Z. J. (2006). Antiviral innate immunity pathways. *Cell Research*, *16*(2), 141−147. Available from https://doi.org/10.1038/sj.cr.7310019.

Sewell, M. A., Millar, R. B., Yu, P. C., Kapsenberg, L., & Hofmann, G. E. (2014). Ocean acidification and fertilization in the antarctic sea urchin *Sterechinus neumayeri*: The importance of polyspermy. *Environmental Science and Technology*, *48*(1), 713−722. Available from https://doi.org/10.1021/es402815s.

Shang, Y., Lan, Y., Liu, Z., Kong, H., Huang, X., Wu, F., Liu, L., Hu, M., Huang, W., & Wang, Y. (2018). Synergistic effects of nano-ZnO and low pH of sea water on the physiological energetics of the thick shell mussel *Mytilus coruscus*. *Frontiers in Physiology*, *9*. Available from https://doi.org/10.3389/fphys.2018.00757.

Sharma, B. R., Martin, M. M., & Shafer, J. A. (1984). Alkaline proteases from the gut fluids of detritus-feeding larvae of the crane fly, tipula abdominalis (say) (diptera, tipulidae). *Insect Biochemistry*, *14*(1), 37−44. Available from https://doi.org/10.1016/0020-1790(84)90081-7.

Shirayama, Y., & Thornton, H. (2005). Effect of increased atmospheric CO_2 on shallow water marine benthos. *Journal of Geophysical Research C: Oceans, 110*(9), 1−7. Available from https://doi.org/10.1029/2004JC002618.

Shi, W., Guan, X., Han, Y., Guo, C., Rong, J., Su, W., Zha, S., Wang, Y., & Liu, G. (2018). Waterborne Cd^{2+} weakens the immune responses of blood clam through impacting Ca^{2+} signaling and Ca^{2+} related apoptosis pathways. *Fish and Shellfish Immunology, 77*, 208−213. Available from https://doi.org/10.1016/j.fsi.2018.03.055.

Shi, W., Han, Y., Guo, C., Su, W., Zhao, X., Zha, S., Wang, Y., & Liu, G. (2019). Ocean acidification increases the accumulation of titanium dioxide nanoparticles ($nTiO_2$) in edible bivalve mollusks and poses a potential threat to seafood safety. *Scientific Reports, 9*(1). Available from https://doi.org/10.1038/s41598-019-40047-1.

Shi, W., Han, Y., Guo, C., Zhao, X., Liu, S., Su, W., Wang, Y., Zha, S., Chai, X., & Liu, G. (2017). Ocean acidification hampers sperm-egg collisions, gamete fusion, and generation of Ca^{2+} oscillations of a broadcast spawning bivalve, *Tegillarca granosa*. *Marine Environmental Research, 130*, 106−112. Available from https://doi.org/10.1016/j.marenvres.2017.07.016.

Shi, W., Zhao, X., Han, Y., Guo, C., Liu, S., Su, W., Wang, Y., Zha, S., Chai, X., & Fu, W. (2017). Effects of reduced pH and elevated pCO_2 on sperm motility and fertilisation success in blood clam, *Tegillarca granosa*. *New Zealand Journal of Marine and Freshwater Research, 51*(4), 543−554.

Stemmer, K., Nehrke, G., & Brey, T. (2013). Elevated CO_2 levels do not affect the shell structure of the bivalve Arctica islandica from the Western Baltic. *PLoS One, 8*(7), e70106.

Stumpp, M., Dupont, S., Thorndyke, M. C., & Melzner, F. (2011). CO_2 induced seawater acidification impacts sea urchin larval development II: Gene expression patterns in pluteus larvae. *Comparative Biochemistry and Physiology - A Molecular and Integrative Physiology, 160*(3), 320−330. Available from https://doi.org/10.1016/j.cbpa.2011.06.023.

Stumpp, M., Hu, M., Casties, I., Saborowski, R., Bleich, M., Melzner, F., & Dupont, S. (2013). Digestion in sea urchin larvae impaired under ocean acidification. *Nature Climate Change, 3*(12), 1044−1049. Available from https://doi.org/10.1038/nclimate2028.

Stumpp, M., Trübenbach, K., Brennecke, D., Hu, M. Y., & Melzner, F. (2012). Resource allocation and extracellular acid-base status in the sea urchin Strongylocentrotus droebachiensis in response to CO_2 induced seawater acidification. *Aquatic Toxicology, 110−111*, 194−207. Available from https://doi.org/10.1016/j.aquatox.2011.12.020.

Stumpp, M., Wren, J., Melzner, F., Thorndyke, M. C., & Dupont, S. T. (2011). CO_2 induced seawater acidification impacts sea urchin larval development I: Elevated metabolic rates decrease scope for growth and induce developmental delay. *Comparative Biochemistry and Physiology - A Molecular and Integrative Physiology, 160*(3), 331−340. Available from https://doi.org/10.1016/j.cbpa.2011.06.022.

Styan, C. A. (1998). Polyspermy, egg size, and the fertilization kinetics of free-spawning marine invertebrates. *American Naturalist, 152*(2), 290−297. Available from https://doi.org/10.1086/286168.

Su, W., Rong, J., Zha, S., Yan, M., Fang, J., & Liu, G. (2018). Ocean acidification affects the cytoskeleton, lysozymes, and nitric oxide of hemocytes: A possible explanation for the hampered phagocytosis in blood clams, *Tegillarca granosa*. *Frontiers in Physiology, 9*. Available from https://doi.org/10.3389/fphys.2018.00619.

Talmage, S. C., & Gobler, C. J. (2010). Effects of past, present, and future ocean carbon dioxide concentrations on the growth and survival of larval shellfish. *Proceedings of the National Academy of Sciences of the United States of America, 107*(40), 17246. Available from https://doi.org/10.1073/pnas.0913804107.

Taylor, J. R., Gilleard, J. M., Allen, M. C., & Deheyn, D. D. (2015). Effects of CO_2-induced pH reduction on the exoskeleton structure and biophotonic properties of the shrimp *Lysmata californica*. *Scientific Reports*, *5*, 10608.

Thomsen, J., Casties, I., Pansch, C., Körtzinger, A., & Melzner, F. (2013). Food availability outweighs ocean acidification effects in juvenile *Mytilus edulis*: Laboratory and field experiments. *Global Change Biology*, *19*(4), 1017−1027. Available from https://doi.org/10.1111/gcb.12109.

Thomsen, J. örn, Gutowska, M., Saphörster, J., Heinemann, A., Trübenbach, K., Fietzke, J., Hiebenthal, C., Eisenhauer, A., Körtzinger, A., & Wahl, M. (2010). Calcifying invertebrates succeed in a naturally CO_2 enriched coastal habitat but are threatened by high levels of future acidification. *Biogeosciences (BG)*, 7(11), 3879−3891.

Timmins-Schiffman, E., Coffey, W. D., Hua, W., Nunn, B. L., Dickinson, G. H., & Roberts, S. B. (2014). Shotgun proteomics reveals physiological response to ocean acidification in *Crassostrea gigas*. *BMC Genomics*, *15*(1), 951.

Todgham, A. E., & Hofmann, G. E. (2009). Transcriptomic response of sea urchin larvae *Strongylocentrotus purpuratus* to CO_2-driven seawater acidification. *Journal of Experimental Biology*, *212*(16), 2579−2594. Available from https://doi.org/10.1242/jeb.032540.

Tosti, E. (1994). Sperm activation in species with external fertilisation. *Zygote*, *2*(4), 359−361. Available from https://doi.org/10.1017/S0967199400002215.

Tosti, E., & Ménézo, Y. (2016). Gamete activation: Basic knowledge and clinical applications. *Human Reproduction Update*, *22*(4), 420−439. Available from https://doi.org/10.1093/humupd/dmw014.

Turvey, S. E., & Broide, D. H. (2010). Innate immunity. *Journal of Allergy and Clinical Immunology*, *125*(2), S24−S32. Available from https://doi.org/10.1016/j.jaci.2009.07.016.

Uthicke, S., Pecorino, D., Albright, R., Negri, A. P., Cantin, N., Liddy, M., Dworjanyn, S., Kamya, P., Byrne, M., & Lamare, M. (2013). Impacts of ocean acidification on early life-history stages and settlement of the coral-eating sea star *Acanthaster planci*. *PLoS One*, *8*(12). Available from https://doi.org/10.1371/journal.pone.0082938.

Ventura, A., Schulz, S., & Dupont, S. (2016). Maintained larval growth in mussel larvae exposed to acidified under-saturated seawater. *Scientific Reports*, *6*. Available from https://doi.org/10.1038/srep23728.

Vidal-Dupiol, J., Zoccola, D., Tambutté, E., Grunau, C., Cosseau, C., Smith, K. M., Freitag, M., Dheilly, N. M., Allemand, D., & Tambutté, S. (2013). Genes related to ion-transport and energy production are upregulated in response to CO_2-driven pH decrease in corals: New insights from transcriptome analysis. *PLoS One*, *8*(3). Available from https://doi.org/10.1371/journal.pone.0058652.

Vihtakari, M., Hendriks, I. E., Holding, J., Renaud, P. E., Duarte, C. M., & Havenhand, J. N. (2013). Effects of ocean acidification and warming on sperm activity and early life stages of the Mediterranean mussel (*Mytilus galloprovincialis*). *Water*, *5*, 1890−1915.

Vogel, H., Czihak, G., Chang, P., & Wolf, W. (1982). Fertilization kinetics of sea urchin eggs. *Mathematical Biosciences*, *58*(2), 189−216. Available from https://doi.org/10.1016/0025-5564(82)90073-6.

Waldbusser, G. G., Hales, B., Langdon, C. J., Haley, B. A., Schrader, P., Brunner, E. L., Gray, M. W., Miller, C. A., & Gimenez, I. (2015). Saturation-state sensitivity of marine bivalve larvae to ocean acidification. *Nature Climate Change*, *5*(3), 273−280. Available from https://doi.org/10.1038/nclimate2479.

Walther, K., Anger, K., & Pörtner, H. O. (2010). Effects of ocean acidification and warming on the larval development of the spider crab *Hyas araneus* from different latitudes (54° vs. 79°N). *Mar Ecol Prog Ser*, *417*, 159−170.

Wang, Q., Cao, R., Ning, X., You, L., Mu, C., Wang, C., Wei, L., Cong, M., Wu, H., & Zhao, J. (2016). Effects of ocean acidification on immune responses of the Pacific oyster *Crassostrea gigas*. *Fish and Shellfish Immunology*, *49*, 24−33. Available from https://doi.org/10.1016/j.fsi.2015.12.025.

Wang, M., Jeong, C. B., Lee, Y. H., & Lee, J. S. (2018). Effects of ocean acidification on copepods. *Aquatic Toxicology*, *196*, 17−24. Available from https://doi.org/10.1016/j.aquatox.2018.01.004.

Wang, Y., Li, L., Hu, M., & Lu, W. (2015). Physiological energetics of the thick shell mussel *Mytilus coruscus* exposed to seawater acidification and thermal stress. *Science of the Total Environment*, *514*, 261−272. Available from https://doi.org/10.1016/j.scitotenv.2015.01.092.

Webster, N. S., Uthicke, S., Botté, E. S., Flores, F., & Negri, A. P. (2013). Ocean acidification reduces induction of coral settlement by crustose coralline algae. *Global Change Biology*, *19*(1), 303−315. Available from https://doi.org/10.1111/gcb.12008.

White, M., McCorkle, D., Mullineaux, L., & Cohen, A. (2013). Early exposure of bay scallops (*Argopecten irradians*) to high CO_2 causes a decrease in larval shell growth. *PLoS One*, *8*, e61065. Available from https://doi.org/10.1371/journal.pone.0061065.

Wikramanayake, A. H., Peterson, R., Chen, J., Huang, L., Bince, J. M., McClay, D. R., & Klein, W. H. (2000). Nuclear beta-catenin-dependent Wnt8 signaling in vegetal cells of the early sea urchin embryo regulates gastrulation and differentiation of endoderm and mesodermal cell lineages. *Genesis*, *39*, 194−205.

Witt, V., Wild, C., Anthony, K. R. N., Diaz-Pulido, G., & Uthicke, S. (2011). Effects of ocean acidification on microbial community composition of, and oxygen fluxes through, biofilms from the great barrier reef. *Environmental Microbiology*, *13*(11), 2976−2989. Available from https://doi.org/10.1111/j.1462-2920.2011.02571.x.

Wood, H. L., Spicer, J. I., & Widdicombe, S. (2008). Ocean acidification may increase calcification rates, but at a cost. *Proceedings of the Royal Society B: Biological Sciences*, *275* (1644), 1767−1773. Available from https://doi.org/10.1098/rspb.2008.0343.

Xu, X., Yang, F., Zhao, L., & Yan, X. (2016). Seawater acidification affects the physiological energetics and spawning capacity of the Manila clam Ruditapes philippinarum during gonadal maturation. *Comparative Biochemistry and Physiology - Part A: Molecular and Integrative Physiology*, *196*, 20−29. Available from https://doi.org/10.1016/j.cbpa.2016.02.014.

Yuan, X., Yuan, T., Huang, H., Jiang, L., Zhou, W., & Liu, S. (2018). Elevated CO_2 delays the early development of scleractinian coral *Acropora gemmifera*. *Scientific Reports*, *8*, 1−10.

Zhan, Y., Hu, W., Zhang, W., Liu, M., Duan, L., Huang, X., Chang, Y., & Li, C. (2016). The impact of CO_2-driven ocean acidification on early development and calcification in the sea urchin *Strongylocentrotus intermedius*. *Marine Pollution Bulletin*, *112* (1−2), 291−302. Available from https://doi.org/10.1016/j.marpolbul.2016.08.003.

Zhao, X., Han, Y., Chen, B., Xia, B., Qu, K., & Liu, G. (2020). CO2-driven ocean acidification weakens mussel shell defense capacity and induces global molecular compensatory responses. *Chemosphere*, *243*. Available from https://doi.org/10.1016/j.chemosphere.2019.125415.

Zhao, X., Shi, W., Han, Y., Liu, S., Guo, C., Fu, W., Chai, X., & Liu, G. (2017). Ocean acidification adversely influences metabolism, extracellular pH and calcification of an economically important marine bivalve, *Tegillarca granosa*. *Marine Environmental Research*, *125*, 82−89. Available from https://doi.org/10.1016/j.marenvres.2017.01.007.

Zha, S., Liu, S., Su, W., Shi, W., Xiao, G., Yan, M., & Liu, G. (2017). Laboratory simulation reveals significant impacts of ocean acidification on microbial community composition and host-pathogen interactions between the blood clam and Vibrio harveyi. *Fish and Shellfish Immunology*, *71*, 393−398. Available from https://doi.org/10.1016/j.fsi.2017.10.034.

CHAPTER TWO

Physiological impacts of ocean acidification on marine vertebrates

Shiguo Li[1,2]
[1]Research Center for Eco-Environmental Sciences, Chinese Academy of Sciences, Beijing, P.R. China
[2]University of Chinese Academy of Sciences, Chinese Academy of Sciences, Beijing, P.R. China

Introduction

In addition to increasing temperature, the rise in atmospheric carbon dioxide (CO_2) concentration driven by human activities has caused another serious ecological problem: ocean acidification (OA). OA is a phenomenon in which seawater absorbs excessive CO_2 from the atmosphere, which results in a decrease in the pH value of the seawater (Doney et al., 2009). The pH value is a quantitative way of seawater acidification. Normal seawater is weakly alkaline and the pH value of surface seawater is about 8.2. A large amount of existing evidence shows that in the past 200 years, the ocean has absorbed 20%−30% of the CO_2 produced due to the activities of human beings, which has reduced the average pH value of the surface seawater from 8.2 at the beginning of the Industrial Revolution to about 8.1 at present (Gruber et al., 2019; Orr et al., 2005). According to the prediction by the Intergovernmental Panel on Climate Change, the average pH value of seawater will drop by about 0.3−0.4 units by 2100 to pH 7.9 or pH 7.8. Seawater acidity will increase by about 100%−150% compared with that at the beginning of the Industrial Revolution. Furthermore, the average pH value of global seawater may drop by about 0.5 units by 2300 (Intergovernmental, 2014). OA, which is caused by the elevated CO_2 concentration, causes changes in seawater chemistry. CO_2 is easily soluble in seawater and combines with water to form carbonic acid (H_2CO_3). A small amount of H_2CO_3 remains in water in its original form, while most of it is dissociated into bicarbonate ions (HCO_3^-)

Ocean Acidification and Marine Wildlife.
DOI: https://doi.org/10.1016/B978-0-12-822330-7.00006-X

and acidic hydrogen ions (H^+), resulting in a continuous drop in pH (Riebesell et al., 2007). The increase in seawater acidity alters the equilibrium state of seawater chemistry, which will threaten the survival of marine organisms and even the health of marine ecosystems that are dependent on the stability of the chemical environment (Mostofa et al., 2016).

At present, the effects of OA on marine organisms involve many biological aspects including the growth, development, and reproduction of marine organisms. The main research object species are invertebrates (such as echinoderm, mollusk and arthropod), phytoplankton, and teleost, while the physiological processes of research focus are calcification and biomineralization (Kroeker et al., 2013; Poloczanska et al., 2016). Calcification and biomineralization are key processes for the formation of the shell and skeleton of these marine organisms. They are also necessary conditions for the formation of shells of coral, mollusk, and some calcified algae, and the basis for the formation of the skeleton and otolith of marine vertebrates. Therefore a decrease in $CaCO_3$ saturation caused by OA directly affects some physiological functions such as respiration, calcification, metabolism, photosynthesis, and regeneration of marine organisms, which can greatly reduce the calcification rate of these species and seriously affect their growth, development, and reproduction, which in turn affects the structural stability and health status of the entire marine ecosystem (Albright et al., 2018; Death et al., 2009; Gazeau et al., 2007; Iglesiasrodriguez et al., 2008).

The direct and indirect effects of OA on marine organisms have been one of the hot research topics in recent years. Although OA is highly concerned by people from all walks of life, including researchers, there are still many problems to be solved. OA mainly revolves around sensitive species, such as shellfish, with calcium carbonate forming the shells, or coral reef groups, with calcium carbonate forming the exoskeleton (Pfister et al., 2014). Compared with these species, marine vertebrates such as teleosts seem to be more adaptable to OA stress (Cattano et al., 2018; Esbaugh, 2018; Munday, Gagliano et al., 2011; Munday, Hernaman et al., 2011). The reason why marine animals respond differently to OA may be that there are fundamental differences in their physiological functions. For example, marine vertebrates, especially marine teleosts, can move freely in seawater and their physiological metabolism would remain active. They can balance the pH decline caused by OA in the blood, whereas it is more difficult for coral or shellfish to do the same. Due to the lack of effective physiological regulation mechanisms, corals and shellfish are unable to balance the pH decline caused by the high level of CO_2 concentration (Hofmann et al., 2010). If the pH value of body fluid is not compensated, the calcification rate will be lower, which will affect the growth and development of these species. As a representative marine vertebrate, the marine teleost has a certain ability to regulate the acid—base imbalance induced by OA, but some physiological aspects are still significantly affected (Esbaugh, 2018). Although we do not fully understand the impacts of OA on marine teleosts, this does not mean that there are no effects. From the perspective of fishery resources, the proportion of large-sized fish living in some areas of the oceans may drop with the worsening global climate change (Munday et al., 2010; Pörtner & Peck, 2010). Scientists compared the results obtained from the most advanced evaluation model with the actual survey, proving that they can simulate the changes in fish resources under dual pressures, both from the fishing industry and environmental factors. They concluded that fishery management objectives based on fish production cannot be achieved without considering the effects of environmental changes such as OA on fish growth (Doney et al., 2020). From the perspective of individual fish, with the gradual deepening of related studies, it has been discovered that the effects of OA on marine teleosts almost involves their entire life history, starting from the sperm and egg stages to the embryonic, larval, juvenile, and adult stages, and most of their physiological processes such as ion and acid—base balance, embryonic development, body growth, behaviors, and metabolism (Espinel-Velasco et al., 2018).

OA is an imminent environmental pressure that all marine organisms are facing, so it has been a burning issue in the field of marine ecology and environment in recent years. Based on the above facts, the physiological impacts of OA on marine vertebrates, especially teleosts, have been interpreted in this chapter. Apart from teleosts, there are a few reports about the responses of other marine vertebrates such as marine reptiles and mammals to OA. Therefore taking teleosts as an example, we summarize the common and specific physiological effects of OA on marine vertebrates (Fig. 2.1 and Table 2.1). The systematic summary in this chapter can provide a comprehensive overview of the current research progress in this field, which will help us to better understand the various effects of OA caused by the global climate change on marine teleosts and create awareness about the effects of OA on marine vertebrates.

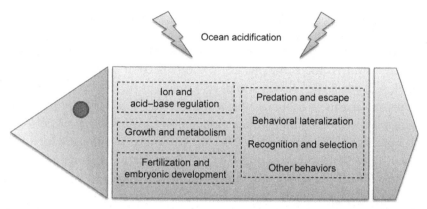

Figure 2.1 Diagrammatic presentation of the effects of ocean acidification on marine teleosts.

Impacts of ocean acidification on acid−base and ion regulation of marine vertebrates

The most direct effect of OA on fish physiology occurs in the respiratory and circulatory systems. In marine fish, the gill is used to absorb the dissolved oxygen from the seawater environment and transfer it to the blood circulation system. At the same time, it discharges CO_2 to the external environment (Wegner, 2015). In most vertebrates, the change in blood pH is generally kept within a relatively narrow range, forming acid−base homeostasis. This process involves the ion and acid−base regulation mechanisms in vertebrates such as marine teleosts. Acid−base and ion regulation occur in many tissues in the teleost, such as those in the gill, which is the most important organ, the kidney, and the intestine (Baker et al., 2015; Hwang, 2009). There are special cytological mechanisms behind the gill acid−base and ion regulation, which are different among freshwater and marine teleosts. The gill tissue of marine teleosts consists of filaments and lamellae. The capillaries in the lamella flow to the seawater in a counter-current manner, which is beneficial for improving the gas and ion exchange efficiencies inside and outside the capillary. A special group of epithelial cells in the gill, called ionocytes, is responsible for the exchange of ion and acid−base regulation across cell membranes. Ionocytes, which are also known as chloride cells, present in the gills of marine teleosts can secrete chloride into the extracellular space (Wegner, 2015; Zadunaisky, 1996).

Table 2.1 List of the published studies on the effects of ocean acidification on marine teleost physiology.

Species	Developmental stage	Acidification level (pH/concentration/pCO_2)	Duration	Effects	Negative/ Positive	References
Ambon damselfish (*Pomacentrus amboinensis* and *Pseudochromis fuscus*)	Larva and adult	pCO_2: 880 µatm	4 days	Predator—prey interaction	Negative	Allan et al. (2013)
Red drum (*Sciaenops ocellatus*)	Sub-adult	pCO_2: 1000, 2000, 5000, 15,000 and 30,000 µatm	16 h	Net H^+ excretion	Negative	Allmon and Esbaugh (2017)
Estuarine fish (*Menidia beryllina*)	Embryos: newly fertilized eggs (<24-h old)	CO_2: 390—1070 ppm	8—13 days	Larval survival, average survival, larval length	Negative	(Baumann et al., 2012)
Cobia (*Rachycentron canadum*)	Larva	pCO_2: 800 µatm pCO_2: 2100 µatm	20 days	Volume, surface area, density, sagittal otoliths	Positive	Bignami et al. (2013)
Cobia (*Rachycentron canadum*)	Larva	EX1 pCO_2: 800 and 2100 µatm EX2 pCO_2: 3500 and 5400 µatm	EX1: 2, 5, 8—12, 15, 17, 22 dph EX1: 2, 6—10, 14 dph	Size-at-age, developmental delay and mortality	Negative	Bignami et al. (2013)
Mahi-mahi (*Coryphaena hippurus*)	Larva	pCO_2: 7700, 1460, 1190, and 2170 µatm	EX1: until 17 dph EX2: until 13 dph EX3: until 21 dph	Otoliths size, hatch rate, standard length, hatched eggs, swimming ability	—	Bignami et al. (2014)

(*Continued*)

Table 2.1 (Continued)

Species	Developmental stage	Acidification level (pH/concentration/pCO_2)	Duration	Effects	Negative/ Positive	References
Painted goby (*Pomatoschistus pictus*)	Larva	pCO_2: 1503 µatm	10, 15, and 25 days	Auditory responses	Negative	Castro et al. (2017)
Damselfish (*Chromis viridis*)	Sub-adult	pCO_2: 900 µatm	5 days	Predator–prey dynamics	Negative	Cattano et al. (2019)
Summer flounder (*Paralichthys dentatus*)	Embryo and larva	pCO_2: 775, 1671 and 4714 µatm	28 dph	Survival of summer flounder embryos, larval size, condition, developmental rates, larval tissues, and craniofacial features	Negative	Chambers et al. (2014)
Rainbow trout (*Oncorhynchus mykiss*)	Fry	380, 1000, and 2000 ppm	30 days	Weight gain of fry, ATP levels, pyruvate kinase, pyruvate metabolism, decomposition of muscle, levels of branched-chain amino acids, pantothenate	Negative	Chen et al. (2020)
European sea bass (*Dicentrarchus labrax*)	Larva and juvenile	pCO_2: 650, 1150, and 1700 µatm	2 dph	Somatic growth rate, swimming capacity	—	Cominassi et al. (2019)
Sea bass (*Dicentrarchus labrax*)	Juvenile	pCO_2: 650 and 1750 µatm	277 days, 367 days	Growth	Negative	Cominassi et al. (2020)
Benthic eelpout (*Zoarces viviparus*)	Adult	CO_2: 10,000 ppm	0, 8, 24, 48 h, 4, 7, 14, 42 days	Acclimation patterns, rate of ion and acid–base regulation	Negative	Deigweiher et al. (2008)

Species	Life stage	CO₂ concentration	Duration	Measured effect	Direction	Reference
Goby (*Paragobiodon xanthosomus* and *Gobiodon histrio*)	adult	pCO_2: 440 and 880 μatm	4 days	Ability to locate coral habitat	Negative	Devine and Munday (2013)
Cardinalfish (*Cheilodipterus quinquelineatus*)	adult	pCO_2: 550, 700, and 950 μatm	4 days	Navigational capabilities and homing behavior	Negative	Devine et al. (2012a, 2012b)
Damselfish (*Pomacentrus amboinensis, Pomacentrus chrysurus,* and *Pomacentrus moluccensis*)	larva	CO_2: 700 and 850 ppm	4 days	Olfactory discrimination	Negative	Devine et al. (2012b)
Orange clownfish (*Amphiprion percula*)	larva	CO_2: 1000 ppm	11 dph	Detect predators by olfactory cues and differentiate	Negative	Dixson et al. (2010)
Reef fish (*Neopomacentrus azysron*)	larva	pCO_2: 880 μatm	4 days	Brain function	Negative	Domenici et al. (2012)
Marine damselfish (*Pomacentrus wardi*)	larva and juvenile	pCO_2: 930 μatm	4 days	Right to left lateralization, neurotransmitter functions	Negative	Domenici et al. (2014)
Gulf toadfish (*Opsanus beta*)	Adult	pCO_2: 750, 1000, and 1900 μatm	24 h	Enzyme activity	Negative	Esbaugh et al. (2012)
Dottyback (*Pseudochromis fuscus*), Damselfish (*Pomacentrus moluccensis, P. amboinensis, P. nagasakiensis, and P. chrysurus*)	Adult, juvenile	pCO_2: 440 and 700 μatm	24 h	Behavioral changes Predation rates and prey selectivity	Negative	Ferrari et al. (2011)

(Continued)

Table 2.1 (Continued)

Species	Developmental stage	Acidification level (pH/concentration/pCO_2)	Duration	Effects	Negative/Positive	References
Two-spotted goby (*Gobiusculus fflavescens*)	Embryo	pCO_2: 1400 µatm	From mating to early larval performance	Embryonic development and sensory responses	Negative	Forsgren et al. (2013)
Atlantic herring (*Clupea harengus L.*)	Egg	pCO_2: 1260, 1859, 2626, 2903, and 4635 µatm	Fertilized and incubated until the main hatch of herring larvae occurred	Metabolism of herring embryos	Negative	Franke and Clemmesen (2011)
Pink salmon (*Oncorhynchus gorbuscha*)	Juvenile	pCO_2: 850, 1500, and 2000 µatm	2 weeks	Hematocrit	Positive	Frommel et al. (2020)
Atlantic cod (*Gadus morhua*)	Larva and juvenile	pCO_2: 1800 and 4200 µatm	32 and 46 dph	Hypercapnia	Positive	Frommel et al. (2020)
Atlantic cod (*Gadus morhua*)	Larva	pCO_2: 1800 and 4200 µatm	From newly fertilized eggs to 7 weeks	Lethal tissue damage	Negative	Frommel et al. (2012)
Atlantic herring (*Clupea harengus*)	Larva	pCO_2: 0.183 and 0.426 kPa	From newly fertilized eggs to 39 dph	Growth and development and tissue damage	Negative	Frommel et al. (2014)
Yellowfin tuna (*Thunnus albacares*)	Embryo and pre-flexion larva	pH: 6.9, 7.3, 7.6, and 8.1	5 and 7 days	Organ damage in the kidney, liver, pancreas, eye, and muscle	Negative	Frommel et al. (2016)
Baltic cod (*Gadus morhua*)	Embryo and larva	pCO_2: 380, 560, 860, and 1120 µatm	From newly fertilized eggs to early nonfeeding larvae	Organ damage	Negative	Frommel et al. (2013)

Species	Life stage	Treatment	Duration	Response	Effect	Reference
Baltic cod (*Gadus morhua*)	Embryo and larva	pH: 8.080 and 7.558	NC	Sperm speed, rate of change of direction, or percent motility	—	Frommel et al. (2010)
Rockfish (genus *Sebastes*), Copper rockfish (*S. caurinus*), Blue rockfish (*S. mystinus*)	Juvenile	pCO_2: 500, 750, 1900, and 2800 μatm	21 weeks	Behavioral lateralization, swimming speed, aerobic scope, metabolic enzyme activity, transcription factors and regulatory genes, muscle structural genes	Negative	Hamilton et al. (2017)
Rockfish (*Sebastes diploproa*)	Juvenile	pCO_2: 1125 ± 100 μatm	12 days	Anxiety	Negative	Hamilton et al. (2014)
Gulf toadfish (*Opsanus beta*)	Adult	pCO_2: 1900 μatm	2–4 weeks	Intestinal HCO_3^- secretion, O_2 consumption rate	Positive	Heuer and Grosell (2016)
Walleye pollock (*Theragra chalcogramma*)	Juvenile	EX1 pCO_2: 414 ± 45, 478 ± 50, 815 ± 167, 1805 ± 212 μatm EX2 pCO_2: 596 ± 178 and 225 ± 35 μatm pCO_2: 828 ± 144 and 386 ± 112 μatm	EX1: 6 weeks EX2: 28 weeks	Growth	—	Hurst et al. (2012)

(*Continued*)

Table 2.1 (Continued)

Species	Developmental stage	Acidification level (pH/concentration/pCO_2)	Duration	Effects	Negative/Positive	References
Coral reef anemonefish (*Amphiprion melanopus*)	Juvenile and adult	pCO_2: 1285 ± 321 and 643 ± 169 µatm pCO_2: 2894 ± 343 and 1543 ± 293 µatm	Adult: 3 months	The effects of diel CO_2 cycles on juveniles	–	Jarrold and Munday (2019)
Atlantic cod (*Gadus morhua*)	Juvenile	pCO_2: 1000 µatm diel-cycling elevated pCO_2: 1000 ± 300 µatm	1 month	Behaviors, shelter, relative lateralization, and absolute lateralization	–	Jutfelt and Hedgärde (2015)
Olive flounder (*Paralichthys olivaceus*)	Egg and larva	pH: 4.0, 6.0, and 8.0	5 days	Fertilized eggs	Negative	Kim et al. (2020)
Atlantic cod (*Gadus morhua*)	Adult	pCO_2: 550, 1200, and 2200 µatm	3–4 weeks	Energy consumption and budget	–	Kreiss et al. (2015)
Stickleback (*Gasterosteus aculeatus*)	Adult	pCO_2: 992.1 ± 119.3 µatm	40 days	Lateralization	Negative	Lai et al. (2015)
Estuarine red drum (*Sciaenops ocellatus*)	Early life stage, embryo, and larva	pCO_2: 1300 and 3000 µatm	72 h	Yolk depletion rate, standard length, and scototaxis behavior	–	Lonthair et al. (2017)

Species	Life stage	pCO_2	Duration	Trait	Effect	Reference
Sand smelt (*Atherina presbyter*)	Larva	pCO_2: 537 µatm	7 and 21 days	Group cohesion, lateralization	Negative	Lopes et al. (2016)
Meagre (*Argyrosomus regius*)	Juvenile	pCO_2: 500 and 1500 µatm	28 days	Exploration and shoal cohesion	Negative	Maulvault et al. (2018)
Pomacentridae (*Pomacentrus wardi*), Pseudochromidae (*Pseudochromis fuscus*)	Juvenile, adult	pCO_2: 925 µatm	6–8 days	Perception	Negative	McCormick et al. (2018)
Ocellated wrasse (*Symphodus ocellatus*)	Adult	pCO_2: 400 and 1100 µatm	16 days	Reproductive behavior	Negative	Milazzo et al. (2016)
Anemonefish (*Amphiprion melanopus*)	Adult	pCO_2: 644 and 1134 µatm	10 months	Reproductive traits	Negative	Miller et al. (2015)
Anemonefish (*Amphiprion melanopus*)	Juvenile and adult	pCO_2: 581 and 1032 µatm	21 days	Performance	Positive	Miller et al. (2012)
Cinnamon anemonefish (*Amphiprion melanopus*)	Adult	pCO_2: 584 and 1032 µatm	9 months	Seeding activity	Negative	Miller et al. (2013)
Atlantic cod (*Gadus morhua L.*)	Juvenile	pCO_2: 1000, 3800, and 8500 µatm	55 days	Weight gain, growth rate, and condition factor	Negative	Moran and Stottrup (2011)
Damselfish (*Dascyllus aruanus* and *Pomacentrus moluccensis*), cardinalfishes (*Apogon cyanosoma* and *Cheilodipterus quinquelineatus*)	Juvenile	pCO_2: 441–998 µatm pCO_2: 346–413 µatm	14 months	Hepatosomatic index, reproductive traits	—	Munday et al. (2014)

(*Continued*)

Table 2.1 (Continued)

Species	Developmental stage	Acidification level (pH/concentration/pCO_2)	Duration	Effects	Negative/Positive	References
Coral reef fishes (*Ostorhinchus doederleini* and *Ostorhinchus cyanosoma*)	Adult	pH: 8.15 and 7.8	1 week	Mortality rate	Negative	Munday et al. (2009)
Orange clownfish (*Amphiprion percula*)	Settlement-stage larva	pH: 8.15, 7.8, and 7.6	11 days	Olfactory	Negative	Munday et al. (2009)
Clownfish (*Amphiprion percula*)	Larva	CO_2: 500, 700, and 850 ppm	10 days	Behavior, mortality	Negative	Munday et al. (2010)
Damselfish (*Acanthochromis polyacanthus*)	Juvenile	pCO_2: 600, 725, and 850 µatm	3 weeks	Growth, survival, size, otolith size, shape, and symmetry	–	Munday et al. (2011)
Clownfish (*Amphiprion percula*)	Larva	pCO_2: 1050 and 1721 µatm	6–8 days	Otolith size, shape, otolith area, maximum length	–	Munday et al. (2011)
Coral trout (*Plectropomus leopardus*)	Juvenile	pCO_2: 490, 570, 700, and 960 µatm	4 weeks	Juvenile performance	Negative	Munday et al. (2013)
Three-spined sticklebacks (*Gasterosteus aculeatus*)	Adult	pCO_2: 400 and 1000 µatm	10 and 20 days	Predator avoidance behaviors	Negative	Näslund et al. (2015)
Clownfish (*Amphiprion percula*), damselfish (*Neopomacentrus azysron*)	Larva	pCO_2: 450 and 900 µatm	Clownfish: 11 days damselfish: 4 days	Neurotransmitter function	Negative	Nilsson et al. (2012)

Species	Life stage	Treatment	Duration	Endpoints measured	Effect	Reference
Sea urchin (*Strongylocentrotus purpuratus*)	Larva	pCO_2: 800 µatm	14 days	Size, metabolic rate, biochemical content, and gene expression	Negative	Pan et al. (2015)
Dolphinfish (*Coryphaena hippurus*)	Larva	pCO_2: 400 and 1600 µatm	3 dph	Oxygen consumption rate, and swimming duration, orientation frequency	Negative	Pimentel et al. (2014)
Gilthead seabream (*Sparus aurata*), meagre (*Argyrosomus regius*)	Larva	pCO_2: 350 and 1400 µatm	15 dph	Hatching success and larval survival, time of swimming, attack and capture rates of prey	Negative	Pimentel et al. (2016)
Sea bass (*Dicentrarchus labrax*)	Juvenile	pCO_2: 450 and 1000 µatm	2, 7, and 48 days	Chances for fish to detect food or predators, olfactory system central brain function	Negative	Porteus et al. (2018)
Meagre (*Argyrosomus regius*)	Juvenile	pH: 8.0 and 7.5	30 days	Hg accumulation	Negative	Sampaio et al. (2018)
Three-spined stickleback (*Gasterosteus aculeatus*)	Juvenile and adult	pCO_2: 400 and 1000 µatm	3 months	Clutch size, juvenile survival and growth rate, body size, and the size of otoliths	Negative	Schade et al. (2014)
Spiny damselfish (*Acanthochromis polyacanthus*)	Adult	pCO_2: 414 and 754 µatm	7 days	Brain molecular phenotype	Negative	Schunter et al. (2016)

(Continued)

Table 2.1 (Continued)

Species	Developmental stage	Acidification level (pH/concentration/pCO_2)	Duration	Effects	Negative/ Positive	References
European sea bass (*Dicentrarchus labrax*)	Juvenile	pCO_2: 400 and 1000 µatm	3, 7, and 21 days	Ammonia excretion rate	Negative	Shrivastava et al. (2019)
Sand smelt (*Atherina presbyter*)	Larva	pCO_2: 1000 and 1800 µatm	15 days	Swimming speed, energetic costs, and morphometric changes	–	Silva et al. (2016)
Orange clownfish (*Amphiprion percula*)	Juvenile	pCO_2: 600, 700, and 900 µatm	17–20 days posthatching	Avoiding the reef noise	Negative	Simpson et al. (2011)
Marine medaka (*Oryzias melastigma*)	Egg and larva	pCO_2: 1080 µatm	30 dph	Histopathological anomalies	Negative	Sun et al. (2019)
Two-spotted gobies (*Gobiusculus flavescens*)	Adult	pCO_2: 1000 µatm	2 years	Strength of the side bias (absolute lateralization)	–	Sundin and Jutfelt (2018)
Seabream (*Sparus aurata*)	Adult	pCO_2: 1000 µatm	4–8 weeks	Olfactory system	Negative	Velez et al. (2019)
Marine medaka (*Oryzias melastigma*)	Egg	pCO_2: 1160 and 1783 µatm	From fertilization until hatching	Yolks of larvae, newly hatched larvae, and C-start escape behavior	Negative	Wang et al. (2017)
Reef damselfish (*Amphiprion percula* and *Acanthochromis polyacanthus*)	Adult	pCO_2: 652 and 912 µatm	9 months	Number of eggs per clutch, reproductive output, survival, hatching success, offspring size, fish reproduction	Negative Positive	Welch and Munday (2016)

EX, Experiment. "–" indicates no significant impact.

During the normal respiration and metabolism processes in the teleost, the CO_2 produced by metabolism will be transported to the ionocytes of the gill epithelia in the form of HCO_3^-. Under the catalysis of carbonic anhydrase (CA), these HCO_3^- are reformed into gaseous CO_2, and then discharged from the body by ionocyte diffusion in the gill epithelium (Tresguerres et al., 2019). In this process, the contents of H^+ and HCO_3^- in ionocytes are in a state of dynamic equilibrium, namely acid—base and ion balance, which is strictly controlled by the teleost's physiology (Brauner et al., 2019). However, this regulation is reversible and the change in CO_2 concentration in the external environment has a significant impact on the physiological balance of the teleost. When the increased CO_2 concentration in the external environment leads to OA, the high concentration of CO_2 will enter the blood through the epithelial cells in the gill through free diffusion and the concentration of CO_2 in the teleost will also increase (Brauner et al., 2019).

OA makes the carbon dioxide partial pressure (pCO_2) levels in the fish blood increase, followed by a rapid decrease in the pH value and finally accumulation of a large amount of bicarbonate in the blood. Because of the differences in living environments, the mechanism of regulating the acid—base balance in fish is quite different from that of terrestrial animals (Heisler, 1986). At the same time, the ability and mechanism for regulating the acid—base balance in teleosts living in distinct environments (such as seawater, freshwater, or saline-alkali water) and even the same teleost species in different development periods are also different. The ability of marine teleosts to regulate acid—base balance is relatively higher than that of freshwater teleosts, and its buffering speed in responding to the change in ion concentration caused by OA is relatively rapid. This is mainly caused by the existence of abundant bicarbonate and other related ions (such as Na^+, Cl^-, etc.) in the seawater environments (Brauner & Baker, 2009). Although acid—base imbalance is the consistent response of the tested marine teleost to OA, it is becoming increasingly clear that the degree of these impacts between different species varies greatly. As an important physicochemical factor of seawater, the pH value is similar to temperature or salinity in that it plays a very important role in the normal survival of marine organisms. After long-term evolution, many marine organisms such as the Cephalopods, Brachypods, and teleosts develop their own relatively complete acid—base balance regulation system (Tresguerres & Hamilton, 2017). Similar to other vertebrates, marine teleosts must maintain their intracellular and extracellular acid—base balance to healthily survive in harsh marine environments.

A large number of adverse effects of OA on teleosts have been observed in previous investigations. The general view is that these effects are mainly the consequence of chemical changes in the blood pCO_2 caused by the imbalance in ion and acid—base regulation (Esbaugh, 2018; Heuer & Grosell, 2016). The most direct effect of OA on marine teleosts is the change in the chemical state of CO_2 in blood, especially in the gill. In a study, two species of the marine teleost, namely *Acanthochromis polyacanthus* and *Amblyglyphidodon curacao*, were exposed to control pCO_2 treatment, two fluctuating pCO_2 treatments, and a stably elevating pCO_2 treatment for 8 hours. The blood draw examination and the hematological parameter test after the treatments indicated that the blood chemistry of these species was significantly influenced by OA (Hannan et al., 2020). It was found that with elevating of pCO_2 in seawater, the hemoglobin—O_2 affinity of the red blood cells of the sea bass *Dicentrarchus labrax* increased, showing stronger tolerance to hypoxia. This further indicated that OA can promote the absorption of O_2 by blood under low concentrations of O_2, thus enhancing tolerance to hypoxia of this marine teleost (Montgomery et al., 2019). Esbaugh et al. (2016) studied a red drum fish *Sciaenops ocellatus* native to the mouth of the Gulf of Mexico. The results showed that under OA conditions (pCO_2: 1000 µatm), the red drum fish showed respiratory plasticity with a slight physiological trade-off (Esbaugh et al., 2016). However, the plasticity was not enough to completely offset the acid—base disturbance caused by OA, so it is unlikely to affect the recovery ability of this species. Heuer and Grosell (2016) also found that the compensation of CO_2 by *Opsanus beta* under OA conditions (pCO_2: 1900 µatm) resulted in continuous alkali (HCO_3^-) loss, which caused energy consumption at the tissue level (Heuer & Grosell, 2016). The continuous increase in standard metabolic rate (SMR) may lead to energy redistribution at the individual level in animals. This change is consistent with the phenomenon of respiratory acidosis in marine teleosts, that is, hypercapnia, followed by metabolic compensation (Esbaugh, 2018). Under OA stress, the higher pCO_2 in the seawater environment changed the gradient of the CO_2 levels between blood and water, which led to the fish having to increase the partial pressure in the blood to keep up the excretion rate of CO_2 (Brauner et al., 2019). As an acidic gas, the increase in pCO_2 in the plasma will reduce the extracellular pH value. Marine teleosts balance this pH change mainly by retaining extra HCO_3^- in the blood. According to statistics, all marine teleosts exposed to OA stress will follow this pattern, and the increased range of plasma HCO_3^- is 30%–60% of the normal concentration (Esbaugh, 2018). When the marine teleost *O. beta* was exposed to pCO_2 stress at 1000–1900 μatm, the acidification of seawater affected the transport of respiratory gas and the acid–base balance in the teleost (Esbaugh et al., 2012). Although the physiological effects caused by the increase in HCO_3^- concentration in the blood are not clear, the long-term increase in HCO_3^- concentration in blood may damage various physiological systems of *O. beta* (Esbaugh et al., 2012). Consistently, when exposed to a similar concentration of acidified seawater, the blood chemical properties of *Scyliorhinus canicula* were changed due to the accumulation of HCO_3, which buffered the blood acidosis. Its swimming pattern at night was also affected significantly and the organism showed a certain tendency for lateralization (Rosa et al., 2017). These behavioral effects may indicate that the neurophysiology of the Elasmobranchii was influenced by higher pCO_2, just like some marine teleosts, leading to the changes in its behavior (Green & Jutfelt, 2014). It is worth noting that elevated plasma pCO_2 and HCO_3^- have a series of potential biochemical effects on marine teleosts, which may change some downstream physiological performances including swimming behavior, metabolism, and osmotic regulation.

When the concentration of CO_2 in seawater increases, OA occurs. Specific CA can catalyze the reaction of CO_2 and H_2O to produce H^+ and HCO_3^-; this process is called hydration. Hydration makes the fish body show signs of acidification, namely hypercapnia (Pörtner, 2008). At this time, the acid–base balance in the fish is impaired. The disorder of acid–base imbalance is disadvantageous to fish physiology. It will change the degree of protonation in the body thus affecting the structure and function of biomacromolecules such as proteins, lipids, nucleic acids, and carbohydrates, as well as the metabolic activity. Besides, the biological activities of some organelles are also significantly affected, because they all have the optimal pH range to maintain their basic functions (Occhipinti & Boron, 2019), which will have significant impacts on the physiological activity and survival of the fish. Therefore in order to restore the acid–base and ion balance so that normal physiological activities are not affected, the ionocytes in the fish gill epithelia will compensate for the gradient of ions and acid–base imbalances in two ways, namely respiratory compensation (by regulating pCO_2 in the blood) and metabolic compensation (by regulating the net acid excretion rate). However, the concentration of CO_2 in the blood cannot be regulated to be lower than that in water just by implementing the respiratory compensation mechanism. It is necessary to alleviate the symptoms of hypercapnia by relying on the metabolic compensation mechanism (Esbaugh, 2018; Kreiss et al., 2015; Tresguerres & Hamilton, 2017). The process of metabolic compensation involves the efflux and absorption of many ions into the ionocytes of the gill epithelia in fish. There are many different types of transport systems that can transport these ions on the membrane ionocytes. A lot of researchers have studied the distribution, physiology, structure, and function of these transporters (Guh et al., 2015; Hwang et al., 2011; Kültz, 2015; Perry et al., 2009), such as Na^+ absorption and H^+ secretion—related transporters (Na^+/H^+ exchangers, NHE; thiazide-sensitive Na/Cl cotransporter, NCC; $Na^+/K^+/Cl^-$ cotransporter, NKCC; Na^+/K^+ ATPase, NKA; Na^+/HCO_3^- cotransporter, NBC; vacuolar H^+ pump, V-type ATPase; and other Na^+ channels), Cl^- absorption and HCO_3^- secretion—related transporters (anion exchanger, AE; and Cl^-/HCO_3^- exchanger, SLC26A transporter family members), and Ca^{2+} absorption—related transporters (epithelial Ca^{2+} channel, ECAC; plasma membrane Ca^{2+} pump, PMCA; and Na^+/Ca^{2+} exchanger NCX). These transporters are mainly distributed in the apical and basolateral gill epithelia of fish (Fig. 2.2). The epithelial cells in the fish gill perform acid efflux function in the form of direct H^+ secretion. This process is mainly accomplished by transporters such as apical NHE and vacuolar-type H^+-ATPase,

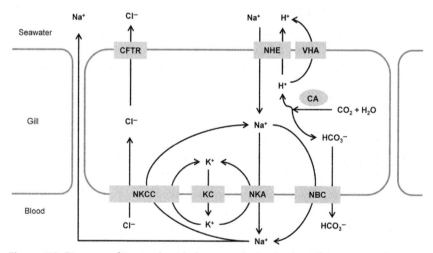

Figure 2.2 Diagram of ion and acid—base regulation in the gill ionocytes of marine teleosts exposed to ocean acidification. *CA*, Carbonic anhydrase; *CFTR*, cystic fibrosis transmembrane conductance regulator; *NBC*, Na^+/HCO_3^- cotransporter; *NHE*, Na^+-H^+ exchangers; *NKA*, Na^+/K^+ ATPase; *NKCC*, $Na^+/K^+/Cl^-$ cotransporter; *VHA*, V-type ATPase; *KC*, K^+ channel.

which is also accompanied by the absorption of Na^+ (Kwong et al., 2014). At the same time, excess HCO_3^- is transported to seawater or blood by the SLC26A transporter on the apical or basolateral membrane, which is often accompanied by Cl^- absorption (Alper & Sharma, 2013). Other transporters can also play an auxiliary role in the secretion. Of course, these forms of metabolic compensation are not endless. Some environmental factors, such as the upper limit threshold of HCO_3^- concentration, ambient temperature, net acid efflux capacity, and HCO_3^- concentration in the cytoplasm, also can limit the extent of compensation (Shartau et al., 2019).

A large number of investigations on the response patterns of the gill epithelial cells under OA conditions have confirmed the conclusion that fish can alleviate the effects of OA through metabolic compensation. Claiborne et al. (1999) confirmed for the first time that NHE1 distributed on the epithelial basolateral membrane played an important role in the compensation process of metabolic acidosis. Subsequent studies have shown that different types of NHE were related to the acid—base balance regulation. After exposure to OA, the NHE1 gene in the basolateral membrane of marine teleosts was found to be downregulated, which reduced the transportation ability of H^+ from the epithelial cells to the blood and induced the occurrence of hypercapnia (Michael et al., 2016). The abundance of NHE3 protein in gills was significantly increased after exposure to OA (pCO_2: 10,000 μatm). Even at a very low level of OA (pCO_2: 1200 μatm), the expression of the NHE3 gene in the gills was also increased in the larval stage. The change in the expression trend is helpful for marine teleosts to excrete H^+, the concentration of which is increased in cells due to OA, and reduce the adverse effects. Following acidosis caused by OA, the post-translational modifications of NHE2 and other ion transport systems (such as NHE3) were found to play an important role in the short-term pH adjustment and H^+ outflow in the gill of the dogfish *Squalus acanthias* (Claiborne et al., 2008). NHE2 may implement a variety of functions in the gill of this species, involving the outflow of H^+ from the acid-secreting cells, the reabsorption of H^+ outside the basolateral membrane, and the production of HCO_3^- in the alkali-secreting cells together with H^+-ATPase (Claiborne et al., 2008). In another marine teleost *Myoxocephalus octodecemspinosus*, there was apical NHE2 distributed in larger ovoid chlorine cells of the gill tissue, which was colocated with NKA in the same type of cells. This result indicates that the acid-secreting cells in the gills of *M. octodecemspinosus* exhibit apical Na^+/H^+ antiporter (NHA) and basolateral NKA activities, suggesting the potential roles of the ion transport systems such as NHE2 in the responses of marine teleost to OA (Catches et al., 2006). In addition, there are many studies conducted on the functions of transport systems such as NHE and NKA in marine teleosts, which constitute the cotransport system of Na^+ and HCO_3^- and outline other metabolic compensation methods adopted by marine teleosts to cope with OA stress. In the case of HCO_3^- transport systems, AE is one of the carriers that are significantly affected by OA. Under OA stress (pCO_2: 7000 μatm), AE1a and AE1b in the basolateral gill of the medaka *Oryzias latipes* were significantly upregulated, which promoted the transport of HCO_3^- from the cytoplasm to the blood (Tseng et al., 2013). Unlike AE1, the AE2 gene expression was significantly reduced in the gill tissue of a toadfish after 72 hours of treatment at 1900 μatm (Brauner et al., 2019; Romero et al., 2004). The decrease in AE2 expression may be a way of limiting HCO_3^- excretion into seawater. For NBC, the current research results showed that its response to OA was diverse, and there was no consistent expression trend for related genes (Deigweiher et al., 2008; Esbaugh et al., 2012). This also indicates that marine teleosts may have a variety of methods to deal with OA, but these need further, in-depth exploration.

When marine teleosts respond to OA stress, they often show some physiological plasticity. Allmon and Esbaugh (2017) studied the response and physiological plasticity of metabolic compensation in the gill tissue of estuarine red drum fish to different pCO_2 stresses. The results showed that the red drum fish can use the NBC regulation mechanism in the gill basolateral cells to quickly compensate for the interference of acid or alkali stress related to environmental changes, and at the same time, it has certain physiological plasticity when dealing with extreme challenges regarding acid or alkali (Allmon & Esbaugh, 2017). When the Atlantic cod *Gadus morhua* was exposed to different levels of OA stresses, it showed certain physiological plasticity and was found to be able to adapt to the acidified environmental conditions within a certain stress level (Michael et al., 2016). In a recent study, the European sea bass *D. labrax* was exposed to the interactive stress of OA, ammonia, and low salinity. By analyzing the changes in gene expression, ion osmoregulation, and other physiological responses, it was found that this species was more vulnerable to OA at a low salinity level, showing a certain degree of physiological plasticity (Shrivastava et al., 2019). The feature of physiological plasticity in the acid–base transport systems of some marine teleosts in response to OA stress can be found through the analyses of gene expression, enzyme activity, and protein abundance in the gill of these species. Early studies on the medaka *O. latipes* have analyzed the expression of a variety of acid—base regulation—related genes, including NHE3, NBCa, NBCb, AE1a, AE1b, ATP1a1a, ATP1a1b, ATP1b1a, Rhag, Rhbg, and Rhcg under the condition of OA stress. The results showed that NHE3 and Rhcg were upregulated both in the embryonic development period and in various tissues, indicating that they played a decisive role in acid—base regulation. In addition, some amino acid metabolism—related genes (ALT1, ALT2, AST1a, AST1b, AST2, and GLUD) were also upregulated in the larval stage, which may provide energy support for the larvae to cope with OA stress. These results indicated that the medaka *O. latipes* has certain molecular plasticity in response to OA (Tseng et al., 2013). Schunter et al. (2016) integrated and analyzed 33 transcriptome and proteome data in the brain tissue of the spiny damselfish, *A. polyacanthus*, after exposure to OA. They screened out a large number of differentially expressed genes and proteins and also studied the molecular phenotype of their offspring individuals. The results showed that the stress tolerance of the parent teleosts was beneficial to the improvement of the offsprings' ability to adapt to OA (Schunter et al., 2016). In general, most marine teleosts usually keep enough proteins in the gill cells to effectively offset the OA stress induced by the rise in the environmental CO_2 concentration.

It is generally accepted by scientists that the regulation of acid—base balance or the transport of ions by teleosts is a process with energy consumption under the condition of OA caused by high pCO_2. These regulatory processes require a large amount of ATP that is catalyzed by various ATPases, which may lead to higher energy costs and change the overall energy distribution pattern of the fish body. These energy allocation changes will have a significant impact on the downstream physiological activities of marine teleosts such as growth, embryonic development, metabolism, and behaviors (Heuer & Grosell, 2014). Given the importance of these fish behaviors, we will summarize and describe these effects in the following paragraphs.

Impacts of ocean acidification on fertilization and embryonic development of marine vertebrates

Similar to other marine organisms, OA has more obvious negative effects on fertilization and embryonic development of marine teleosts.

The early developmental stage of marine teleosts is more sensitive to the change in seawater pH due to the weak adaption ability of individuals in this period (Cattano et al., 2018; Wang et al., 2017). Specifically, there are two main reasons for this phenomenon. (1) The ion regulation ability of marine teleosts is closely related to their gills. In the early developmental stage, the gill of a marine teleost is not yet mature and its function may not be fully implemented. The larval gill does not have strong acid—base and ion regulation ability when compared to an adult gill. (2) Even if the gill and its functions are well developed, compared with that of the adult, the marine teleost has a larger gill surface area in its early developmental stage and it needs more energy to maintain pH homeostasis (Fu et al., 2010; Melzner et al., 2009). The early developmental stage of a marine teleost is therefore more susceptible to the changes in different marine environmental factors (Ross et al., 2011).

The formation and motility of the sperm and the egg in marine teleosts are both affected by OA. Studies have shown that the fecundity of *Amphiprion percula* and *Gasterosteus aculeatus* increased with the increase in pCO_2 in seawater (Munday, Gagliano et al., 2011; Munday, Hernaman et al., 2011; Schade et al., 2014). On the contrary, this increasing trend does not occur in *Gobiusculus flavescens* (Forsgren et al., 2013; Miller et al., 2013; Schade et al., 2014). The sperm motility of *Limanda yokohamae* was significantly inhibited by OA stress (Inaba et al., 2003). Olive flounder (*Paralichthys olivaceus*) was exposed to the different levels of OA conditions and the results showed that the hatching and osmolality of the fertilized eggs were negative affected by OA (Kim et al., 2020). The sperm fluidity of *G. morhua* also deteriorated after exposure to OA (Frommel et al., 2010). Contrarily, *A. melanopus* showed an increasing trend in sperm motility (Miller et al., 2013). OA usually leads to a decline in teleosts' fertilization ability or even the delay of the fertilization period. These effects are closely associated with the swimming ability of these species. Decreased swimming ability under OA conditions can affect the mating quality of marine teleosts (gamete emission), which leads to failure in completing the mating. Moreover, the delayed fertilization in marine teleosts exposed to OA was found to disturb regular gonadal development and maturation, leading to mating failure or mating in the nonmating season (Servili et al., 2020). All these changes will result in the inability of the insemination or hatching process of marine teleosts to proceed normally. Under high pCO_2, the food intake of the marine teleost sea bass *D. labrax* decreased, and the gonadal growth and development were also affected (Cominassi et al., 2020). The gamete activity of Baltic cod (*G. morhua*) was also found to be insufficient, which affected the gamete combination, development, and hatching (Frommel et al., 2010). Most marine teleosts fertilize in vitro and the sperms are activated by seawater. OA could soften the egg membrane, inducing yolk sac collapse and even rupture (Stoss, 1983). It was found that the sperm of *Oncorhynchus keta* could not be fertilized successfully under OA conditions, which affected the reproduction of this species and the continuation of their populations, leading to a new round of survival of the fittest (Mathis et al., 2015). It was found that the volume of yolk sac in *Clupea pallasii* eggs decreased under high pCO_2, which indicated that the decrease in protein synthesis would affect the hatching rate of eggs and the survival rate of their larvae (Frommel et al., 2012; Love et al., 2018). In addition, under OA conditions, a low pH value led to the imbalance in the male and female ratio of some sensitive marine teleosts. This imbalance will be further enlarged after long-term reproduction and even endanger the health and stability of the whole population (Cattano et al., 2018; Frommel, Carless et al., 2020; Frommel, Hermann et al., 2020).

Investigations of the effects of OA on marine teleosts also focus on early embryonic development, which is generally considered to be the most sensitive stage to CO_2-driven OA owing to the fact that marine teleosts lack pH-mediated intracellular regulation mechanisms during this stage. Studies have shown that OA has serious negative impacts on the embryonic development of the marine teleost *Menidia beryllina*, with a 73% decrease in the 10-day survival rate after exposure to a pCO_2 level of 780 μatm (Baumann et al., 2012). Although subsequent studies found that this species is an abnormal teleost in terms of sensitivity, one sure conclusion is that OA can significantly reduce the survival rate of many marine teleosts during the rearing period. The average survival rate and length of *M. beryllina* embryos exposed to OA with a pCO_2 level of 1000 μatm for 7 days after hatching were reduced by 74% and 18%, respectively. Teleosts in the egg stage were also more susceptible to high CO_2 concentrations than those in the larval stage after hatching (Baumann et al., 2012). A series of studies by one research team confirmed that OA caused embryonic retardation and extensive tissue damage in the gilthead seabream (*Sparus aurata*) and the meagre (*Argyrosomus regius*) (Frommel et al., 2012, 2014, 2016). Lipid metabolism disorder and histopathological anomalies were both observed in a study on the marine medaka *O. melastigma*, in which individuals that were 30 days of age were exposed to OA

(Sun et al., 2019). On the contrary, it is also clear that some teleosts are highly tolerant to high pCO_2 at their larval stage. The responses of two commercially important teleost species, *S. aurata* and *A. regius*, to a predicted OA condition with pH 7.6 were investigated. Hypercapnia caused by OA reduced the hatching success rate by 26.4% and 14.3% in these two species, respectively, while the survival rate decreased by 50%; although the length of the newly hatched larvae of *S. aurata* and *A. regius* was not significantly affected (Pimentel et al., 2016). Some contradictory results have been observed in the study on marine teleosts. When *S. ocellatus* was exposed to pCO_2 of 1300 and 3000 μatm, OA significantly reduced the survival rate of its embryo (Lonthair et al., 2017), but it also had a considerable number of individuals with CO_2 tolerance. Juvenile individuals of the Atlantic cod (*G. morhua*) were exposed to pCO_2 stress for 46 hours, and the expression levels of the genes related to glycogen and fatty acid synthesis (including CPTA1, PPAR1b, ACoA, GYS2, 6PGL, and FAS) were not significantly affected, confirming a higher tolerance of *G. morhua* to OA (Frommel, Carless et al., 2020; Frommel, Hermann et al., 2020). In contrast to the abovementioned adverse effects, a series of studies on coral reef fish have shown that some marine teleosts possessed elastic reproduction ability under OA conditions, and a few species even exhibited significant increases in the reproductive output (Miller et al., 2013; Welch & Munday, 2016). In these studies, the hatching success rate of the embryo and the size of larvae did not decrease when the teleosts were exposed to OA. The offspring of some teleosts had significantly higher egg survival rate and hatching success rate under the same OA level. Branch et al. (2013) conducted an experiment on several fish species to test the effects of OA on reproduction and early life stages. The results showed that although there were some differences among species, OA had no significant effects on sperm motility and embryonic development, even at a CO_2 concentration of 2300 μatm (Branch et al., 2013). Hamilton et al. (2017) conducted a similar experiment on two species of rock teleosts within the genus *Sebastes*, aiming to illustrate the importance of differences between species. The results showed that the swimming speed and aerobic respiration capacity of one teleost species decreased with the decrease in pH, while the same properties of the other species were not affected (Hamilton et al., 2017). These pieces of evidence indicate that the responses of marine teleosts to OA are different in their early embryonic developmental stages. Taken together, it is necessary to consider many factors and not to give any single result too much weight when comprehensively evaluating the effects of OA on the embryonic development of marine teleosts.

It has been hypothesized that the sensitivity during the embryonic stage of marine teleosts to OA may be induced by the lack of a mature functional gill in these species during this stage. A previous investigation showed that the eggs and early-stage larvae of G. morhua in the Baltic Sea were well adapted to a high level of pCO_2 (3200 μatm), which may be due to a specialized acid–base and ion regulation ability despite the lack of a mature gill (Frommel et al., 2013). Tseng et al. (2013) carried out a test on O. latipes and the obtained results provided detailed gene expression patterns in response to OA-induced hypercapnia throughout the ontogenetic process of this species. The results showed that the sensitivity of O. latipes embryos to OA was higher at the early embryonic stage when the epithelial cells in the functional gill are not yet developed (Tseng et al., 2013). OA-induced damage in critically important organs such as the smaller functional gill can be observed in the Atlantic cod (G. morhua) (Frommel, Carless et al., 2020; Frommel, Hermann et al., 2020). Under normal conditions, marine teleost larvae can excrete H^+ through their skin (Liu et al., 2013), but the mechanism of actual acid excretion in the gill issue of their larvae under hypercapnia is still unclear.

Another important consideration in the discussion of OA tolerance by the marine teleosts in the early developmental stage is transgenerational plasticity. Intergenerational effects have been shown to be able to improve the effects of OA on the embryonic development of several marine teleosts. Miller et al. (2012) studied the effects of OA on the growth and development of different generations of A. melanopus. The results showed that OA increased the metabolic rate of the parent A. melanopus and decreased the length, weight, and survival rate of the offspring (Miller et al., 2012). However, when the parents were exposed to higher pCO_2, the effects of CO_2-driven OA on the larvae no longer existed or were reversed, showing that the nonhereditary intergenerational effects could significantly alter the responses of marine teleosts to OA (Miller et al., 2012). Some marine teleosts can adapt to OA by adopting the powerful mechanism of intergenerational plasticity. In some studies, transgenerational treatment partially or completely reduced a series of negative effects of OA on marine teleosts. For example, after the coral reef anemonefish A. melanopus was exposed to stably elevated pCO_2 stress, the negative effects of OA on the growth of their offspring were ameliorated (Jarrold & Munday, 2019). The mechanisms of this transgenerational adaptation are still unclear, but the continuous development in modern molecular biotechnology could provide a powerful tool for exploring these mechanisms. Judging from current evidence, the stress memory of parents can affect the phenotype of their offspring through various nongenetic mechanisms. This mechanism involves egg nutrition supply, hormone or protein transfer to the egg, and epigenetic markers. When teleost parents experience environmental stress, the adaptive behavior of their offspring will be improved in a similar environment. A typical case is the adaptive ability of marine teleosts to OA (Munday et al., 2014).

Impacts of ocean acidification on metabolism and growth of marine vertebrates

Assessment of the effects of OA on fish metabolism can provide us with new insights into the impacts of this environmental stress from the perspectives of ecological and physiological mechanisms. When studying the effects of OA on the metabolism of teleosts, two indicators are generally considered, namely the SMR and the rest metabolic rate (RMR). Due to the limited number of relevant investigations, there is still a lack of data on the effects of metabolism on marine teleosts in the early developmental stage. However, for adult marine teleosts, OA often leads to a significant increase in their RMR values (Cattano et al., 2018). After being exposed to OA stress, marine teleosts will activate their acid−base and ion regulation mechanisms to balance the excess CO_2 and increase the metabolic intensity in vivo, which is the possible reason for the increase in RMR. This conclusion is based on the results reported in many works in the literature, but it is inconsistent with the results of an earlier *meta*-analysis. Lefevre (2016) suggested that OA had no significant effect on RMR and the metabolic aspects of marine teleosts and attributed this phenomenon to the high variability in the metabolic response of fish. Other studies on the marine teleost metabolism obtained negative results. Under long-term high-level pCO_2 stress, the RMR of some marine teleosts were found to have decreased significantly (Heuer & Grosell, 2016; Rummer & Brauner, 2015). These results indicate that the metabolic effects of OA on marine teleosts are not single, but rather diverse, and maybe species-specific.

Some researchers believe that the effects of OA on acid−base homeostasis, respiration, and subsequent osmoregulation in marine teleosts usually lead to an increase in body SMR (Pörtner & Peck, 2010). However, the available experimental data so far do not fully support this conclusion. These metabolic costs are likely to be offset by reduced energy requirements elsewhere in marine teleosts. This type of metabolic redistribution has been explored in juvenile invertebrates (Pan et al., 2015). Importantly, there are some data on marine teleosts that support the notion that low pH levels can induce the metabolic inhibition of gill cells (Stapp et al., 2015). When we consider the overall body metabolism, data obtained by using fewer tissues as the experimental materials cannot be used to comprehensively assess the impacts of OA on the metabolism costs of marine teleosts. Therefore SMR could not be used as the only indicator to reflect the impacts of OA on fish metabolism. It is an indisputable fact that hormones and other regulatory factors such as seasonal variation and circadian rhythm can also affect fish metabolism (Fabbri et al., 1998; Fujisawa et al., 2019), so the above differences may be due to the lack of consideration of these factors.

Marine teleosts and other marine animals must ingest food, digest food, absorb nutrients, and metabolize energy to maintain their life activities. However, OA can alter the energy metabolism of marine teleosts, thus indirectly affecting their survival and growth. Photosynthesis of aquatic organisms is the main source of the energy in water environments (Liu et al., 2016). Photosynthesis of marine algae and coral and photoautotrophic organisms is highly sensitive to pCO_2, which is mainly due to the change in ribulose diphosphate carboxylase under high pCO_2 and low O_2 conditions (Gao et al., 2019). However, marine teleosts and heterotrophic organisms acquire nutrition mainly by ingesting these autotrophic organisms. On the one hand, OA affects the food chain of marine ecosystems by altering the photosynthesis of marine autotrophs; on the other hand, OA directly affects the nutrient metabolism of marine animals, including marine teleosts, and changes their metabolic rate. In order to cope with OA by achieving acid—base compensation and ion homeostasis, marine animals usually transfer most of the energy to enabling key physiological processes such as growth, immunity, and reproduction so as to improve their energy demand for responding to environmental stress (Fabry et al., 2008; Kroeker et al., 2010). However, when the OA level exceeds the maximum regulation range that the animal body can bear, it will cause a pathological reaction. For example, intracellular acidosis and environmental hypercapnia were found to inhibit marine teleost metabolism (Chen et al., 2020; Esbaugh, 2018). This kind of inhibition is an internal adaptive strategy used by many marine organisms to prolong the survival time under the conditions of short-term hypercapnia, hypoxia, and starvation. The typical metabolism inhibition method is stopping energy consumption, such as protein synthesis and ion conversion, which could therefore influence the growth and reproduction of marine teleosts (Esbaugh, 2018).

OA significantly affects the activities of various enzymes involved in protein metabolism. Esbaugh et al. (2012) reported that when *O. beta* was exposed to pCO_2 of 1900 μatm for 24 hours, the expression level of the CA gene was downregulated in the gill, and the activity of Na^+-K^+ ATPase was also decreased (Esbaugh et al., 2012). The superoxide dismutase and catalase enzyme activities were both affected in the larvae of sand smelt (*Atherina presbyter*) when they were exposed to CO_2-driven OA conditions (Silva et al., 2016). In addition, OA also can affect nucleic acid metabolism, which has been proved by many exposure experiments. For instance, Franke and Clemmesen (2011) found that the acidified seawater did not affect the embryo formation and hatching rate of *C. harengus*, but there was an obvious negative linear relationship between acidified levels and the ratio of RNA/DNA (Franke & Clemmesen, 2011).

According to the literature reports, the effects of OA on the growth of marine teleosts are more complex. These studies mainly focused on marine teleosts in the juvenile stage and various conclusions have been drawn, among which the most common results are inhibitory impacts (Frommel et al., 2016; Hancock et al., 2020; Miller et al., 2012; Noor & Das, 2019). This growth inhibition was found in estuarine inland silverside (*M. beryllina*) (pCO_2: 780 μatm; Gobler et al., 2018) and cinnamon anemonefish (*A. melanopus*) (pCO_2: 1032 μatm, Jarrold & Munday, 2019). In the study on juvenile Atlantic cod (*G. morhua*), the growth rate even decreased in a concentration-dependent manner (Baumann et al., 2012; Miller et al., 2015; Moran & Støttrup, 2011). However, the growth rate of juvenile individuals of *Solea senegalensis* was significantly reduced after OA exposure (Pimentel et al., 2014). Studies on the growth of the sea bass *D. labrax* larvae also drew the same conclusion, which was that the growth rate of marine teleosts exposed to pCO_2 of 1750 μatm is significantly lower than those reared in conventional seawater (Cominassi et al., 2020). At present, some studies have shown that OA has no significant effects on marine teleost growth. For example, OA (pCO_2: 1150 and 1700 μatm) had little effect on the growth of juvenile European sea bass (*D. labrax*) (Cominassi et al., 2019). Both the body growth and swimming ability of the sergeant fish *Rachycentron canadum* were found to be insignificantly changed when exposed to OA stress with pCO_2 of 800 and 2100 μatm (Bignami, Enochs et al., 2013; Bignami, Sponaugle et al., 2013). Under extreme OA conditions with pCO_2 of 3500 and 5400 μatm, the growth of juvenile R. canadum was delayed by 2−3 days compared with the control incubated in normal seawater. This strong OA resistance of R. canadum may be attributed to the fact that they usually live in coastal environments and their development usually experiences changing environmental conditions, resulting in the strong adaptive ability of this species (Bignami, Enochs et al., 2013; Bignami, Sponaugle et al., 2013). Franke and Clemmesen (2011) observed that OA did not affect the hatching rate of fertilized eggs in the marine teleost C. harengus. There was no linear relationship between pCO_2 levels and the total length, dry weight, yolk sac area, and otolith size of the newly hatched larvae (Franke & Clemmesen, 2011). Generally speaking, the growth-related parameters of marine teleosts such as Baltic cod, Atlantic herring, spiny damselfly fish, coral trout, and cobia have no significant changes when the pCO_2 level is lower than 2000 μatm (Bignami, Enochs et al., 2013; Bignami, Sponaugle et al., 2013; Franke & Clemmesen, 2011; Frommel et al., 2012; Heuer & Grosell, 2014; Hurst et al., 2012; Munday, Gagliano et al., 2011; Munday, Hernaman et al., 2011; Munday et al., 2013). In contrast to the above results, some studies also reported the positive effects of OA on the growth of juvenile marine teleosts (Bignami et al., 2014; Munday, Crawley et al., 2009; Munday, Dixson et al., 2009). Upon exposure to medium and higher CO_2 levels, the early body length of the marine teleost P. dentatus was found to be longer and their growth rate was faster than that of the control group (Chambers et al., 2014). When exposed to OA with pCO_2 of 770 and 2170 μatm, the marine teleost Coryphaena hippurus showed significantly higher growth and development rates than expected. The larval growth rates of the big-eye cod, the orange clownfish (A. percula), and the cinnamon anemonefish (A. melanopus) also increased after being exposed to seawater with pCO_2 of 1032 μatm (Hurst et al., 2012; Miller et al., 2012; Munday, Crawley et al., 2009; Munday, Dixson et al., 2009). Some researchers believe that the growth-promoting phenomenon may have resulted from the strong metabolic compensation mechanism mastered by some marine teleosts, which can improve the adaptability of marine teleosts to OA by reducing the conventional swimming speed and frequency. In summary, the existing research results cannot explain the complex responses of teleost to OA.

Other physiological impacts of ocean acidification on marine vertebrates

As can be seen from a large number of reports, OA is likely to have a wide range of impacts on the behavior of marine teleosts, threatening the stability and function of marine ecosystems. The related studies mainly involve functionally important behaviors, such as predation, escape behavior, behavioral lateralization, habitat identification, and selection. Moreover, a small number of studies focus on the behaviors of taxis, reproduction, collective behavior, and courage level (or anxiety level).

Effects of ocean acidification on predation and escape behavior

Predation is a series of behaviors undertaken by predators to hunt for food, mainly involving catching, processing, searching, chasing, and ingestion. However, through food chasing analysis, Dixson et al. (2015) found that OA exposure made the adult *Mustelus canis* slow to respond to food smell, which shortened the chase time, reducing the chase behavior. This means that OA will weaken the ability of marine teleosts to perceive food and eventually damage their predatory behavior. The prey does not wait to be killed in the same place during the process of hunting. When they feel the danger coming, they take various measures, such as escape and mimicry, to avoid predation. Even if they are injured or preyed on, they also release chemical signals to warn other individuals in the population to take defensive measures. Although the group cohesion was not changed, a risky behavior was observed in the damselfish *Chromis viridis* exposed to CO_2-driven OA conditions (Cattano et al., 2019). However, Munday et al. (2010) found that after 4 days of OA treatment, the sensitivity of *Pomacentrus wardi* larvae to predatory danger signals decreased and their residence time in an environment with predatory threat was prolonged, indicating that OA can weaken the ability of marine teleosts to perceive the danger of predation and render them impotent to take actions quickly and effectively to escape.

In recent years, in order to more truly reflect the effects of OA on predator—prey dynamics, many experiments have tried to study the prey and the predator simultaneously. For example, Allan et al. (2013) carried out OA exposure and predation experiments on *Pseudochromis fuscus* and *P. amboinensis* at the same time. They found that the ability of

P. amboinensis to escape from the predation of *P. fuscus* was seriously weakened and the predation ability of *P. fuscus* against *P. amboinensis* did not change significantly, which increased the final predation rate (Allan et al., 2013). Similarly, McCormick et al. (2018) found that *P. fuscus* had lower predatory enthusiasm, shorter chase distance, and less attack speed. Meanwhile, the ability of *P. wardi* to escape from *P. fuscus* was also reduced, which kept the final predation rate unchanged (McCormick et al., 2018). These studies suggest that although OA does not consequentially influence the final rate of predatory success, it will inevitably alter the predator–prey dynamics. It is very important to study the influences of OA on the dynamics of predator–prey interactions because OA can further alter the food chain structure of marine organisms, thus threatening the health and stability of the entire marine ecosystem (Domenici & Seebacher, 2020).

There are many kinds of prey for predators in natural ecosystems. Due to the difficulty of predation and other restrictions, predators often choose a specific prey and this is the food selectivity of that predator. The difficulty of predation is closely related to the predator's preying ability and the prey's ability to avoid predation. The same is true for marine teleosts. Therefore it is not difficult to speculate that the changes in the predatory behavior and the escape behavior of marine teleosts caused by OA are likely to be related to the change in food selectivity of these species. Ferrari et al. (2011) confirmed this hypothesis by conducting OA exposure and predation experiments on *P. fuscus* and four food sources of this marine teleost: *P. moluccensis*, *P. amboinensis*, *P. nagasakiensis*, and *P. chrysurus*. The results showed that the food selectivity of *P. fuscus* to these four kinds of teleosts changed from having a preference toward *P. nagasakiensis* and *P. chrysurus* to having a preference toward *P. moluccensis* and *P. amboinensis* after exposure to OA (Ferrari et al., 2011).

Effect of ocean acidification on behavioral lateralization

Behavioral lateralization is a phenomenon in which animals prefer to use one side of the body or sensory organs when they conduct a certain behavior. It is an external manifestation of asymmetry of brain function. Behavioral lateralization can effectively shorten the decision-making time of animals and improve the fitness of animals to the changing environment. The T-maze test confirmed that OA can significantly change the intensity of behavioral lateralization in marine teleosts. For example, the absolute index of behavioral lateralization of *Neopomacentrus azysron* larvae was significantly lower than that of the control group after 4 days of OA exposure (Domenici et al., 2012; Nilsson et al., 2012). However, in another study, there was no significant difference in the absolute index of behavioral lateralization between juvenile Atlantic cod (*G. morhua*) and the control group after exposure for 30 days (Jutfelt & Hedgärde, 2015). The same conclusion was also drawn in a study on sailfin molly (*Poecilia latipinna*), which showed that this teleost was tolerant to OA. However, the author also stressed that more studies are needed to confirm the lateralization tolerance of marine teleosts (Remnitz, 2018). Unlike the above results, Green and Jutfelt (2014) found that the absolute index of behavior lateralization of juvenile *S. canicula* was significantly increased by OA (Green & Jutfelt, 2014).

The T-maze experiment confirmed that OA can change the direction of fish lateralization. For example, OA exposure not only reduced the absolute indices of behavioral lateralization of the three-spined stickleback (*G. aculeatus*) but also made this species exhibit particularly left lateralization at the population level (Lai et al., 2015; Näslund et al., 2015). The conclusion is supported by more experimental results. The lateralization behavior of juvenile *P. wardi* shifted from right to left after 7 days of OA exposure (Domenici et al., 2014). Similarly, the lateralization behavior of adult *G. flavescens* changed from right to left under OA conditions (Sundin and Jutfelt). Behavioral lateralization of marine teleosts is closely related to their behaviors of predation, escape behavior, positioning, and clustering behaviors, which means that the intensity and direction of behavioral lateralization caused by OA would have profound impacts on marine teleosts.

Effects of ocean acidification on habitat recognition and selection

Marine fish usually recognize, locate, and select suitable survival habitats by using chemical signals, sound signals, and light signals through their olfactory, auditory, and visual senses, respectively. As an important source of population replenishment, juvenile fish need to find suitable environments to settle after the completion of the planktonic period, which is directly related to the maintenance of the fish population structure. After exposure to OA for 11 days, the ability of juvenile *A. percula* to recognize habitats using chemical signals decreased, while the residence time in the reefs was also shortened (Munday, Crawley et al., 2009;

Munday, Dixson et al., 2009), implying that it was more difficult for the juvenile individuals of this species to complete the settlement behavior due to the influence of serious OA conditions. In the experiments on the larvae of *P. amboinensis*, *P. chrysurus*, and *P. moluccensis*, it was found that OA not only weakened the teleosts' abilities to seek habitats using chemical signals but also changed the time rhythm of their settlement (Devine et al., 2012a, 2012b). Studies have shown that OA also can affect marine teleosts' ability to recognize habitats through sound signals. For instance, the juveniles of *Pennahia argentata* and *Pomatoschistus pictus* were no longer sensitive to the sound signals of local habitats after OA exposure, resulting in the synchronous decrease in the recognition ability and the success rate of settlement (Castro et al., 2017). Moreover, OA can cause marine teleost juveniles to be attracted by unrelated sound signals. Rossi et al. (2018) found that the ability of juvenile *Lates calcarifer* to recognize sound signals in their living habitats decreased after exposure to OA and that they were inversely attracted by unrelated sound signals (Rossi, Pistevos, Connell, & Nagelkerken, 2018).

Similar to juveniles, OA also alters the ability of adult teleosts to recognize their habitats and then affects the subsequent habitat positioning and settlement behavior. The sensitivity and recognition ability of *Cheilodipterus quinquelineatus* adults to habitat chemical signals decreased and the final successful homing rate was only 69%−78% of that of the control group after 4 days of OA exposure (Devine et al., 2012a, 2012b). Not only that, OA also even changes the habitat preference of marine teleosts. For example, adult *Paragobiodon xanthosoma*, which originally preferred to inhabit around the coral *Seriatopora hystrix*, was found to no longer prefer to inhabit around *S. hystrix* after exposure to OA for 4 days (Devine & Munday, 2013). OA can impair teleosts' ability to detect odors. This, in turn, makes it harder for them to find food, avoid predators, or find suitable spawning areas. In addition, the electrophysiological activities of the nervous system of marine teleosts were recorded in a study. The results showed that the expressions of odorant sensing/processing genes in the olfactory organs and the brain were decreased significantly, which suggests that OA may cause the loss of olfactory ability. Their abilities to detect and respond to certain odors associated with food and threat situations were significantly negatively affected (Porteus et al., 2018).

The triggering of fish behaviors includes three consecutive processes: first, external signal recognition and conduction; second, nerve center processing the signal and giving instructions; and third, organs making actions, which is the effect. Any problem in these processes will lead to abnormal behaviors in marine teleosts. However, the current research on this aspect mainly focuses on two points: sensory function and neural function. OA can cause abnormal behaviors in marine teleosts by interfering with a variety of sensory and neural functions (Ashur et al., 2017; Cattano et al., 2018; Esbaugh, 2018).

Like many marine animals, fish also have olfactory, visual, auditory, and other sensory functions. It is with the help of these sensory functions that external signals can be recognized by marine teleosts, which then trigger various behaviors. Olfactory-mediated behavior is the main research object of current studies. The olfactory sensitivity was found to be decreased in gilthead seabream (*S. aurata*) exposed to short- and medium-term OA conditions (Velez et al., 2019). It is the OA stress that hinders olfactory function leading to these behavioral abnormalities. The protonation and deprotonation of some chemical substances including purines, pterins, sterols, amino acids, and polypeptides acting as chemical signals are very sensitive to pH changes. Therefore OA is likely to alter the conformation of these chemical substances by affecting the processes of protonation and deprotonation, resulting in abnormal chemical signal recognition in marine teleosts (Ashur et al., 2017; Munday, Crawley et al., 2009; Munday, Dixson et al., 2009). Although there were no reports directly confirming this view, this mechanism has long been confirmed to be one of the reasons for marine invertebrates' abnormal behavior caused by OA. For example, a study confirmed that OA changed the conformation of the signal peptide glycyl-glycyl-L-arginine, which affected the recognition of this signal peptide, resulting in the reduction in ventilation times of *Carcinus maenas* eggs (Roggatz et al., 2016). At the same time, a similar mechanism has also been observed in freshwater fish and has been widely recognized as one of the reasons for their abnormal behavior (Leduc et al., 2013). These results show that this mechanism may be widespread in marine teleosts. It can be concluded that the disturbance in the recognition of chemical signals is likely the reason for abnormal olfactory function and behaviors of marine teleosts.

In addition to olfaction, OA also interferes with auditory function, resulting in some abnormal behaviors in marine teleosts. For example, OA was found to impair the auditory avoidance behavior of *A. percula* and reduce the ability of this species to evade coral reef noise (Simpson et al., 2011). The ability of a variety of marine teleost juveniles to recognize their habitats through sound declined after exposure to OA

(Castro et al., 2017; Rossi et al., 2016, 2018). It was found that OA affected the development of teleost otolith, resulting in changes in otolith morphology and density (Bignami, Enochs et al., 2013; Bignami, Sponaugle et al., 2013; Di Franco et al., 2019). The hearing function of marine teleosts is closely related to the otolith, so it is generally believed that changes in otolith may be the cause of the abnormal hearing function. For example, the otolith volume and density in R. *canadum* larvae increased under OA conditions, which enhanced their auditory ability, making this species able in recognizing more sounds according to the model prediction (Bignami, Enochs et al., 2013; Bignami, Sponaugle et al., 2013). However, for marine teleosts, an elevation in auditory function may not be good news, as it can lead to interferences by irrelevant sounds and increase the difficulty of target sound reorganization. For example, OA increased the abilities of the otolith and hearing in A. *japonicus* but decreased its ability to recognize its habitat by sound and its colonization rate (Rossi, Nagelkerken, Pistevos, & Connell, 2016). In addition, the ability of L. *calcarifer* to recognize the sounds of the habitat decreased, and they were subsequently attracted by unrelated sounds (Rossi, Pistevos, Connell, & Nagelkerken, 2018). This evidence reinforces our conclusion about the effects of OA on marine teleosts.

It is speculated that OA may affect the function of the γ-aminobutyric acid type A (GABA$_A$) receptor and then change the behaviors of marine teleosts (Heuer et al., 2019). This conclusion is mainly drawn based on some experiments and is supported by the following results. For example, the sensitivity of A. *percula* to the chemical signals of predators decreased after exposure to OA, while the residence time in an area with threat from predators was prolonged, and the behavior of fleeing from predators was also impaired (Dixson et al., 2010). The behavioral lateralization ability of N. *azysron* decreased under high pCO$_2$ conditions (Domenici et al., 2012). Both of these behavioral damages could be reversed by gabazine, a specific antagonist of the GABA$_A$ receptor (Masiulis et al., 2019). This is the first time that scientists explained the behavioral changes in marine teleosts caused by OA from the perspective of neural function at the molecular level. Using similar methods, more and more studies have confirmed that this mechanism was ubiquitous in marine teleosts (Esbaugh, 2018). The GABA$_A$ receptor is an inhibitory receptor of the central nervous system. It is an ionic receptor and a kind of ligand-gated ion channel. After binding with the endogenous ligand neurotransmitter GABA, the ion channel is opened and a large amount of Cl$^-$ and a small amount of

HCO_3^- flow from outside the cell to inside the cell, thus causing the hyperpolarization of neurons, which plays an inhibitory role in marine teleost behaviors (Tresguerres & Hamilton, 2017). In order to maintain acid—base balance under OA conditions, marine teleosts must excrete H^+ and Cl^- from extracellular spaces into the seawater and accumulate intracellular HCO_3^-, which will change the intracellular and extracellular concentration gradients of Cl^- and HCO_3^-. After the GABA receptor is activated, the ion channel is opened, and the Cl^{-1} and HCO_3^- then flow from the inside to outside of the cells. This change causes the depolarization of neurons, playing an exciting role in the nervous system (Tresguerres & Hamilton, 2017). In this case, the function of the $GABA_A$ receptor changes from "inhibition" to "excitability", which will cause alterations in the sensory function and even the following behaviors of marine teleosts.

Influence of ocean acidification on other behaviors

In addition to the abovementioned behaviors, a few studies have also reported the effects of OA on teleosts' tropism, reproductive behavior, shoaling behavior, and boldness. For example, it was found that OA enhanced the sensitivity of G. flavescens larvae to the light source, altering the number of phototactic individuals and the positive phototaxis (Forsgren et al., 2013). On the contrary, the anxiety level and cautious behavior of juvenile S. diplopora increased while the courage level decreased after exposure to OA. Meanwhile, the individuals of this species exposed to OA preferred to stay in the dark (Hamilton et al., 2014). In any case, OA can significantly change the phototaxis behavior and courage level of marine teleosts in the planktonic period. It is not difficult to speculate that these effects may further alter the dispersal of juvenile marine teleosts. OA also can affect the group behavior of marine teleosts. Lopes et al. (2016) found that the colony strength of A. presbyter significantly decreased when compared with the control group after 7 days of OA exposure. Similarly, OA significantly shortened the colony formation time of A. regius (Maulvault et al., 2018; Sampaio et al., 2018). Another significant impact of OA on marine teleosts is that it can damage reproductive behavior, thus threatening the reproduction of the population. Milazzo et al. (2016) detected that OA significantly reduced the efficiencies of courtship and mating of Symphodus ocellatus, prolonging the duraions of courtship and mating.

Case studies

Case study 1—Ion and acid—base regulation

Scientists from the University of Gothenburg and Helmholtz Center for Polar and Marine Research investigated the response of key ions and acid—base regulatory genes in the gill of Atlantic cod (*G. morhua*), a marine teleost with important ecological and commercial values, to increase in seawater pCO_2 levels (Fig. 2.3; Michael et al., 2016). The cod *G. morhua* is widely distributed across the Atlantic Ocean. In this study, *G. morhua*

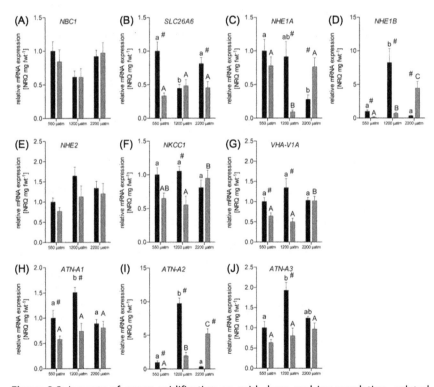

Figure 2.3 Impacts of ocean acidification on acid—base and ion regulation—related genes in the gills of the Atlantic cod. *From Michael, K., Kreiss, C., Hu, M., Koschnick, N., Bickmeyer, U., Dupont, S., Pörtner, H., & Lucassen, M. (2016). Adjustments of molecular key components of branchial ion and pH regulation in Atlantic cod* (Gadus morhua) *in response to ocean acidification and warming. Comparative Biochemistry and Physiology Part - B: Biochemistry and Molecular Biology, 193, 33—46. https://doi.org/ 10.1016/j.cbpb.2015.12.006.*

collected from the Skagerrak/Kattegat population in Sweden was exposed to acidified seawater, covering current and future pCO_2 levels, namely 550 µatm (low), 1200 µatm (medium), and 2200 µatm (high). All the exposure experiments were conducted at both 10°C and 18°C. The functional relationships between Na^+/K^+-ATPase, NKCC1, and NBC1 in the ionocytes of G. morhua were accessed by analyzing the cell localizations of these ion transporters. The expression changes with different ion transporters such as NKCC1, NBC1, NHE1, and NHE2; SLC26A6, a member of the anion transporter family, in the gill tissue of G. morhua exposed to OA was characterized. In order to reveal the possible effects of OA on energy budget in the gills of G. morhua, the activities of ATP-dependent ion pumps including Na^+/K^+-ATPase and V-type H^+-ATPase were also measured at the mRNA, protein, and function levels.

The results obtained from this study showed that the mRNA expression levels of some ion transporters were significantly higher in medium pCO_2 than those in low pCO_2 and high pCO_2 at 10°C. The expression level of NHE1B under the medium pCO_2 condition increased nearly 10-fold than that in low and high pCO_2 conditions. Similarly, the expression levels of all the three $Na + /K + $-ATPase α subunits (ATNA1, ATNA2, and ATNA3) increased under the medium pCO_2 condition. On the contrary, the expression level of the anion transporter SLC26A6 decreased nearly twice under the same condition. Under the medium pCO_2 condition, the mRNA expression level of NBC1 decreased significantly. In addition, the expressions of NHE1A mRNA showed different reaction patterns, because the expression level of NHE1A in the gill tissue of G. morhua exposed to high pCO_2 decreased nearly fourfold compared with the other two pCO_2 levels.

When the pCO_2 level was higher, the mRNA expression levels of H^+-ATPase subunit V1A (VHA-V1A) and NKCC1 increased nearly twice as much as those under low and medium pCO_2 conditions at 18°C, (Fig. 2.3). Compared with the G. morhua exposed to low and high pCO_2, the expression level of NHE1A decreased about eightfold under medium pCO_2. Under the same exposure condition, the mRNA expression level of NBC1 decreased obviously. Moreover, the expression difference was the largest between NHE1B and ATNA2. Further analysis found that there was a positive correlation between gene expression levels and the increase in pCO_2, which increased about 10-fold for NHE1B and 55-fold for ATNA2 from low pCO_2 to high pCO_2. It is worth noting that the abundance of NHE1B in the gill tissue of G. morhua was about 400-fold lower than that of NHE1A. For ATNA2, the difference was even greater. The average expression level of ATNA2 was 1000-fold lower than that of ATNA1, which showed the highest cell subtype abundance.

The analysis of the enzyme activity also showed that OA has a significant impact on ion and acid—base regulation. The results showed that Na^+/K^+-ATPase activity was not related to the pCO_2 level. However, the activity of this enzyme at 18°C was nearly twice as high as that at 10°C. The activity of H^+-ATPase was inhibited by medium pCO_2 stress at 10°, C but it was not affected by medium pCO_2 stress at 18°C. However, the temperature sensitivity of H^+-ATPase decreased after thermal acclimation. The authors found that this decrease appeared to be dependent on the pCO_2 level. Under medium and high pCO_2 conditions, the activity of F1F0-ATP synthase increased significantly (about twice). Similar to Na^+/K^+-ATPase, the Q10 value of F1F0-ATP synthase remained unchanged among the exposure groups. The correlation between Na^+/K^+-ATPase and F1F0-ATP synthetase activities was the strongest at both 10°C and 18°C.

In this study, the authors compared the regulation modes of some important gene components related to ion and acid—base homeostasis in the gill of G. morhua after exposure to different pCO_2 levels to reveal the basic mechanisms involved in the responses of marine teleosts to OA. The results showed that the expression levels of most transporter genes increased under medium pCO_2 conditions. The mRNA and protein expression levels of different ion transporters are closely related, which confirmed the relationship between their functions. The results also revealed the importance of the coordination between transporters involved in acid—base and ion regulation in the gill of G. morhua when the pCO_2 was increased. The functional activity of F1F0-ATP synthetase was usually related to the activities of Na^+/K^+-ATPase and H^+-ATPase. These expression patterns therefore suggested that the genes related to acid—base and ion regulation in the gills of marine teleosts were easily affected by OA. It can be inferred from this study that in order to adapt to the changes in marine environments, especially OA, the gills of marine teleosts often show certain physiological plasticity.

Case study 2—Metabolism

The Atlantic cod G. morhua was also taken as the research object to study the effects of OA on marine teleost metabolism. Frommel, Hermann et al. (2020) investigated the effects of OA on the expressions of 18 lipid metabolism-related genes in this species, hoping to provide the scientific basis for clarifying the potential mechanisms of intestinal, kidney, pancreatic, and liver disease occurrences caused by elevated pCO_2. The fertilized eggs of *G. morhua* were raised in seawater under three acidification levels (control: 380 μatm, medium: 1800 μatm, and high: 4200 μatm). After exposure for 32 hours (larvae) and 46 hours (juvenile), the fish samples were collected for subsequent gene expression analysis.

The results showed that after 32 hours of exposure, the gene expression patterns of 18 lipid metabolism-related genes in some cod larvae were similar to those of the control, while other larvae showed highly upregulated or downregulated gene expression patterns (Fig. 2.4). Particularly, the 6PGL, ACoA, and PPAR1b genes were significantly upregulated under medium and high pCO_2 conditions, while the CPTA1, FAS, and GYS2 genes were significantly upregulated under the medium pCO_2 condition. Compared with the control, the expression level of the PPAR1B gene encoding a transcription factor increased threefold at the medium pCO_2 level and increased twofold at the high pCO_2 level. The expression level of the GYS2 gene, which plays an important role in glycogen synthesis, increased threefold when compared to the control at the medium pCO_2 level, but its expression level was similar to the control at the high pCO_2 condition. The expression of the 6PGL gene involved in the pentose phosphate pathway increased twofold at the medium pCO_2 level and increased by threefold at the high pCO_2 level. The ACoA, FAS, and CPTA1 genes play important roles in fatty acid synthesis. The expression of the ACoA gene increased by 2.5-fold at the medium pCO_2 level and only the ACoA gene was significantly upregulated at the high pCO_2 level. In addition, the genes related to ion regulation (NKA), glycolysis (GAPDH), glycogen synthesis (GYS1), glucose oxidation (GDH), β oxidation, and digestion (TRP1A, BAL) showed complex upregulation and downregulation patterns. The expressions of genes that are important for the process of gluconeogenesis (PEPCK) and growth (GH) were upregulated at both the medium and high pCO_2 levels. Unlike the larvae, the cod juveniles showed a similar gene expression pattern after 46 hours of OA exposure, and their response to pCO_2 was almost unchanged. Compared with the control, the expressions of most genes were slightly downregulated under these two pCO_2 levels, but there was no significant difference in the average value.

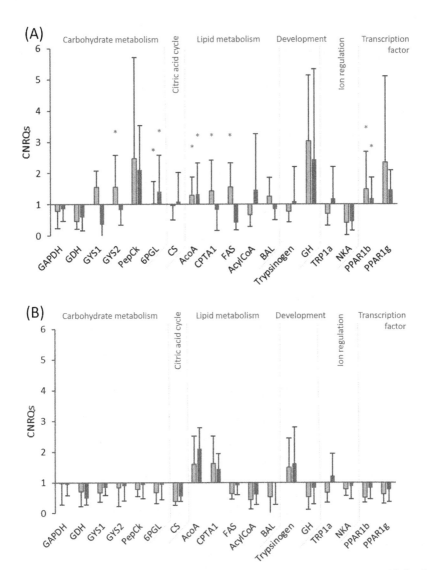

Figure 2.4 Impacts of ocean acidification on the expression patterns of lipid metabolism-related genes in the Atlantic cod. A, juveniles at 32 dph. B, juveniles at 46 dph. *From Frommel, A., Hermann, B., Michael, K., Lucassen, M., Clemmesen, C., Hanel, R., & Reusch, T. (2020). Differential gene expression patterns related to lipid metabolism in response to ocean acidification in larvae and juveniles of Atlantic cod. Comparative Biochemistry and Physiology Part A: Molecular & Integrative Physiology, 247, 110740. https://doi.org/10.1016/j.cbpa.2020.110740.*

Among the six genes related to carbohydrate metabolism, two genes were significantly upregulated in *G. morhua* exposed to high $p\mathrm{CO_2}$, namely GYS2 and 6PGL. GYS2 encodes a protein subtype of glycogen synthase associated with fat and muscle tissue. The upregulation of GYS2 may lead to the accumulation of glycogen in the form of energy storage and send out stress response signals. Similarly, the overexpression of 6PGL has been proven to play a role in the overproduction of proteins regarding stress response, thus causing metabolic burden. The 6PGL gene can catalyze the first two steps of the oxidation phase of the pentose phosphate pathway, which is connected with other important metabolic pathways. The enhanced activity of pentose phosphate therefore provides a mechanistic explanation for abnormal lipid deposition in *G. morhua* larvae at 32 dph. The increased demand for NADPH indicated the enhancement in oxidative stress as it is necessary to reduce glutathione disulfide to glutathione. The synchronous increase in the expression level of PPAR1b in this study further supported this conclusion, as this transcription factor is closely related to apoptosis and interception of oxidative stress. Moreover, the results also indicated that the expressions of the ACoA and fatty acid synthase (FAS) genes in medium and high $p\mathrm{CO_2}$ conditions were twice as high as those in the control. ACoA is mainly expressed in adipogenic tissues, such as the liver, which is very important for the biosynthesis of long-chain fatty acids catalyzed by FAS. Long-chain fatty acids can combine with triglycerides and phospholipids to form a main energy storage mode in animals. However, the accumulation of triglycerides in liver cells can lead to liver steatosis, which may be a way in which liver injury occurs in *G. morhua* larvae under high $p\mathrm{CO_2}$ conditions. After the exposure of *G. morhua* larvae to high $p\mathrm{CO_2}$, the expression of ACoA increased, while the expression of FAS decreased. This result suggested that the production of malonyl CoA in larvae was increased under high $p\mathrm{CO_2}$ conditions. The authors believe that the increase in malonyl CoA content can further stimulate glycolysis, resulting in more carbohydrates being converted into lipids in the larvae of *G. morhua*.

From the point of view of development, the expressions of some genes involved in transcription and energy metabolism increased in *G. morhua* in their larval stage (32 dph), especially the genes involved in the synthesis of glycogen and fatty acids, while the corresponding gene expressions in juvenile individuals (46 dph) were basically unaffected. These findings may reflect that the ion and acid–base balance in *G. morhua* larvae was destroyed, and the oxidative stress and metabolic capacity decreased significantly under OA.

Taken together, the authors of this study believe that the expressions of the genes related to major metabolic pathways of marine teleosts represented *by G. morhua* changed under OA conditions, especially the upward expression trend of the genes participating in fatty acid and glycogen metabolisms, which can support the previously reported results about the pathological changes caused by lipid metabolism disorder in marine teleosts exposed to OA.

Case study 3—Escape behavior

The marine medaka *O. melastigma* is a model organism that has been used in recent years for marine ecotoxicology research. The authors from Shanghai Ocean University studied the effects of OA on the escape behavior of juvenile *O. melastigma* (Wang et al., 2017). The fertilized eggs of *O. melastigma* were cultured in three containers until they hatched. The acidification degrees of the seawater in the experimental containers were set as three gradients: the control pCO_2, 450 µatm; medium pCO_2, 1160 µatm; and high pCO_2, 1783 µatm. A total of 350 individuals were cultured in each container and exposed to OA conditions at 28°C ± 1°C. Every day from 1700 to 1800, a total of 40% of the treating water in the container was replaced by fresh acidified seawater. The pCO_2 level of seawater in each container was maintained by using a CO_2 concentrator. The concentrator aerates seawater through a fine pore stone at a flow rate of 0.6 L/min so as to keep the seawater carbonate system at a stable level. The top of each container was sealed with transparent plastic film to limit the exchange of CO_2 with the atmosphere.

In this study, the escape behavior of the marine medaka *O. melastigma* in a Petri dish under OA conditions was observed and recorded. The detailed experimental methods are given as follows: the culture dish wall was tapped with a plastic ball with a diameter of 1.2 cm and the plastic ball was dropped from the same height (16 cm) each time, so as to induce the C-start escape behavior of *O. melastigma*. The C-start behavior of *O. melastigma* was recorded by a high-speed digital camera and was used to characterize the escape behavior. In order to avoid the influence of surrounding obstacles on the escape behavior of the fish, the authors only analyzed the behavior of *O. melastigma* at the center of each culture dish. A successful C-start of *O. melastigma* was detected from the captured image.

The results showed that *O. melastigma* was prone to C-start escape from mechanical stimulation (Fig. 2.5). The occurrence of the C-start

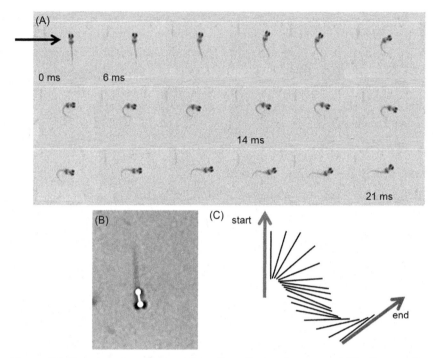

Figure 2.5 C-start escape for assessing the effects of ocean acidification on the escape behavior of the marine medaka. A, experimental setup for the stimulation. B, the C-start behaviour. C, frame analysis. *From Wang, X., Song, L., Chen, Y., Ran, H., & Song, J. (2017). Impact of ocean acidification on the early development and escape behavior of marine medaka (Oryzias melastigma). Marine Environmental Research, 131, 10–18. https://doi.org/10.1016/j.marenvres.2017.09.001.*

behavior of *O. melastigma* in the medium and high pCO_2 conditions was significantly lower than that in the control, but there were no significant differences in the latency of the initiation of the movement, the maximum bending angle, and the duration of the first stage of the escape behavior of *O. melastigma* exposed to the two levels of pCO_2 stresses when compared to the control.

The authors of this study believe that the behavior and sensory systems of juvenile teleosts are easily affected by OA. Vision is very important for daytime marine life to capture prey and avoid predators. Therefore any defects in the visual system may have serious negative impacts on the survival of teleost individuals. In this study, the eye defects and impairment in brain development of newly hatched

O. melastigma larvae were very obvious under high *p*CO$_2$ conditions. Abnormal eye development will affect some behaviors that need good binocular vision in marine teleosts, such as the escape behavior. The activation of the Mauthner cells usually leads to the C-start escape behavior. The Mauthner cell is a command-like neuron that can receive a large number of sensory input signals. In this study, the medium and high *p*CO$_2$ conditions affected the probability of the C-start escape behavior of *O. melastigma*, which may indicate that an increase in *p*CO$_2$ has a significant negative impact on otolith development and sensory input. The otolith of many marine teleosts has been proved to be altered by an increase in *p*CO$_2$. The changes in otolith size, density, and mass will affect the mechanics of the otolith and the auditory function of marine teleosts. Therefore the authors concluded that the effect of OA on the C-start behavior may be related to changes in the otolith of *O. melastigma* exposed to high *p*CO$_2$. The effects of OA on the eye position of the marine medaka *O. melastigma* may change the binaural hearing and further affect the sound-induced escape behavior under high *p*CO$_2$ conditions. In addition, the authors also believe that the regulatory changes in HCO$_3^-$ and Cl$^-$ concentrations during CO$_2$ exposure will interfere with the neurotransmitter function of *O. melastigma*, thus affecting some behaviors and causing drastic changes in sensory preferences of the organisms of this species. These effects may indirectly result in changes in the escape behavior of *O. melastigma*. Although these results have been obtained, the authors pointed out that in order to fully understand the relationship between OA and the abnormal behaviors of marine teleosts, more in-depth research works are needed.

Summary

As mentioned above, serious OA caused by increasing seawater CO$_2$ concentration has become another major problem threatening marine ecosystems and endangering marine ecological security. Recently, more and more attention has been paid to the influences of OA on various physiological functions of marine vertebrates, especially marine teleosts, including acid—base and ion regulation, larval growth, embryonic development, reproduction, and behaviors. Existing research results showed that OA can affect acid–base homeostasis, growth speed, fertilization rate, embryonic development process, predation, escape behavior, lateralization, habitat recognition, and the selection of marine teleosts by interfering with their physiological states, nerve conduction, and external signal perception and recognition. Because of the important fishery value and ecological function of marine teleosts, we need to be vigilant against the subsequent risks of OA. Obviously, we should deeply understand the effects of OA on marine teleosts and reveal the underlying physiological and molecular mechanisms behind these impacts, which is the prerequisite for scientifically assessing the threat of OA. Although most studies have shown that OA would damage the physiology of marine teleosts, some studies have also shown the presence of no significant effects, and even positive effects. This means that different marine vertebrates or different life stages of the same species possess diverse and complex tolerances to OA. In addition, a large number of studies confirmed that OA has serious impacts on the behaviors of marine teleosts. However, scientists have recently tested the repeatability of these obtained results and have drawn a contradictory conclusion. In their 3-year study, Clark et al. examined more than 900 wild and farmed marine teleosts from six different species. They analyzed their predator avoidance, activity levels, and behavioral lateralization. It was found that increased OA did not affect the key behaviors of coral reef teleosts. Compared with previous studies, the future impacts of pCO_2 levels at the end of this century on key behaviors of coral reef teleosts were found to be negligible. It was also found by data simulation that the huge effects and the small intragroup differences caused by OA reported in previous studies are impossible (Clark et al., 2020). In conclusion, this study suggests that the reported effects of OA on the behaviors of coral reef teleosts are unrepeatable, suggesting that behavioral disturbance is not the main consequence of the effects of OA on marine teleosts. This puts forward new requirements for the scientific evaluation of the impacts of OA on marine teleosts.

The effects of OA on marine vertebrates are complex, so more attention should be paid to this field in future research. In particular, we should carry out more long-term, intergenerational exposure studies to determine whether marine vertebrates can adapt to future OA and analyze the combined effects of various marine environmental stresses. We should focus on population biology studies, so as to accurately predict the potential impacts of OA on the global marine economy, especially marine fisheries. In order to understand the underlying mechanisms of the effects of OA on marine verte-brates from the perspectives of the population, individual, physiological, and molecular levels, we should conduct deeper basic biological research as well.

References

Albright, R., Takeshita, Y., Koweek, D. A., Ninokawa, A., Wolfe, K., Rivlin, T., Nebuchina, Y., Young, J., & Caldeira, K. (2018). Carbon dioxide addition to coral reef waters suppresses net community calcification. *Nature*, *555*(7697), 516−519.

Allan, B. J. M., Domenici, P., McCormick, M. I., Watson, S., & Munday, P. L. (2013). Elevated CO_2 affects predator-prey interactions through altered performance. *PLoS One*, *8*(3), e58520.

Allmon, E. B., & Esbaugh, A. J. (2017). Carbon dioxide induced plasticity of branchial acid-base pathways in an estuarine teleost. *Scientific Reports*, *7*(1), 45680.

Alper, S. L., & Sharma, A. K. (2013). The SLC26 gene family of anion transporters and channels. *Molecular Aspects of Medicine*, *34*(2), 494−515.

Ashur, M. M., Johnston, N. K., & Dixson, D. L. (2017). Impacts of ocean acidification on sensory function in marine organisms. *Integrative and Comparative Biology*, *57*(1), 63−80.

Baker, D. W., Sardella, B., Rummer, J. L., Sackville, M., & Brauner, C. J. (2015). Hagfish: Champions of CO_2 tolerance question the origins of vertebrate gill function. *Scientific Reports*, *5*(1), 11182.

Baumann, H., Talmage, S. C., & Gobler, C. J. (2012). Reduced early life growth and sur-vival in a fish in direct response to increased carbon dioxide. *Nature Climate Change*, *2* (1), 38−41.

Bignami, S., Enochs, I. C., Manzello, D. P., Sponaugle, S., & Cowen, R. K. (2013). Ocean acidification alters the otoliths of a pantropical fish species with implications for sensory function. *Proceedings of the National Academy of Sciences of the United States of America*, *110*(18), 7366.

Bignami, S., Sponaugle, S., & Cowen, R. K. (2013). Response to ocean acidification in larvae of a large tropical marine fish, *Rachycentron canadum*. *Global Change Biology*, *19* (4), 996−1006.

Bignami, S., Sponaugle, S., & Cowen, R. K. (2014). Effects of ocean acidification on the larvae of a high-value pelagic fisheries species, mahi-mahi *Coryphaena hippurus*. *Aquatic Biology*, *21*(3), 249−260.

Branch, T. A., DeJoseph, B. M., Ray, L. J., & Wagner, C. A. (2013). Impacts of ocean acidification on marine seafood. *Trends in Ecology & Evolution*, *28*(3), 178−186.

Brauner, C. J., & Baker, D. W. (2009). Patterns of acid−base regulation during exposure to hypercarbia in fishes. In M. L. Glass, & S. C. Wood (Eds.), *Cardio-respiratory control in vertebrates: Comparative and evolutionary aspects* (pp. 43−63). Berlin, Heidelberg: Springer Berlin Heidelberg.

Brauner, C. J., Shartau, R. B., Damsgaard, C., Esbaugh, A. J., Wilson, R. W., & Grosell, M. (2019). Acid-base physiology and CO_2 homeostasis: Regulation and compensation in response to elevated environmental CO_2. In M. Grosell, P. L. Munday, A. P. Farrell, & C. J. Brauner (Eds.), *Fish physiology* (pp. 69−132). Academic Press.

Castro, J. M., Amorim, M. C. P., Oliveira, A. P., Gonçalves, E. J., Munday, P. L., Simpson, S. D., & Faria, A. M. (2017). Painted goby larvae under high-CO_2 fail to recognize reef sounds. *PLoS One*, *12*(1), e170838.

Catches, J. S., Burns, J. M., Edwards, S. L., & Claiborne, J. B. (2006). Na^+/H^+ antiporter, $V\text{-}H^+\text{-}ATPase$ and $Na^+/K^+\text{-}ATPase$ immunolocalization in a marine teleost (*Myoxocephalus octodecemspinosus*). *The Journal of Experimental Biology*, *209*(17), 3440.

Cattano, C., Claudet, J., Domenici, P., & Milazzo, M. (2018). Living in a high CO_2 world: A global *meta*-analysis shows multiple trait-mediated fish responses to ocean acidification. *Ecological Monographs, 88*(3), 320−335.

Cattano, C., Fine, M., Quattrocchi, F., Holzman, R., & Milazzo, M. (2019). Behavioural responses of fish groups exposed to a predatory threat under elevated CO_2. *Marine Environmental Research, 147*, 179−184.

Chambers, R. C., Candelmo, A. C., Habeck, E. A., Poach, M. E., Wieczorek, D., Cooper, K. R., Greenfield, C. E., & Phelan, B. A. (2014). Effects of elevated CO_2 in the early life stages of summer flounder, *Paralichthys dentatus*, and potential consequences of ocean acidification. *Biogeosciences, 11*(6), 1613−1626.

Chen, Y., Bai, Y., Hu, X., Yang, X., & Xu, S. (2020). Energy metabolism responses in muscle tissue of rainbow trout *Oncorhynchus mykiss* fry to CO_2-induced aquatic acidification based on metabolomics. *Aquatic Toxicology, 220*, 105400.

Claiborne, J. B., Blackston, C. R., Choe, K. P., Dawson, D. C., Harris, S. P., Mackenzie, L. A., & Morrison-Shetlar, A. I. (1999). A mechanism for branchial acid excretion in marine fish: Identification of multiple Na^+/H^+ antiporter (NHE) isoforms in gills of two seawater teleosts. *The Journal of Experimental Biology, 202*(3), 315.

Claiborne, J. B., Choe, K. P., Morrison-Shetlar, A. I., Weakley, J. C., Havird, J., Freiji, A., Evans, D. H., & Edwards, S. L. (2008). Molecular detection and immunological localization of gill Na^+/H^+ exchanger in the dogfish (*Squalus acanthias*). *American Journal of Physiology Regulatory Integrative and Comparative Physiology, 294*(3), 1092−1102.

Clark, T. D., Raby, G. D., Roche, D. G., Binning, S. A., Speers-Roesch, B., Jutfelt, F., & Sundin, J. (2020). Ocean acidification does not impair the behaviour of coral reef fishes. *Nature, 577*(7790), 370−375.

Cominassi, L., Moyano, M., Claireaux, G., Howald, S., Mark, F. C., Zambonino-Infante, J., Le Bayon, N., & Peck, M. A. (2019). Combined effects of ocean acidification and temperature on larval and juvenile growth, development and swimming performance of European sea bass (*Dicentrarchus labrax*). *PLoS One, 14*(9), e221283.

Cominassi, L., Moyano, M., Claireaux, G., Howald, S., Mark, F. C., Zambonino-Infante, J., & Peck, M. A. (2020). Food availability modulates the combined effects of ocean acidification and warming on fish growth. *Scientific Reports, 10*(1), 2338.

Death, G., Lough, J. M., & Fabricius, K. E. (2009). Declining coral calcification on the great barrier reef. *Science, 323*(5910), 116−119.

Deigweiher, K., Koschnick, N., Pörtner, H., & Lucassen, M. (2008). Acclimation of ion regulatory capacities in gills of marine fish under environmental hypercapnia. *American Journal of Physioloy Regulatory Integrative and Comparative Physiology, 295*(5), 1660−1670.

Devine, B. M., & Munday, P. L. (2013). Habitat preferences of coral-associated fishes are altered by short-term exposure to elevated CO_2. *Marine Biology, 160*(8), 1955−1962.

Devine, B. M., Munday, P. L., & Jones, G. P. (2012a). Homing ability of adult cardinalfish is affected by elevated carbon dioxide. *Oecologia, 168*(1), 269−276.

Devine, B. M., Munday, P. L., & Jones, G. P. (2012b). Rising CO_2 concentrations affect settlement behaviour of larval damselfishes. *Coral Reefs, 31*(1), 229−238.

Di Franco, A., Calò, A., Sdiri, K., Cattano, C., Milazzo, M., & Guidetti, P. (2019). Ocean acidification affects somatic and otolith growth relationship in fish: Evidence from an in situ study. *Biology Letters, 15*(2), 20180662.

Dixson, D. L., Jennings, A. R., Atema, J., & Munday, P. L. (2015). Odor tracking in sharks is reduced under future ocean acidification conditions. *Global Change Biology, 21*(4), 1454−1462.

Dixson, D. L., Munday, P. L., & Jones, G. P. (2010). Ocean acidification disrupts the innate ability of fish to detect predator olfactory cues. *Ecology Letters, 13*(1), 68−75.

Domenici, P., Allan, B., McCormick, M. I., & Munday, P. L. (2012). Elevated carbon dioxide affects behavioural lateralization in a coral reef fish. *Biology Letters, 8*(1), 78−81.

Domenici, P., Allan, B. J. M., Watson, S., McCormick, M. I., & Munday, P. L. (2014). Shifting from right to left: The combined effect of elevated CO_2 and temperature on behavioural lateralization in a coral reef fish. *PLoS One, 9*(1), e87969.

Domenici, P., & Seebacher, F. (2020). The impacts of climate change on the biomechanics of animals. *Conservation Physiology, 8*(1), coz102.

Doney, S. C., Busch, D. S., Cooley, S. R., & Kroeker, K. J. (2020). The impacts of ocean acidification on marine ecosystems and reliant human communities. *Annual Review of Environment and Resources, 45*(1), 83—112.

Doney, S. C., Fabry, V. J., Feely, R. A., & Kleypas, J. A. (2009). Ocean acidification: The other CO_2 problem. *Annual Review of Marine Science, 1*(1), 169—192.

Esbaugh, A. J. (2018). Physiological implications of ocean acidification for marine fish: Emerging patterns and new insights. *Journal of Comparative Physiology B, 188*(1), 1—13.

Esbaugh, A. J., Ern, R., Nordi, W. M., & Johnson, A. S. (2016). Respiratory plasticity is insufficient to alleviate blood acid—base disturbances after acclimation to ocean acidification in the estuarine red drum, *Sciaenops ocellatus. Journal of Comparative Physiology B, 186*(1), 97—109.

Esbaugh, A. J., Heuer, R., & Grosell, M. (2012). Impacts of ocean acidification on respiratory gas exchange and acid-base balance in a marine teleost, *Opsanus beta. Journal of Comparative Physiol. B, 182*(7), 921—934.

Espinel-Velasco, N., Hoffmann, L., Agüera, A., Byrne, M., Dupont, S., Uthicke, S., Webster, N. S., & Lamare, M. (2018). Effects of ocean acidification on the settlement and metamorphosis of marine invertebrate and fish larvae: A review. *Marine Ecology Progress Series, 606*, 237—257.

Fabbri, E., Capuzzo, A., & Moon, T. W. (1998). The role of circulating catecholamines in the regulation of fish metabolism: An overview. *Comparative Biochemistry Physiolgy Part C, 120*(2), 177—192.

Fabry, V. J., Seibel, B. A., Feely, R. A., & Orr, J. C. (2008). Impacts of ocean acidification on marine fauna and ecosystem processes. *ICES Journal of Marine Science, 65*(3), 414—432.

Ferrari, M. C. O., McCormick, M. I., Munday, P. L., Meekan, M. G., Dixson, D. L., Lonnstedt, Ö., & Chivers, D. P. (2011). Putting prey and predator into the CO_2 equation-qualitative and quantitative effects of ocean acidification on predator-prey interactions. *Ecology Letters, 14*(11), 1143—1148.

Forsgren, E., Dupont, S., Jutfelt, F., & Amundsen, T. (2013). Elevated CO_2 affects embryonic development and larval phototaxis in a temperate marine fish. *Ecology amd Evoluion, 3*(11), 3637—3646.

Franke, A., & Clemmesen, C. (2011). Effect of ocean acidification on early life stages of Atlantic herring (*Clupea harengus*). *Biogeosciences, 8*(12), 3697—3707.

Frommel, A. Y., Carless, J., Hunt, B. P. V., & Brauner, C. J. (2020). Physiological resilience of pink salmon to naturally occurring ocean acidification. *Conservation Physiology, 8*(1), coaa059.

Frommel, A. Y., Hermann, B. T., Michael, K., Lucassen, M., Clemmesen, C., Hanel, R., & Reusch, T. B. H. (2020). Differential gene expression patterns related to lipid metabolism in response to ocean acidification in larvae and juveniles of Atlantic cod. *Comparative Biochemistry and Physiology. Part A, Molecular & Integrative Physiology, 247*, 110740.

Frommel, A. Y., Maneja, R., Lowe, D., Malzahn, A. M., Geffen, A. J., Folkvord, A., Piatkowski, U., Reusch, T. B. H., & Clemmesen, C. (2012). Severe tissue damage in Atlantic cod larvae under increasing ocean acidification. *Nature Climate Change, 2*(1), 42—46.

Frommel, A. Y., Maneja, R., Lowe, D., Pascoe, C. K., Geffen, A. J., Folkvord, A., Piatkowski, U., & Clemmesen, C. (2014). Organ damage in Atlantic herring larvae as a result of ocean acidification. *Ecological Applications, 24*(5), 1131—1143.

Frommel, A. Y., Margulies, D., Wexler, J. B., Stein, M. S., Scholey, V. P., Williamson, J. E., Bromhead, D., Nicol, S., & Havenhand, J. (2016). Ocean acidification has lethal and sub-lethal effects on larval development of yellowfin tuna, *Thunnus albacares*. *Journal of Experimental Marine Biology and Ecology, 482*, 18−24.

Frommel, A. Y., Schubert, A., Piatkowski, U., & Clemmesen, C. (2013). Egg and early larval stages of Baltic cod, *Gadus morhua*, are robust to high levels of ocean acidification. *Marine Biology, 160*(8), 1825−1834.

Frommel, A. Y., Stiebens, V., Clemmesen, C., & Havenhand, J. (2010). Effect of ocean acidification on marine fish sperm (Baltic cod: *Gadus morhua*). *Biogeosciences, 7*(12), 3915−3919.

Fu, C., Wilson, J. M., Rombough, P. J., & Brauner, C. J. (2010). Ions first: Na$^+$ uptake shifts from the skin to the gills before O$_2$ uptake in developing rainbow trout, *Oncorhynchus mykiss*. *Proceedings of the Royal Society B: Biological Sciences, 277*(1687), 1553−1560.

Fujisawa, K., Takami, T., Shintani, H., Sasai, N., Matsumoto, T., Yamamoto, N., & Sakaida, I. (2019). Evaluation of the impact of seasonal variations in photoperiod on the hepatic metabolism of medaka (*Oryzias latipes*). *bioRxiv*, 745646.

Gao, K., Beardall, J., Häder, D., Hall-Spencer, J. M., Gao, G., & Hutchins, D. A. (2019). Effects of ocean acidification on marine photosynthetic organisms under the concurrent influences of warming, UV radiation, and deoxygenation. *Frontiers in Marine Science, 6*, 322.

Gazeau, F., Quiblier, C., Jansen, J. M., Gattuso, J., Middelburg, J. J., & Heip, C. H. R. (2007). Impact of elevated CO$_2$ on shellfish calcification. *Geophysical Research Letters, 34*(7), L07603.

Gobler, C. J., Merlo, L. R., Morrell, B. K., & Griffith, A. W. (2018). Temperature, acidification, and food supply interact to negatively affect the growth and survival of the forage fish, *Menidia beryllina* (Inland Silverside), and *Cyprinodon variegatus* (Sheepshead Minnow). *Frontiers in Marine Science, 5*, 86.

Green, L., & Jutfelt, F. (2014). Elevated carbon dioxide alters the plasma composition and behaviour of a shark. *Biology Letters, 10*(9), 20140538.

Gruber, N., Clement, D., Carter, B. R., Feely, R. A., van Heuven, S., Hoppema, M., Ishii, M., Key, R. M., Kozyr, A., Lauvset, S. K., Lo Monaco, C., Mathis, J. T., Murata, A., Olsen, A., Perez, F. F., Sabine, C. L., Tanhua, T., & Wanninkhof, R. (2019). The oceanic sink for anthropogenic CO$_2$ from 1994 to 2007. *Science, 363*(6432), 1193.

Guh, Y., Lin, C., & Hwang, P. (2015). Osmoregulation in zebrafish: Ion transport mechanisms and functional regulation. *Excli Journal, 14*, 627−659.

Hamilton, S. L., Logan, C. A., Fennie, H. W., Sogard, S. M., Barry, J. P., Makukhov, A. D., Tobosa, L. R., Boyer, K., Lovera, C. F., & Bernardi, G. (2017). Species-specific responses of juvenile rockfish to elevated pCO$_2$: From behavior to genomics. *PLoS One, 12*(1), e169670.

Hamilton, T. J., Holcombe, A., & Tresguerres, M. (2014). CO$_2$-induced ocean acidification increases anxiety in rockfish via alteration of GABAA receptor functioning. *Proceedings of the Royal Society B: Biologcal Sciences, 281*(1775), 20132509.

Hancock, A. M., King, C. K., Stark, J. S., McMinn, A., & Davidson, A. T. (2020). Effects of ocean acidification on Antarctic marine organisms: A *meta*-analysis. *Ecology and Evolution, 10*(10), 4495−4514.

Hannan, K. D., Munday, P. L., & Rummer, J. L. (2020). The effects of constant and fluctuating elevated pCO$_2$ levels on oxygen uptake rates of coral reef fishes. *The Science of the Total Environment, 741*, 140334.

Heisler, N. (1986). Mechanisms and limitations of fish acid-base regulation. In S. Nilsson, & S. Holmgren (Eds.), *Fish physiology: Recent advances* (pp. 24−49). Dordrecht: Springer.

Heuer, R. M., & Grosell, M. (2014). Physiological impacts of elevated carbon dioxide and ocean acidification on fish. *American Journal of Physiological-Regulatory Integrative and Comparative Physiology, 307*(9), 1061−1084.

Heuer, R. M., & Grosell, M. (2016). Elevated CO_2 increases energetic cost and ion movement in the marine fish intestine. *Scientific Reports, 6*(1), 34480.

Heuer, R. M., Hamilton, T. J., & Nilsson, G. E. (2019). The physiology of behavioral impacts of high CO_2. In M. Grosell, P. L. Munday, A. P. Farrell, & C. J. Brauner (Eds.), *Fish physiology* (pp. 161−194). San Diego, CA: Academic Press.

Hofmann, G. E., Barry, J. P., Edmunds, P. J., Gates, R. D., Hutchins, D. A., Klinger, T., & Sewell, M. A. (2010). The effect of ocean acidification on calcifying organisms in marine ecosystems: An organism-to-ecosystem perspective. *Annual Review of Ecology Evoluionand Systematics, 41*(1), 127−147.

Hurst, T. P., Fernandez, E. R., Mathis, J. T., Miller, J. A., Stinson, C. M., & Ahgeak, E. F. (2012). Resiliency of juvenile walleye pollock to projected levels of ocean acidification. *Aquatic Biology, 17*(3), 247−259.

Hwang, P. (2009). Ion uptake and acid secretion in zebrafish (*Danio rerio*). *The Journal of Experimental Biology, 212*(11), 1745.

Hwang, P., Lee, T., & Lin, L. (2011). Ion regulation in fish gills: Recent progress in the cellular and molecular mechanisms. *American Journal of Physiology-Regulatory Integrative and Comparative Physiology, 301*(1), 28−47.

Iglesiasrodriguez, M. D., Halloran, P. R., Rickaby, R. E. M., Hall, I., Colmenerohidalgo, E., Gittins, J. R., Green, D. R. H., Tyrrell, T., Gibbs, S. J., & Von Dassow, P. (2008). Phytoplankton calcification in a high-CO_2 world. *Science, 320*(5874), 336−340.

Inaba, K., Dréanno, C., & Cosson, J. (2003). Control of flatfish sperm motility by CO_2 and carbonic anhydrase. *Cell Mot Cytoskel, 55*(3), 174−187.

Intergovernmental, P. O. C. C. (2014). *Climate change 2014-impacts, adaptation and vulnerability: Part A: glsobal and sectoral aspects. Working group II contribution to the IPCC fifth assessment report: Volume 1: Global and sectoral aspects.* Cambridge: Cambridge University Press.

Jarrold, M. D., & Munday, P. L. (2019). Diel CO_2 cycles and parental effects have similar benefits to growth of a coral reef fish under ocean acidification. *Biology Letters, 15*(2), 20180724.

Jutfelt, F., & Hedgärde, M. (2015). Juvenile Atlantic cod behavior appears robust to near-future CO_2 levels. *Frontiers in Zoology, 12*(1), 11.

Kim, K., Moon, H., Noh, Y., & Yeo, I. (2020). Influence of osmolality and acidity on fertilized eggs and larvae of olive flounder (*Paralichthys olivaceus*). *Development and Reproduction, 24*(1), 19−30.

Kreiss, C. M., Michael, K., Lucassen, M., Jutfelt, F., Motyka, R., Dupont, S., & Pörtner, H. O. (2015). Ocean warming and acidification modulate energy budget and gill ion regulatory mechanisms in Atlantic cod (*Gadus morhua*). *Journal of Comparative Physiology B, 185*(7), 767−781.

Kroeker, K. J., Kordas, R. L., Crim, R., Hendriks, I. E., Ramajo, L., Singh, G. S., Duarte, C. M., & Gattuso, J. (2013). Impacts of ocean acidification on marine organisms: Quantifying sensitivities and interaction with warming. *Global Change Biology, 19*(6), 1884−1896.

Kroeker, K. J., Kordas, R. L., Crim, R. N., & Singh, G. G. (2010). Meta-analysis reveals negative yet variable effects of ocean acidification on marine organisms. *Ecology Letters, 13*(11), 1419−1434.

Kültz, D. (2015). Physiological mechanisms used by fish to cope with salinity stress. *The Journal of Experimental Biology, 218*(12), 1907.

Kwong, R. W. M., Kumai, Y., & Perry, S. F. (2014). The physiology of fish at low pH: The zebrafish as a model system. *The Journal of Experimental Biology, 217*(5), 651.

Lai, F., Jutfelt, F., & Nilsson, G. E. (2015). Altered neurotransmitter function in CO_2-exposed stickleback (*Gasterosteus aculeatus*): A temperate model species for ocean acidification research. *Conservation Physiology, 3*(1), cov018.

Leduc, A. O. H. C., Munday, P. L., Brown, G. E., & Ferrari, M. C. O. (2013). Effects of acidification on olfactory-mediated behaviour in freshwater and marine ecosystems: A synthesis. *Philosophical Transactions of the Royal Society B: Biological Sciences, 368*(1627), 20120447.

Lefevre, S. (2016). Are global warming and ocean acidification conspiring against marine ectotherms? A *meta*-analysis of the respiratory effects of elevated temperature, high CO_2 and their interaction. *Conservation Physiology, 4*(1), cow009.

Liu, C., Colón, B. C., Ziesack, M., Silver, P. A., & Nocera, D. G. (2016). Water splitting−biosynthetic system with CO_2 reduction efficiencies exceeding photosynthesis. *Science, 352*(6290), 1210.

Liu, S., Tsung, L., Horng, J., & Lin, L. (2013). Proton-facilitated ammonia excretion by ionocytes of medaka (*Oryzias latipes*) acclimated to seawater. *American Journal of Physiology-Regulatory Integrative and Comparative Physiology, 305*(3), 242−251.

Lonthair, J., Ern, R., & Esbaugh, A. J. (2017). Early life stages of an estuarine-dependent fish are tolerant of ocean acidification. *ICES Journal of Marine Science, 74*(4), 1042−1050.

Lopes, A. F., Morais, P., Pimentel, M., Rosa, R., Munday, P. L., Gonçalves, E. J., & Faria, A. M. (2016). Behavioural lateralization and shoaling cohesion of fish larvae altered under ocean acidification. *Marine Biology, 163*(12), 243.

Love, B., Villalobos, C., & Olson, M. B., (2018). Interactive effects of ocean acidification and ocean warming on Pacific herring (*Clupea pallasi*) early life stages.

Masiulis, S., Desai, R., Uchański, T., Serna Martin, I., Laverty, D., Karia, D., Malinauskas, T., Zivanov, J., Pardon, E., Kotecha, A., Steyaert, J., Miller, K. W., & Aricescu, A. R. (2019). GABAA receptor signalling mechanisms revealed by structural pharmacology. *Nature, 565*(7740), 454−459.

Mathis, J. T., Cooley, S. R., Lucey, N., Colt, S., Ekstrom, J., Hurst, T., Hauri, C., Evans, W., Cross, J. N., & Feely, R. A. (2015). Ocean acidification risk assessment for Alaska's fishery sector. *Progress in Oceanograpy, 136*, 71−91.

Maulvault, A. L., Santos, L. H. M. L., Paula, J. R., Camacho, C., Pissarra, V., Fogaça, F., Barbosa, V., Alves, R., Ferreira, P. P., Barceló, D., Rodriguez-Mozaz, S., Marques, A., Diniz, M., & Rosa, R. (2018). Differential behavioural responses to venlafaxine exposure route, warming and acidification in juvenile fish (*Argyrosomus regius*). *The Science of the Total Environment, 634*, 1136−1147.

McCormick, M. I., Watson, S., Simpson, S. D., & Allan, B. J. M. (2018). Effect of elevated CO_2 and small boat noise on the kinematics of predator-prey interactions. *Proceedings of the Royal Society B: Biological Sciences, 285*(1875), 20172650.

Melzner, F., Gutowska, M. A., Langenbuch, M., Dupont, S., Lucassen, M., Thorndyke, M. C., Bleich, M., & Pörtner, H. O. (2009). Physiological basis for high CO_2 tolerance in marine ectothermic animals: Pre-adaptation through lifestyle and ontogeny? *Biogeosciences, 6*(10), 2313−2331.

Michael, K., Kreiss, C. M., Hu, M. Y., Koschnick, N., Bickmeyer, U., Dupont, S., Pörtner, H., & Lucassen, M. (2016). Adjustments of molecular key components of branchial ion and pH regulation in Atlantic cod (*Gadus morhua*) in response to ocean acidification and warming. *Comparative Biochemistry and Physiology. B, Comparative Biochemistry, 193*, 33−46.

Milazzo, M., Cattano, C., Alonzo, S. H., Foggo, A., Gristina, M., Rodolfo-Metalpa, R., Sinopoli, M., Spatafora, D., Stiver, K. A., & Hall-Spencer, J. M. (2016). Ocean acidification affects fish spawning but not paternity at CO_2 seeps. *Proceedings of the Royal Society B: Biological Sciences, 283*(1835), 20161021.

Miller, G. M., Kroon, F. J., Metcalfe, S., & Munday, P. L. (2015). Temperature is the evil twin: Effects of increased temperature and ocean acidification on reproduction in a reef fish. *Ecological Applications, 25*(3), 603−620.

Miller, G. M., Watson, S., Donelson, J. M., McCormick, M. I., & Munday, P. L. (2012). Parental environment mediates impacts of increased carbon dioxide on a coral reef fish. *Nature Climate Change*, *2*(12), 858−861.

Miller, G. M., Watson, S., McCormick, M. I., & Munday, P. L. (2013). Increased CO_2 stimulates reproduction in a coral reef fish. *Global Change Biology*, *19*(10), 3037−3045.

Montgomery, D. W., Simpson, S. D., Engelhard, G. H., Birchenough, S. N. R., & Wilson, R. W. (2019). Rising CO_2 enhances hypoxia tolerance in a marine fish. *Scientific Reports*, *9*(1), 15152.

Moran, D., & Støttrup, J. G. (2011). The effect of carbon dioxide on growth of juvenile Atlantic cod *Gadus morhua* L. *Aquatic Toxicology*, *102*(1), 24−30.

Mostofa, K. M. G., Liu, C. Q., Zhai, W., Minella, M., Vione, D., Gao, K., Minakata, D., Arakaki, T., Yoshioka, T., Hayakawa, K., Konohira, E., Tanoue, E., Akhand, A., Chanda, A., Wang, B., & Sakugawa, H. (2016). Reviews and syntheses: Ocean acidification and its potential impacts on marine ecosystems. *Biogeosciences*, *13*(6), 1767−1786.

Munday, P. L., Cheal, A. J., Dixson, D. L., Rummer, J. L., & Fabricius, K. E. (2014). Behavioural impairment in reef fishes caused by ocean acidification at CO_2 seeps. *Nature Climate Change*, *4*(6), 487−492.

Munday, P. L., Crawley, N. E., & Nilsson, G. E. (2009). Interacting effects of elevated temperature and ocean acidification on the aerobic performance of coral reef fishes. *Marine Ecology Progress Series*, *388*, 235−242.

Munday, P. L., Dixson, D. L., Donelson, J. M., Jones, G. P., Pratchett, M. S., Devitsina, G. V., & Døving, K. B. (2009). Ocean acidification impairs olfactory discrimination and homing ability of a marine fish. *Proceedings of the National Academy of Sciences of the United States of America*, *106*(6), 1848.

Munday, P. L., Dixson, D. L., McCormick, M. I., Meekan, M., Ferrari, M. C. O., & Chivers, D. P. (2010). Replenishment of fish populations is threatened by ocean acidification. *Proceedings of the National Academy of Sciences of the United States of America*, 201004519.

Munday, P. L., Gagliano, M., Donelson, J. M., Dixson, D. L., & Thorrold, S. R. (2011). Ocean acidification does not affect the early life history development of a tropical marine fish. *Marine Ecology Progress Series*, *423*, 211−221.

Munday, P. L., Hernaman, V., Dixson, D. L., & Thorrold, S. R. (2011). Effect of ocean acidification on otolith development in larvae of a tropical marine fish. *Biogeosciences*, *8*(6), 1631−1641.

Munday, P. L., Pratchett, M. S., Dixson, D. L., Donelson, J. M., Endo, G. G. K., Reynolds, A. D., & Knuckey, R. (2013). Elevated CO_2 affects the behavior of an ecologically and economically important coral reef fish. *Marine Biology*, *160*(8), 2137−2144.

Näslund, J., Lindström, E., Lai, F., & Jutfelt, F. (2015). Behavioural responses to simulated bird attacks in marine three-spined sticklebacks after exposure to high CO_2 levels. *Marine and Freshwater Research*, *66*(10), 877−885.

Nilsson, G. E., Dixson, D. L., Domenici, P., McCormick, M. I., Sørensen, C., Watson, S., & Munday, P. L. (2012). Near-future carbon dioxide levels alter fish behaviour by interfering with neurotransmitter function. *Nature Climate Change*, *2*(3), 201−204.

Noor, N. M., & Das, S. K. (2019). Effects of elevated carbon dioxide on marine ecosystem and associated fishes. *Thalassas*, *35*(2), 421−429.

Occhipinti, R., & Boron, W. F. (2019). Role of carbonic anhydrases and inhibitors in acid-base physiology: Insights from mathematical modeling. *International Journal of Molecular Sciences*, *20*(15), 3841.

Orr, J. C., Fabry, V. J., Aumont, O., Bopp, L., Doney, S. C., Feely, R. A., Gnanadesikan, A., Gruber, N., Ishida, A., Joos, F., Key, R. M., Lindsay, K., Maier-Reimer, E., Matear, R., Monfray, P., Mouchet, A., Najjar, R. G., Plattner, G., Rodgers, K. B.,

... Yool, A. (2005). Anthropogenic ocean acidification over the twenty-first century and its impact on calcifying organisms. *Nature, 437*(7059), 681−686.

Pan, T. C. F., Applebaum, S. L., & Manahan, D. T. (2015). Experimental ocean acidification alters the allocation of metabolic energy. *Proceedings of the National Academy of Sciences of the United States of America, 112*(15), 4696.

Perry, S. F., Esbaugh, A., Braun, M., & Gilmour, K. M. (2009). Gas transport and gill function in water-breathing fish. In M. L. Glass, & S. C. Wood (Eds.), *Cardiorespiratory control in vertebrates: Comparative and evolutionary aspects* (pp. 5−42). Berlin, Heidelberg: Springer Berlin Heidelberg.

Pfister, C. A., Esbaugh, A. J., Frieder, C. A., Baumann, H., Bockmon, E. E., White, M. M., Carter, B. R., Benway, H. M., Blanchette, C. A., Carrington, E., McClintock, J. B., McCorkle, D. C., McGillis, W. R., Mooney, T. A., & Ziveri, P. (2014). Detecting the unexpected: A research framework for ocean acidification. *Environmental Science & Technology, 48*(17), 9982−9994.

Pimentel, M., Pegado, M., Repolho, T., & Rosa, R. (2014). Impact of ocean acidification in the metabolism and swimming behavior of the dolphinfish (*Coryphaena hippurus*) early larvae. *Marine Biology, 161*(3), 725−729.

Pimentel, M. S., Faleiro, F., Marques, T., Bispo, R., Dionísio, G., Faria, A. M., Machado, J., Peck, M. A., Pörtner, H., Pousão-Ferreira, P., Gonçalves, E. J., & Rosa, R. (2016). Foraging behaviour, swimming performance and malformations of early stages of commercially important fishes under ocean acidification and warming. *Climatic Change, 137*(3), 495−509.

Poloczanska, E. S., Burrows, M. T., Brown, C. J., García Molinos, J., Halpern, B. S., Hoegh-Guldberg, O., Kappel, C. V., Moore, P. J., Richardson, A. J., Schoeman, D. S., & Sydeman, W. J. (2016). Responses of marine organisms to climate change across oceans. *Frontiers in Marine Science, 3*, 62.

Porteus, C. S., Hubbard, P. C., Uren Webster, T. M., van Aerle, R., Canário, A. V. M., Santos, E. M., & Wilson, R. W. (2018). Near-future CO_2 levels impair the olfactory system of a marine fish. *Nature Climate Change, 8*(8), 737−743.

Pörtner, H. (2008). Ecosystem effects of ocean acidification in times of ocean warming: A physiologist's view. *Marine Ecology Progress Series, 373*, 203−217.

Pörtner, H. O., & Peck, M. A. (2010). Climate change effects on fishes and fisheries: Towards a cause-and-effect understanding. *Journal of Fish Biology, 77*(8), 1745−1779.

Remnitz, A. (2018). Behavioral lateralization and scototaxis unaltered by near future ocean acidification conditions in *Poecilia latipinna* (*Sailfin Molly*) (P1.468). *Neurology, 90*(15 Supplement)), P1−P468.

Riebesell, U., Schulz, K. G., Bellerby, R. G. J., Botros, M., Fritsche, P., Meyerhöfer, M., Neill, C., Nondal, G., Oschlies, A., Wohlers, J., & Zöllner, E. (2007). Enhanced biological carbon consumption in a high CO_2 ocean. *Nature, 450*(7169), 545−548.

Roggatz, C. C., Lorch, M., Hardege, J. D., & Benoit, D. M. (2016). Ocean acidification affects marine chemical communication by changing structure and function of peptide signalling molecules. *Global Change Biology, 22*(12), 3914−3926.

Romero, M., Fulton, C., & Boron, W. (2004). The SLC4 family of HCO_3^- transporters. *Pflügers Archiv European Journal of Physiology, 447*(5), 495−509.

Rosa, R., Rummer, J. L., & Munday, P. L. (2017). Biological responses of sharks to ocean acidification. *Biology Letters, 13*(3), 20160796.

Ross, P. M., Parker, L., O Connor, W. A., & Bailey, E. A. (2011). The impact of ocean acidification on reproduction, early development and settlement of marine organisms. *Water, 3*(4), 1005−1030.

Rossi, T., Nagelkerken, I., Pistevos, J. C. A., & Connell, S. D. (2016). Lost at sea: ocean acidification undermines larval fish orientation via altered hearing and marine soundscape modification. *Biology Letters, 12*, 20150937.

Rossi, T., Pistevos, J. C. A., Connell, S. D., & Nagelkerken, I. (2018). On the wrong track: ocean acidification attracts larval fish to irrelevant environmental cues. *Scientific Reports, 8*(1), 5840.

Rummer, J. L., & Brauner, C. J. (2015). Root effect haemoglobins in fish may greatly enhance general oxygen delivery relative to other vertebrates. *PLoS One, 10*(10), e139477.

Sampaio, E., Lopes, A. R., Francisco, S., Paula, J. R., Pimentel, M., Maulvault, A. L., Repolho, T., Grilo, T. F., Pousão-Ferreira, P., Marques, A., & Rosa, R. (2018). Ocean acidification dampens physiological stress response to warming and contamination in a commercially-important fish (*Argyrosomus regius*). *The Science of the Total Environment, 618*, 388−398.

Schade, F. M., Clemmesen, C., & Mathias Wegner, K. (2014). Within- and transgenerational effects of ocean acidification on life history of marine three-spined stickleback (*Gasterosteus aculeatus*). *Marine Biology, 161*(7), 1667−1676.

Schunter, C., Welch, M. J., Ryu, T., Zhang, H., Berumen, M. L., Nilsson, G. E., Munday, P. L., & Ravasi, T. (2016). Molecular signatures of transgenerational response to ocean acidification in a species of reef fish. *Nature Climate Change, 6*(11), 1014−1018.

Servili, A., Canario, A. V. M., Mouchel, O., & Muñoz-Cueto, J. A. (2020). Climate change impacts on fish reproduction are mediated at multiple levels of the brain-pituitary-gonad axis. *General and Comparative Endocrinology, 291*, 113439.

Shartau, R. B., Damsgaard, C., & Brauner, C. J. (2019). Limits and patterns of acid-base regulation during elevated environmental CO_2 in fish. *Comparative Biochemistry and Physiology. A, Comparative Physiology, 236*, 110524.

Shrivastava, J., Ndugwa, M., Caneos, W., & De Boeck, G. (2019). Physiological trade-offs, acid-base balance and ion-osmoregulatory plasticity in European sea bass (*Dicentrarchus labrax*) juveniles under complex scenarios of salinity variation, ocean acidification and high ammonia challenge. *Aquatic Toxicology, 212*, 54−69.

Silva, C. S. E., Novais, S. C., Lemos, M. F. L., Mendes, S., Oliveira, A. P., Gonçalves, E. J., & Faria, A. M. (2016). Effects of ocean acidification on the swimming ability, development and biochemical responses of sand smelt larvae. *The Science of the Total Environment, 563−564*, 89−98.

Simpson, S. D., Munday, P. L., Wittenrich, M. L., Manassa, R., Dixson, D. L., Gagliano, M., & Yan, H. Y. (2011). Ocean acidification erodes crucial auditory behaviour in a marine fish. *Biology Letters, 7*(6), 917−920.

Stapp, L. S., Kreiss, C. M., Pörtner, H. O., & Lannig, G. (2015). Differential impacts of elevated CO_2 and acidosis on the energy budget of gill and liver cells from Atlantic cod, *Gadus morhua*. *Comparative Biochemistry and Physiology. A, Comparative Physiology, 187*, 160−167.

Stoss, J. (1983). Fish gamete preservation and spermatozoan physiology. In W. S. Hoar, D. J. Randall, & E. M. Donaldson (Eds.), *Fish physiology* (pp. 305−350). San Diego, CA: Academic Press.

Sun, L., Ruan, J., Lu, M., Chen, M., Dai, Z., & Zuo, Z. (2019). Combined effects of ocean acidification and crude oil pollution on tissue damage and lipid metabolism in embryo−larval development of marine medaka (*Oryzias melastigma*). *Environmental Geochemistry and Health, 41*(4), 1847−1860.

Sundin, J., & Jutfelt, F. (2018). Effects of elevated carbon dioxide on male and female behavioural lateralization in a temperate goby. *Royal Society Open Science, 5*(3), 171550.

Tresguerres, M., & Hamilton, T. J. (2017). Acid−base physiology, neurobiology and behaviour in relation to CO_2-induced ocean acidification. *The Journal of Experimental Biology, 220*(12), 2136.

Tresguerres, M., Milsom, W. K., & Perry, S. F. (2019). CO_2 and acid-base sensing. In M. Grosell, P. L. Munday, A. P. Farrell, & C. J. Brauner (Eds.), *Fish physiology* (pp. 33−68). New York: Academic Press.

Tseng, Y. C., Hu, M. Y., Stumpp, M., Lin, L. Y., Melzner, F., & Hwang, P. P. (2013). CO_2-driven seawater acidification differentially affects development and molecular plasticity along life history of fish (*Oryzias latipes*). *Comparative Biochemistry and Physiology Part A: Molecular & Integrative Physiology, 165*(2), 119−130.

Velez, Z., Roggatz, C. C., Benoit, D. M., Hardege, J. D., & Hubbard, P. C. (2019). Short- and medium-term exposure to ocean acidification reduces olfactory sensitivity in gilthead seabream. *Frontiers in Physiology, 10*, 731.

Wang, X., Song, L., Chen, Y., Ran, H., & Song, J. (2017). Impact of ocean acidification on the early development and escape behavior of marine medaka (*Oryzias melastigma*). *Marine Environmental Research, 131*, 10−18.

Wegner, N. C. (2015). Elasmobranch gill structure. In R. E. Shadwick, A. P. Farrell, & C. J. Brauner (Eds.), *Fish physiology* (pp. 101−151). New York: Academic Press.

Welch, M. J., & Munday, P. L. (2016). Contrasting effects of ocean acidification on reproduction in reef fishes. *Coral Reefs, 35*(2), 485−493.

Zadunaisky, J. A. (1996). Chloride cells and osmoregulation. *Kidney International, 49*(6), 1563−1567.

CHAPTER THREE

Behavioral impacts of ocean acidification on marine animals

Youji Wang and Ting Wang

International Research Center for Marine Biosciences at Shanghai Ocean University, Ministry of Science and Technology, Shanghai, P.R. China

Introduction

The rapid increase in atmospheric carbon dioxide (CO_2) concentration and the subsequent ocean acidification (OA) have been reported to have a broad range of biological impacts on marine animals, including effects on physiology, growth and development, calcification, and overall survival (see Cattano et al., 2018; Kroeker, Kordas et al., 2013; Zunino et al., 2017) for *meta*-analysis reviews. Given the critical biological significance of behaviors and significant concerns about the potential ecological risks of acidification, behavioral consequences on marine animals have been of great research interest (Clements & Hunt, 2015; Nagelkerken & Munday, 2016). The current research showed that acidification could not only interfere with the sensory functions of marine animals (including olfactory, hearing, and visual) but also adversely affect neurophysiological functions and cell signaling processes, and subsequently change animal behaviors (Ashur et al., 2017; Melzner et al., 2020; Tresguerres & Hamilton, 2017).

Animal behaviors not only regulate the overall welfare and status of specific species and their populations (Sih et al., 2004) but also have the potential evolutionary ability to affect ecosystems (Fabry et al., 2008). For example, changes in foraging or feeding behavior have a specific impact on the survival and reproduction of animal populations. The impairments of the prey's resistance or evasion to predators can further threaten the stability and function of marine ecosystems (Brose et al., 2019). Marine environments are frequently in the state of dynamic changes, and animals have behavioral strategies for optimizing their fitness to cope with

Ocean Acidification and Marine Wildlife.
DOI: https://doi.org/10.1016/B978-0-12-822330-7.00002-2

environmental pH variations. At present, related studies mainly involve forging, antipredation, habitat selection, and social hierarchy (Fig. 3.1). The research conclusions are mainly based on the adverse effects of acidification on behavioral responses (see below). It can be seen that in the future, global oceanic change is likely to have a broad impact on the behavior of marine animals, thereby threatening the stability and function of marine ecosystems. This chapter aims to summarize and understand how, in general, the behaviors of marine animals. Throughout this chapter, some case studies are used to demonstrate marine animal behavioral responses to OA, which is one of the climate-induced stressors in marine environments.

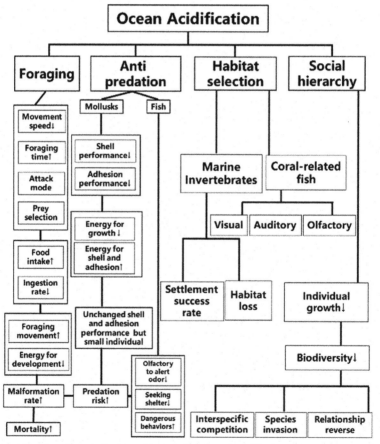

Figure 3.1 Flowchart of the effects of ocean acidification on marine animal behaviors.

Impacts of ocean acidification on foraging

With the increasingly severe environmental concerns of OA, numerous studies have been conducted on the impacts of OA on marine organisms, the first of which is the impact of OA on the predation of organisms (Table 3.1). Bivalves are an essential part of the coastal ecosystem and sentinel species in marine pollution investigations. As far as the current research results are concerned, OA has two main effects on marine bivalves. Firstly, when exposed to acidic conditions, the feeding process of marine bivalves is affected, indicated by the delayed start of feeding, slower foraging movement, change in foraging selection, prolonged foraging time, and decreased foraging success rate, and even the initial feeding of the larvae can be affected (Horwitz et al., 2020; Leung et al., 2015; Li et al., 2020). For example, since gills mediate water flow for feeding, OA exposure impairs the cilia function of blue mussels *Mytilus edulis* and affects its filtration and feeding rates (Meseck et al., 2020). Secondly, as they are affected by acidification, bivalves need to eat more to obtain more energy to adjust their physiological conditions and then adapt to the new living environment (Sadler et al., 2018).

Another vital member of mollusks, cephalopods, has also been reported to be affected by OA. Under OA, the attack latency and attack patterns of cephalopods on prey are changed. The predation rate of the two-toned pygmy squid (*Idiosepius pygmaeus*) decreased by 20%, the attack distance increased, and the predation changed from the tail to the side of the prey (Spady et al., 2018). The activity level of the bigfin reef squid (*Sepioteuthis lessoniana*) (moving distance, moving speed, active time, active degree) was found to be increased in a high pCO_2 environment (Spady et al., 2018). However, it has also been reported that cephalopods, being advanced mollusks, did not change their predatory behavior under high pCO_2 levels (Moura et al., 2019). Other mollusks were also found to be affected by high pCO_2 levels, such as decreased feeding and success rate of predation in the cone snail *Conus marmoreus* (Watson et al., 2017); a higher standard metabolic rate in the whelk *Tenguella marginalba* (Wright et al., 2018); increased food intake in the Olympia oysters *Ostrea lurida* and the Atlantic oyster drill *Urosalpinx cinereal* (Sanford et al., 2014); and poor foraging performance (i.e., prolonged foraging process, reduced foraging success rate, and prolonged feeding time) along with a significant proportion of individuals retracting into the shell in the scavenging gastropod *Nassarius festivus* (Leung et al., 2015).

Table 3.1 A summary of the impacts of ocean acidification on foraging.

Study	Species	pH/pCO$_2$	Control pH/pCO$_2$	Behaviors	Response	Exposure time
Cnidaria						
Bamber et al. (2018)	*Actinia equina*	pH 7.6, 7.9	pH 8.1	Feeding behavior	—	96 h
Mollusca						
Leung et al. (2015)	*Nassarius festivus*	pH 7.0, 7.5	pH 8.0	Foraging performance	→	5 days
Saavedra et al. (2018)	*Perumytilus purpuratus*	750, 1200 µatm	380 µatm	Ingestion rate	↑	2–3 days
Horwitz et al. (2020)	*Stylocheilus striatus*	pH 7.85, 7.65	pH 8.10–8.15	Foraging performance	→	5 weeks
Spady et al. (2018)	*Idiosepius pygmaeus*	737, 934 µatm	438 µatm	Predatory behaviors	→	5 days
	Sepioteuthis lessoniana	935 µatm	435 µatm	Predatory behaviors	→	28 days
Sadler et al. (2018)	*Mytilus edulis*	1000 ppm	400 ppm	Feeding rate	↑	8 weeks
Li et al. (2020)	*Reishia clavigera*	950, 1250 µatm	380 µatm	Foraging performance	→	1 month
Moura et al. (2019)	*Sepia officinalis*	pH 7.7	pH 8.1	Foraging performance	—	15–20 days
Crustaceans						
Wu et al. (2017)	*Charybdis japonica*	pH 7.3	pH 8.1	Foraging performance	→	2 days
Menu-Courey et al. (2019)	*Homarus americanus*	600, 800, 1000, 1200 µatm 2000, 3000 µatm	400 µatm	Ingestion rate	→	40 days
Wang et al. (2017)	*Cancer pagurus*	1200, 2300 µatm	390 µatm	Foraging performance	→	2 weeks
Lord et al. (2019)	*Carinus maenas*	pH 7.8	pH 8.1	Foraging performance	→	5 months

Echinoderms

Reference	Species					
Jellison and Gaylord (2019)	*Leptasterias hexactis*	pH 6.9	pH 8.0	Ingestion rate	↑	7 days

Fish

Park et al. (2020)	*Gondogeneia antarctica*	pH 7.6	pH 8.0	Foraging performance	↓	26 days
Maneja et al., 2015	*Clupea harengus*	1800, 4200 µatm	370 µatm	Foraging performance	—	34 days
Andrade et al. (2018)	*Citharichthys stigmaeus*	900, 1500 µatm	400 µatm	Foraging performance	—	4–6 weeks
Nadermann et al. (2019)	Goldfish	pH 5.5, 6.5	pH 7.5	Foraging performance	↓	2 weeks
Rong et al. (2018)	*Acanthopagrus schlegelii*	pH 7.4	pH 7.8	Foraging performance	↓	15 days
Dixson et al. (2015)	*Mustelus canis*	741, 1064 µatm	405 µatm	Foraging performance	↓	5 days
Ferrari et al. (2012)	*Pseudochromis fuscus*	700 µatm	440 µatm	Foraging performance	↓	4 days
Allan et al. (2013)	*Pseudochromis fuscus*	880 µatm	440 µatm	Foraging performance	↓	4 days
Nowicki et al. (2012)	*Amphiprion melanopus*	530, 960 µatm	420 µatm	Foraging performance	—	21 days
Laubenstein et al. (2019)	*Acanthochromis polyacanthus*	1000 µatm	500 µatm	Food intake	↓	60–66 days

Note: Hereinafter negative effect indicated by "↓"; positive effect indicated by "↑"; no effect indicated by "—".

Crustaceans are one of the most critical organisms of marine ecosystems and they also have high economic value. In a high pCO_2 environment, the time spent by the Japanese stone crab (*Charybdis japonica*), the brown crab (*Cancer pagurus*), and the green crab (*Carcinus maenas*) on searching, crushing, eating, and processing food was found to be increased significantly compared with those in a normal pCO_2 environment, and the predation rate and food conversion rate were observed to be decreased significantly (Wang et al., 2017; Wu et al., 2017). Under OA conditions, the blue crab (*Callinectes sapidus*) ate only part of the prey (Glaspie et al., 2017). The hermit crab *Pagurus bernhardus* and *P. tanneri* showed a lower antenna swing rate (one of the indicators used to measure the "olfactory" response of Decapoda), an inaccurate orientation to the odor source, and decreased activity (Kim et al., 2016; Lecchini et al., 2017). The sensitivity of the Antarctic amphipod *Gondogeneia antarctica* to food was decreased under acidic conditions (Park et al., 2020). The feeding rate of the American lobster (*Homarus americanus*) was decreased, and subsequently, the mortality was increased.

As known, many crustaceans feed on mollusks, but OA can have a significant impact on both. Taking the oyster *Crassostrea virginica* and the mud crab *Panopeus herbstii* as experimental subjects, Dodd et al. (2015) investigated whether OA harms predators or preys when both oysters and crabs are in the same acidic conditions at the same time. The experimental results showed that both oysters and crabs were affected by OA, which is manifested by the decrease in the calcification rate of the oysters and the decreases in prey consumption, handling time, and the duration of failed predation attempt in the crabs. However, in the case of interaction, the adverse effects of OA on crabs entirely offset any benefits from oysters, and crabs were found to have no benefit at all in the predation relationship (Dodd et al., 2015).

In terms of fish, most studies focus on coral reef fish. Under acidic conditions, the predation amount, predation selectivity, and predation success rate of these coral reef fish all decreased to different degrees (Allan et al., 2013; Ferrari et al., 2015; Laubenstein et al., 2019; Nowicki et al., 2012). OA expressed different effects on other fish; for example, reduced feeding activity in the sanddab *Citharichthys stigmaeus* and the goldfish (Andrade et al., 2018; Nadermann et al., 2019); increased response time to food and changes in the feeding movement in the black sea bream (*Acanthopagrus schlegelii*) (Rossi et al., 2018); reduced aggressive behavior toward prey in the smooth dogfish *Mustelus canis* (Dixson et al., 2015); reduced attack distance and speed and changes in attack choices in the brown dottyback (*Pseudochromis fuscus*) (Ferrari et al., 2012; McCormick et al., 2018); and signs of lethargy (reduced activity levels) and a decrease in the frequency of eating and breathing in the seahorse *Hippocampus guttulatus* (Faleiro et al., 2015). The ability to find food through smell was reduced in the shark *Heterodontus portusjacksoni* (Pistevos et al., 2015), and the chance to attack and obtain prey in the larvae of the gilthead seabream *Sparus aurata* and the meagre *Argyrosomus regius* was observed to be reduced (Pimentel et al., 2016).

Besides, OA shows the most negative effects on some marine animals, and there are also anecdotal reports of the effects of OA on other marine organisms. The sea star *Leptasterias hexactis*, under acidified conditions, moved less but ate three times more than it would in normal conditions (Jellison & Gaylord, 2019). The foraging time was significantly longer due to the effects of OA in the sea urchin *Strongylocentrotus fragilis* (Barry et al., 2014). Some species such as the beadlet anemone *Actinia equine* have specific adaptability to OA; its feeding time and competition behavior under OA did not change significantly (Bamber et al., 2018). In a scavenging gastropod *N. festivus*, after short-term acidification treatment and returning back into the normal conditions, the foraging performance was able to return to a normal level (Leung et al., 2015). The foraging behaviors of larval Atlantic herring *Clupea harengus* and the epaulet shark *Hemiscyllium ocellatum* were not affected by OA (Heinrich et al., 2016; Maneja et al., 2015).

Impacts of ocean acidification on antipredation

OA affects not only the predation of marine animals but also the antipredation of organisms (Table 3.2). Mussels have also been a typical research object for research on antipredation behavior. The antipredation response of mussels is mainly affected by three aspects, given as follows. First, seawater with a lower pH may dissolve the mussel shell and reduce the strength, thickness, and cross-sectional area of the shell, thus increasing the risk of mussel predation (Sadler et al., 2018; Yuan et al., 2020). Second, the main properties of the byssus used for attachment, such as the radius, number, length, and strength, are significantly reduced under acidic conditions, rendering mussels susceptible to be captured and eaten by predators (Babarro et al., 2018; Kong et al., 2019; Shang et al., 2019; Sui et al., 2015; Zhao, Guo, et al., 2017; Zhao, Shi, et al., 2017). Third, to

Table 3.2 A summary of the impacts of ocean acidification on antipredation.

Study	Species	pH/pCO_2	Control pH/pCO_2	Behaviors	Response	Exposure time
Mollusca						
Froehlich and Lord (2020)	*Tritia obsoleta*	pH 7.1	pH 8.6	Antipredatory behavior	→	4 weeks
Glaspie et al. (2017)	*Mya arenaria*	pH 7.2	pH 7.8	Shell performance	→	30 days
Alma et al. (2020)	*Crassadoma gigantea*	1050 µatm	365 µatm	Shell performance	→	6 weeks
Kong et al. (2019)	*Mytilus coruscus*	pH 7.7	pH 8.1	Adhesion performance	→	7 days
	Mytilus edulis	pH 7.7	pH 8.1	Adhesion performance	→	7 days
Sui et al. (2015)	*Mytilus coruscus*	pH 7.3, 7.7	pH 8.1	Adhesion performance	→	3 days
Zhao et al. (2020)	*Musculista senhousia*	pH 7.7	pH 8.1	Adhesion performance	→	236 days
Sadler et al. (2018)	*Mytilus edulis*	400 µatm	1000 µatm	Shell performance	→	8 weeks
Shang et al. (2019)	*Mytilus coruscus*	pH 7.7	pH 8.1	Adhesion performance	→	3 days
Lassoued et al. (2019)	*Mytilus galloprovincialis*	800, 1200 µatm	500 µatm	Shell performance	→	3 weeks
Barclay et al. (2019)	*Tegula funebralis*	pH 7.4	pH 7.9	Shell performance	→	6 months
	Nucella ostrina	pH 7.4	pH 7.9	Shell performance	—	6 months
Jellison and Gaylord (2019)	*Tegula funebralis*	pH 6.9	pH 8.0	Antipredatory behavior	→	7 days
Moura et al. (2019)	*Sepia officinalis*	pH 7.7	pH 8.1	Seeking shelter	—	15–20 days
Lord et al. (2019)	*Nucella lapillus*	pH 7.8	pH 8.1	Shell performance	→	5 months
	Nucella ostrina	pH 7.8	pH 8.1	Shell performance	→	5 months
	Urosalpinx cinerea	pH 7.8	pH 8.1	Shell performance	→	5 months

Fish

Reference	Species	CO₂ treatment	Control	Behavior		Duration
Park et al. (2020)	*Gondogeneia antarctica*	pH 7.6	pH 8.0	Staying in the shelter	↓	26 days
Andrade et al. (2018)	*Citharichthys stigmaeus*	900, 1500 µatm	400 µatm	Sensitivity to predators	—	4–6 weeks
Cattano et al. (2019)	*Chromis viridis*	pH 7.88	pH 8.10	Staying in the shelter	→	5 days
Laubenstein et al. (2019)	*Acanthochromis polyacanthus*	1000 µatm	500 µatm	Sensitivity to predators	→	60–66 days
Raby et al. (2018)	*Cephalopholis cyanostigma*	945 µatm	406 µatm	Seeking shelter	—	8–9 days
Dixson et al. (2010)	*Amphiprion percula*	pH 7.8	pH 8.15	Sensitivity to predators	→	11 days
Clark et al. (2020)	*Acanthochromis polyacanthus*	1000 µatm	400 µatm	Sensitivity to predators	—	4 days
	Chromis atripectoralis	1000 µatm	400 µatm	Sensitivity to predators	—	4 days
	Dascyllus aruanus	1000 µatm	400 µatm	Sensitivity to predators	—	4 days
	Dischistodus perspicillatus	1000 µatm	400 µatm	Sensitivity to predators	—	4 days
	Pomacentrus amboinensis	1000 µatm	400 µatm	Sensitivity to predators	—	4 days
	Pomacentrus moluccensis	1000 µatm	400 µatm	Sensitivity to predators	—	4 days
Steckbauer et al. (2018)	*Dicentrarchus labrax*	1000 µatm	380 µatm	Seeking shelter	→	4 days
Davis et al. (2018)	Genus *Sebastes*	1600 µatm	600 µatm	Staying in the shelter	→	3 weeks
Cattano et al. (2017)	*Symphodus ocellatus*	750, 800 µatm	400 µatm	Sensitivity to predators	—	5 min

resist the loss of the dissolved shells under acidic conditions, some mussels consume more energy to produce thicker shells, which in turn reduces the energy used for growth, reduces the individual size, and increases the risk of predation (Kroeker et al., 2014). The effects of OA on the antipredation of other mollusks are mainly manifested in the form of shell dissolution (Alma et al., 2020; Barclay et al., 2019; Glaspie et al., 2017; Lord et al., 2019; Zhao, Shi, et al., 2017), slow response (Froehlich & Lord, 2020; Wright et al., 2018), formation of unshelled larva (Wessel et al., 2018), and reduced time to jumping out of the water (Jellison & Gaylord, 2019).

One interesting experiment showed that the oyster *C. virginica* has two flexible mechanisms to fend off attacks from predators in adverse environments, such as under OA conditions. When external predation pressure is not strong, they add inexpensive calcium carbonate to their shells to get rid of the growing risk quickly. When faced with high risks, the oysters may increase the production of expensive organic materials to increase the strength of their shells. Oysters exhibit a two-tier mechanism in that they can escape quickly from the predator without nearly any cost, but, if necessary, they can also produce thicker shells at a more expensive cost. Understanding this two-tier mechanism expressed by oysters against a predator is very helpful for us to understand the antipredation strategy of marine animals under OA (Schram et al., 2014).

In low pH seawater, the avoidance of the Caribbean spiny lobster (*Panulirus argus*) to their predator (the stone crab) is inhibited and its sensitivity to suitable shelter is reduced (Ross & Behringer, 2019). The sea urchins *Diadema africanum* and *Paracentrotus lividus* are more vulnerable to the predator in low pH seawater than they are in the normal pH seawater because of the bone structure changes. Morover, the resistance of *D. africanum* to OA is stronger than the *P. lividus* (Rodríguez et al., 2017). The incidence of escape behavior in the Indian Ocean medaka *Oryzias melastigma* can be significantly reduced by OA (Wang et al., 2017). The black turban snail *Tegula funebralis* spends less time in shelters under OA conditions, which is irreversible damage (Jellison et al., 2016). It was found that, when compared to normal circumstances, more two-toned pygmy squid (*I. pygmaeus*) change to choose the jet escape responses under OA instead of defensive arm postures or using ink as a defense strategy (Spady et al., 2018).

Unlike mollusks, OA primarily impairs the perception and daily behavior of fish, thereby increasing the risk of predation. The goldsinny wrasse *Ctenolabrus rupestris*, the clownfish *Amphiprion percula*, the damselfish *Pomacentrus wardi* and *P. moluccensis*, the brown dottyback *P. fuscus*, the juvenile spiny chromis *Acanthochromis polyacanthus*, and the coral reef fish *P. wardi* and *A. percula* all have a reduced ability to sense predators or alert odors under acidic conditions, thus increasing the risk of predation (Briffa et al., 2012; Cattano et al., 2018; Dixson et al., 2010; Laubenstein et al., 2019; McCormick et al., 2018). Further physiological studies have been carried out to study the changes in the olfactory sensation of fish under OA. It has been found that elevated environmental CO_2 can change the olfactory sensation of fish by altering the function of the γ-aminobutyric acid type A ($GABA_A$) receptor in the fish brain (Heuer & Grosell, 2014; Heuer et al., 2016). Regan et al. (2016) explored the antipredation behavior of fish living in a typical, high carbonic acid environment exposed to low CO_2 levels and verified the effect of CO_2 on the olfactory-related $GABA_A$ receptor in the fish brain on the contrary. The Antarctic amphipod *G. antarctica*, the gregarious damselfish *Chromis viridis*, and the European seabass *Dicentrachus labrax* spent significantly less time in the shelter under acidic conditions and performed more dangerous behaviors (Cattano et al., 2019; Park et al., 2020; Steckbauer et al., 2018).

However, many studies have pointed out that fish have some adaptability to OA and that they even make some behavioral and physiological changes to adapt to OA. The perception of the predators of the speckled sanddab *C. stigmaeus* and the coral reef fish (such as *A. polyacanthus*, *C. atripectoralis*, *Dascyllus aruanus*, *Dischistodus perspicillatus*, *P. amboinensis*, and *P. moluccensis*) was not found to be significantly changed under acidic conditions (Andrade et al., 2018; Clark et al., 2020). Basic functions of vision, swimming, and decision-making in the adult blue-spotted rock cod *Cephalopholis cyanostigma* for sensing location were also unaffected by OA (Raby et al., 2018). The amount of time spent in the shelter was also not affected by OA in the Atlantic cod *Gadus morhua* (Jutfelt & Hedgärde, 2015). After 3 weeks of domestication under acidic conditions, juvenile rockfish's (genus *Sebastes*) metabolism and behavior were moderately compensated, showing certain adaptability to OA. The gregarious damselfish *C. viridis* and *Symphodus ocellatus* were able to resist the predation of predators by living in groups and increasing the sizes of individuals (Cattano et al., 2017, 2019). The three-spined stickleback *Gasterosteus aculeatus* remained resistant to birds even after exposure to acidic seawater (Nasuchon et al., 2016).

In addition to these direct effects, it has also been reported that high levels of CO_2 in the seawater may indirectly affect the antipredatory response of marine organisms, such as mollusks, by exacerbating other adverse physiological reactions (Schram et al., 2014; Sui et al., 2017). However, in the case of fish, Nasuchon et al. (2016) draw a contrary conclusion that an acidic environment does not exacerbate the adverse effects of warm seawater on the antipredatory behavior of the Japanese anchovy *Engraulis japonicas*. In general, fish are more evolutionarily advanced than mollusks. They have a more comprehensive range of activities than most mollusks, which may be one reason why fish are more resilient to OA than mollusks.

All of the above results are based on the prediction of stable seawater pH in the future. However, we know that the pH of seawater, especially the intertidal zone, fluctuates within a specific range every day. Therefore in order to explore the influence of the diel CO_2 cycle in seawater on the antipredation effect of marine organisms, Michael et al. chose juvenile coral reef fish as the experimental object, simulated the conditions of the diel CO_2 cycle in seawater in the laboratory, and carried out an interesting research. The results showed that in the case of fluctuating acidic seawater, which represents the natural seawater condition, juveniles exposed at stable, elevated pCO_2 spent more time in predator cue water compared to those in normal seawater. Compared with the individuals in the stable acidic seawater, the individuals in the fluctuating acidic seawater showed recovery of some predator-resistant behaviors (Jarrold et al., 2017).

Impacts of ocean acidification on habitat selection

For most marine animals, population supplementation and persistence depend on dispersed larvae, finding a suitable habitat, and surviving until reproductive stages (Lecchini et al., 2017; Rodriguez et al., 1993). Under OA, both physical condition and behavioral activity are affected, and the first response of animals to environmental change is mainly change in behavior, especially habitat selection (Clements & Hunt, 2015; Nagelkerken & Munday, 2016) (Table 3.3).

Different marine organisms choose habitats by using different mechanisms and have different sensitivity to OA (Byrne & Przeslawski, 2013). Upper marine larvae rely on environmental cues to make behavioral

Table 3.3 A summary of the impacts of ocean acidification on habitat selection.

Study	Species	pH/pCO_2	Control pH/pCO_2	Behaviors	Response	Exposure time
Cnidaria						
Doropoulos et al. (2012)	*Acropora millepora*	800, 1300 µatm	401 µatm	Settlement	↓	6 days
Doropoulos & Diaz–Pulido (2013)	*Acropora selago*	705, 1214 µatm	447 µatm	Settlement	↓	3 days
Webster et al. (2013)	*Acropora millepora*	822, 1187, 1638 µatm	464 µatm	Settlement	↓	6 weeks
	Acropora tenuis	822, 1187, 1638 µatm	464 µatm	Settlement	↓	6 weeks
Viyakarn et al. (2015)	*Pocillopora damicornis*	pH 7.6, 7.9	pH 8.1	Settlement	↓	24 h
Crustacean						
Lecchini et al. (2017)	*Stenopus hispidus*	pH 7.84, 7.96	pH 8.11	Recruitment	↓	3 h
Fish						
Munday et al. (2010)	*Amphiprion percula*	pH 7.6, 7.8	pH 8.15	Settlement	↓	11 days
Devine et al. (2012a)	*Cheilodipterus quinquelineatus*	550, 700, 950 ppm	390 ppm	Habitat choice	↓	4 days
Devine et al. (2012b)	*Pomacentrus amboinensis*	700, 850 ppm	440 ppm	Habitat choice	↓	4 days
	Pomacentrus chrysurus	700, 850 ppm	440 ppm	Habitat choice	↓	4 days
	Pomacentrus moluccensis	700, 850 ppm	440 ppm	Habitat choice	↓	4 days
Devine and Munday (2013)	*Paragobiodon xanthosomus*	pH 7.89	pH 8.15	Habitat preference	↓	4 days
	Gobiodon histrio	pH 7.89	pH 8.15	Habitat preference	↓	4 days

decisions, and chemical information plays an essential role in the settlement and habitat selection (Lecchini et al., 2017). Barnacles also use various substrate-related chemical cues to select a suitable microhabitat for attachment (Thiyagarajan, 2010). Sampaio et al. (2016) discovered that when exposed to ocean warming and acidification, and polluted conditions, cryptic flatfish lose the ability to choose a habitat. As for migratory fish, the survival rate of glass eels and their ability to migrate to the estuary appears to be diminished, which also affects their search for habitat and growth in freshwater (Borges et al., 2019).

Currently, there are many examples of habitat selection in marine organisms. However, habitat selection studies mostly focus on coral-related fish and some marine invertebrates. Especially for coral-related fish, with the current climate conditions changing faster than ever and thus exceeding the resilience of coral reefs, their habitat could be destroyed to some extent (Hoegh-Guldberg et al., 2017; Przeslawski et al., 2008). The coral ecosystem may be destroyed, and the fish sensation may also be affected. Early juvenile reef fish can use a variety of sensory cues that enable the efficient selection of new habitats (Huijbers et al., 2012; Leis et al., 2011). OA may impair their ability to choose their preferred place of residence (Lecchini et al., 2005). Besides, OA has some negative effects that may affect the auditory behavior of fish, which also affects the hearing-based habitat selection of many fish (Simpson et al., 2011). For example, Castro et al. (2017) reported that the larvae of the painted goby could not distinguish the sound of reefs at high CO_2 concentrations. Apart from hearing, larval reef fish also use the olfactory sense to find habitat (Atema et al., 2002). The olfactory sense has also been shown to be affected by OA, which affects the ability to home (Munday et al., 2010) Some studies have shown that visual information is also influenced. Ferrari et al. (2012) found that when CO_2 concentrations are elevated, the ability of organisms to use visual information to locate settlements has been significantly affected. However, one study reported that larvae could rely on other sensory information such as visual cues to compensate for the effects of the impaired olfactory ability within a small space to select their habitat. Huijbers et al. (2012) also reported that habitat selection might be related to the order of using different senses.

Most invertebrates have planktonic larval stages. As most marine invertebrates move from the planktonic stage to the benthic, they assess the environmental quality to choose a suitable environment that is beneficial for individual survival and reproduction (Hadfield, 2011; Hunt & Scheibling, 1997; Kendall et al., 2013; Rodriguez et al., 1993). Many marine invertebrates need to live with corals to get food or settle through them (Stella et al., 2011). Under OA, the process of metamorphosis can be hindered in the coral *Acropora digitifera*. However, if larvae survive, the settlement and recruitment of corals are indirectly affected (Nakamura et al., 2011). Concrete surfaces provide suitable conditions for settlement due to their high roughness with pores and microcracks, and many marine invertebrate larvae could use biofilm components as clues to find appropriate settlement sites (Hadfield, 2011). However, OA can affect the settlement of the larvae marine invertebrates on biofilms and indirectly affect their settlement choices, such as echinoderms, foraminifera, nematodes, polychaetes, mollusks, crustaceans, and chaetognaths (Cigliano et al., 2010; Fabricius et al., 2014; García et al., 2018; Maboloc & Chan, 2017; Nelson et al., 2020; Pecquet et al., 2017; Uthicke et al., 2013; Wang & Wang, 2020). Investigation of the European oysters found significant differences in larvae settlement on different substrates. Studies have also shown that bryozoans and barnacles have different preferences for settlement on artificial materials.

Above all, under the condition of seawater acidification, the destruction of habitat leads to the change or loss of some clues, which is very unfavorable to the choice of marine life habitat. Also, the study of the settlement status of marine animals under OA is crucial for the conservation and restoration of species, which is also related to the distribution of organisms and the continuation of ethnic groups in the future marine ecosystem.

Impacts of ocean acidification on the social hierarchy

As the climate changes, species antagonism and interactions are also affected (Tylianakis et al., 2008). In ecosystems, there is a balanced in-group competition, and OA can interrupt that balance, leading to changes in the community (Kroeker, Micheli, et al., 2013) (Table 3.4). Byers claimed that changes in or loss of habitats might make some communities more vulnerable to invasive species (Byers, 2002). In particular, the relationship between species reversed due to OA may change the process of shaping communities and ecosystems (McCormick et al., 2018). Besides, under OA, the competitive advantage is expanded, and the density of

Table 3.4 A summary of the impacts of ocean acidification on the social hierarchy.

Study	Species	pH/pCO_2	Control pH/pCO_2	Behaviors	Response	Exposure time
Reef fish						
McMahon et al. (2019)	*Amphiprion percula*	489 μatm	1022 μatm	Individuals' growth	↓	50 days
				Social hierarchy	—	50 days
Community						
Hale et al. (2011)	Marine benthic communities	pH 6.7, 7.3, 7.7	pH 8.0	Biodiversity	↓	60 days

Note: negative effect indicated by "↓"; positive effect indicated by "↑"; no effect indicated by "—".

species in quantity and competitive disadvantage is reduced (Nagelkerken & Munday, 2016). However, there is also the possibility of CO_2 leakage in the natural environment in addition to the air diffusion; although this may occur only in parts of the region, it may also affect many animal communities (Widdicombe et al., 2015).

An increase in CO_2 leads to carbonate changes in specific habitats, such as coastal zone, which then influence the organisms (i.e., bivalves) that exist there (Gazeau et al., 2013). In recent years, there have been many kinds of research on the community structure of the coral reef ecosystem and the microbial community. Effects of OA on habitats further regulate biodiversity (Sunday et al., 2017). Horwitz reported that coral reef ecosystems are seriously threatened with an increasing threat of acidification (Horwitz et al., 2017). In coral reef ecosystems, acidification leads to reduced coral cover and habitat complexity, which seriously affects invertebrate communities there (Fabricius et al., 2014). Moreover, an interspecific competition between corals will likely change the composition, structure, and function of coral reef communities. OA coupled with some climatic changes result in loss of coral cover and indirectly affect invertebrate communities inhabiting coral reefs, thus reducing the biodiversity (Fabricius et al., 2014; Hale et al., 2011).

As for microbial communities, existing studies have found that OA has a direct effect on heterotrophic planktonic bacteria in the warm ocean at low latitudes and therefore may affect the global biogeochemical cycle (Xia et al., 2019). Nevertheless, there are also some opposing views. Roy et al. (2013) claimed that the effects of OA on bacterial communities in the upper middle area of the coastal zone are negligible. Under OA, the responses of different classes in marine ecosystems vary (Wittmann & Pörtner, 2013). Here, we mention some typical marine animals affected by OA. The shell thickness of calcifying protists, such as foraminifera, was significantly impacted (Moy et al., 2009). In California, the dissolution of the shell of the pelagic snail *Limacina helicina* was used as a sign of decline in habitat adaptability under OA. Additionally, it is expected that the incidence of shell dissolution will triple by 2050 (Bednaršek et al., 2014).

As for calcified zooplankton, pteropods of pelagic mollusks are vulnerable to OA, although the extent of their effects remains to be determined (Roberts et al., 2011). Little is known about the potential impacts of OA in benthic ecosystems, though some studies have proven that the skeletons would be damaged (McClintock et al., 2009; Venn et al., 2013). In addition to the sensitivity of marine invertebrates to acidification, fish are also vulnerable to acidification. In marine ecosystems, fish play an important role in maintaining ecological and interspecific balance. However, the physiological status of fish and the replenishment of fish stocks may be threatened by OA (Heuer & Grosell, 2014; Ishimatsu et al., 2008). At high concentrations of CO_2, the larvae of coral-related fish are more active and perform riskier actions (Munday et al., 2010; Pankhurst & Munday, 2011). Nilsson et al. (2012) predicted that fish behavior would be interfered with by changes in neurotransmitter function under the elevated CO_2 shortly. Elevated levels of CO_2 in seawater can cause anxiety in rockfish by altering $GABA_A$ receptor function (Tresguerres & Hamilton, 2017). However, some people hold different opinions. McMahon et al. noted that high levels of CO_2 affect the growth of the individuals of the orange clownfish *A. percula* but have no significant effect on the size-based social hierarchy (McMahon et al., 2019). Hence, the mechanistic understanding of the underlying physiological effects of OA on marine organisms needs to be clarified urgently, such as the calcification process in invertebrates and the reproductive physiology of fish.

It is now widely believed that elevated CO_2 levels in the air lead to OA, and of course, human activity also contributes significantly to seawater acidification. Marine organisms, especially those with carbonated shells, struggle to survive in such conditions (Ridgwell & Schmidt, 2010). We can see that in the future, marine organisms may not be able to adapt quickly to OA, and people do not have the ability to change oceanic pH values. Currently, the views of each hierarchy in marine ecosystems under OA are not inconsistent, so more experimental data and conclusions are needed to support the underlying mechanisms of OA for identifying the changes in each community.

Case studies

Case study 1—foraging

Foraging behavior is a series of actions that predators take to hunt for food and it includes searching, chasing, catching, processing, and ingesting. It is an excellent indicator for evaluating individual performance as marine organisms need food to supply vital energy for basic life functions. To obtain food from natural marine environments, many animals mainly rely on chemoreception to detect chemical cues originating from available food sources. An abundance of studies claim that decreased pH could interfere with such sensory modality, leading directly to a decline in food consumption, preying ability, and the subsequent adaptability and survival of marine animals, including fish, crabs, and shellfish (Leung et al., 2015; Wang et al., 2017; Wu et al., 2017).

A study conducted by Leung et al. (2015) demonstrated that the foraging performance of a scavenging gastropod *N. festivus* could be diminished by OA. However, such performance was restored when *N. festivus* was returned to normal pH conditions. The gastropods *N. festivus* are mainly distributed across the mid-intertidal zones of sandy shores, where they are regularly exposed to pH fluctuations. The authors designed the following experiment to explain how acidification affects the foraging performance of *N. festivus* and their capability to recover from acidified conditions. Three pH levels (pH 8.0, 7.5, and 7.0) and three distances (2.5, 10, and 15 cm) from the bait were set as a 3×3 factorial design. After 1 hour of exposure, they found that the foraging performance (travel speed, foraging success, and consumption rate) of the individuals were significantly decreased at low pH, and *N. festivus* spent more time on feeding compared to those in the normal pH (Leung et al., 2015). The authors attributed such a reduction in foraging performance to the direct injury to the chemoreceptors (i.e., olfactory and neurotransmitter receptors) due to the increased concentration of hydrogen ions. The reduced consumption rate and longer feeding time under low pH indicated that *N. festivus* spends more effort meeting energy requirements under stress (Leung et al., 2015). However, all the foraging performances returned to normal level after a 48-hour recovery period, indicating that *N. festivus* can regulate body function after short-term acidic stress and that this species can still populate sandy shores with pH fluctuations within a certain range. If *N. festivus* is adaptable to such changing environments and capable of coping with the rapid influences of the acidic marine environment, then it is supposed that some shallow-coastal organisms may have the compatibility to adapt to the gradual changes in marine environmental chemistry than presently anticipated.

OA can also alter foraging behavior by impairing sensory functions. Olfaction is essential for food localization in elasmobranch fish, and acidic seawater may interfere with the chemical cues and make fish exhibit sensory and behavioral abnormalities. The smooth dogfish (*M. canis*) is one of the coastal elasmobranch fishes, which populates in the North-Western

Atlantic Ocean. With various baits (such as invertebrates and small fish), the dogfish *M. canis* was considered to be an ideal species to investigate the foraging behavior changes influenced by OA. In a study by Dixson et al. (2015), adult *M. canis* were exposed to three pCO_2 conditions (current: 405 ± 26 μatm, mid: 741 ± 22 μatm, and high: 1064 ± 17 μatm) for 5 days. The results showed that sharks in the high CO_2 treatments avoided the odor cues of food. In contrast, the sharks in the other two treatments could maintain the usual odor tracking behavior. Additionally, the sharks exposed to high CO_2 tended to spend less time in the water stream with food stimulus than those in the control treatment (Fig. 3.2). Moreover, sharks in elevated CO_2 conditions reduced their attack behaviors compared with the control treatment. The results of this study imply that changes in the chemical cues can influence the shark's foraging behavior in acidic seawater. This experiment was performed in the laboratory, and such mimic conditions might not wholly evaluate the probable influences of OA on the foraging performance of *M. canis* in their natural environment. Nevertheless, the present observation claims that the outcome of predatory abilities and vulnerabilities could be affected by acidification in a natural marine environment. Besides, the present study was a short-term test, equivalent to acute exposure, and the acclimated and adaptable possibility was not involved. Whether such marine organisms are capable of adapting to the long-term OA exposure remains unknown. The outcome of adaptability to changing ocean CO_2 might have

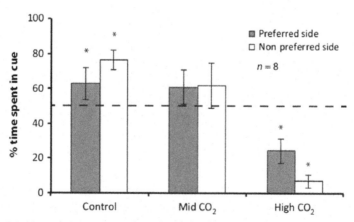

Figure 3.2 Mean time spent on the side of the food odor when presented on either the preferred or nonpreferred side for sharks exposed to control and elevated CO_2 conditions.

significant influences on the evolutional probability to salvage populations from the demographical consequences of increasing $p\mathrm{CO}_2$.

Being a vital part of biodiversity, intraspecific trait variation involves shaping communities via altering ecological interactions. Predator trait variation can potentially influence food webs, and environmental changes such as temperature elevation and pH reduction can modify the foraging traits of predators in a marine environment. Contolini et al. (2020) evaluated how local environmental factors contribute to variation in prey selectivity among marine intertidal populations. This study demonstrated that the dog whelk *Nucella lapillus* (one intertidal drilling predator) could drill larger mussels *M. californianus* at the sample location with less variable pH. More stable pH conditions could reduce the marine organism's exposure to a repetitive stressor, thus offering more chances for active foraging. Under low pH conditions, *N. lapillus* tends to select smaller prey; which may be attributed to the less handling time in acidic water for consuming smaller mussels. Their results highlight the role of oceanic change in shaping marine predator selectivity in a keystone species. Low pH could also affect the mussel (prey) traits in terms of size and shell thickness, so net changes in the *Nucella—Mytilus* interaction eventually are a tradeoff for both the predator and the prey responding to the changing environment. With the constant changes in oceanic conditions, a tremendous amount of marine organisms are facing growing degenerative abiotic conditions, which to a large extent is based on the interactions among different regional climate dynamics. Since every population faces unique habituating conditions, marine organisms will make responses accordingly by altering behavioral or physiological traits, which can make alternations in population interactions and community dynamics. It has been confirmed that OA can change the predatory traits of some marine organisms, but only a few researchers have addressed if different predator populations react differently to elevated CO_2 conditions. Contolini et al. (2020) sampled three dog whelk *N. lapillus* populations in California to test if they could alter the consumption of prey in acidified seawater. In general, it was observed that elevated CO_2 increased the consumption time variation among the three populations. Compared with the populations that were exposed to acidified conditions less frequently, those that experienced more frequent acidification in their habitat were observed to have reduced the consumption time significantly. In terms of prey searching time, whelks from the location with higher temperature and pH enhanced their time when exposed to acidification. However, the populations naturally living in acidified conditions did not. This phenomenon suggests that prior exposure to elevated CO_2 could help whelks alleviate the adverse effects caused by acidification by adjusting their physiology or behaviors.

The population-level difference should be a nonnegligible source of variation for organisms responding to climate change, but previous works have rarely considered such factors in OA research. The above case found that exposure to acidification altered the predator performance (i.e., consumption time) depending on the population (Contolini et al., 2020). Population-specific responses are involved with the environmental regimes of source populations, such as elevated CO_2 conditions. Populations from different pH preexposure conditions seem to produce different responses to acidification that could help ease the adverse impacts of acidification; the predation responses to acute acidification could be population specific. By understanding the impacts of intraspecific variation in reacting to OA, we may understand how organisms would respond to environmental changes and provide more accurate predictions about the coming changes in ecological communities. This case also highlights how population-specific responses to oceanic change can result in changes in ecological effects, and such differences may restructure prey communities in the local area.

Case study 2—antipredation

To reflect the impact of OA on predator–prey dynamics, many scholars have tried to experiment on prey and predator at the same time in recent years. The prey can exhibit antipredation behaviors for self-defense against predators, and these behaviors are demonstrated to be susceptible to OA. To investigate the predator–prey interactions under acidic conditions, Froehlich and Lord et al. assessed the escape response of the mud snail *Tritia obsoleta* to its predator the mud crab *P. herbstii* under acidified conditions (pH 7.6) (Froehlich & Lord, 2020). The mud snails *T. obsoleta* populate the intertidal and subtidal zones of the western Atlantic shoreline, and they can escape from the predator by thrashing around using their foot, fast crawling away, and self-burial (Froehlich & Lord, 2020). The authors found that OA has negative impacts on the escape responses of the mud snail *T. obsoleta* to its predator's chemical cues. The mud snail *T. obsoleta* primarily crawls away in the presence of predators under normal pH 8.1, but they prefer to bury under low pH. Such different responses were found probably because *T. obsoleta* can bury faster than crawl out of the water. The authors speculated that low pH could increase the intensity of danger cues and make *T. obsoleta* bury more under OA. In addition to perceiving threats from predators, marine animals also need to maintain acid−base balance, which may change energy allocation under acidification exposure and thus affect the antipredation behavior (Briffa et al., 2012; Dodd et al., 2015). Another underlying mechanism that explains the behavioral changes is the altered perception of threat cues.

On the one hand, OA can cause disturbance in the ability of specific receptors binding to the signal molecules, such as GABA receptors; on the other hand, it can interfere with the transmission of signals from the receptors to the olfactory lobe of the brain (Rong et al., 2018). The latter hypothesis involves the upregulation of genes that prevent signal transduction pathways from receptors (Rong et al., 2018), which gives the possible interpretation of the behavioral changes in mud snails under acidification exposure. If snails are not able to receive signals from their receptors that typically sense threat cues, then they may decrease their movement or bury in the sediments. The burial behavior helps mud snails to be less susceptible to predation. However, if they bury more frequently, this may potentially change the prey−predator dynamics. This may trigger bottom-up changes across trophic levels that reduce mud snail consumption and lead to decreases in predator numbers as well as changes in the functioning of the ecosystem. If consumers like *T. obsoleta* are not capable of sensing the chemical cues of potential predators, such diminished chemosensory abilities under OA will substantially impact the predator−prey relationships in shaping ecosystems. Further studies should focus on the multiple trophic levels and responses to OA at the community level to better understand and forecast the outcomes of OA. In general, Froehlich and Lord (2020) demonstrated that acidified conditions threaten marine mollusks not only by hindering shell formation but also by meddling with their behavior and changing predator−prey dynamics.

For some sessile organisms, such as mussels, the chemical cues released from the predator can trigger their antipredator behavior (i.e., increasing byssal thread production and mechanical properties) to avoid being prey. Byssal production and adhesion are not only just mediated by chemical reactions but also biologically controlled processes. The synthesis and secretion of byssal proteins are energy-consuming processes; if the energy budget is decreased due to OA, mussels might produce fewer, shorter, and weaker byssal threads compared with those in normal seawater conditions. The differential byssal thread performance of the mussel *Musculista*

High pH (8.1)   Single-generational   Trans-generational
(control)   exposure to low pH (7.7)   exposure to low pH (7.7)

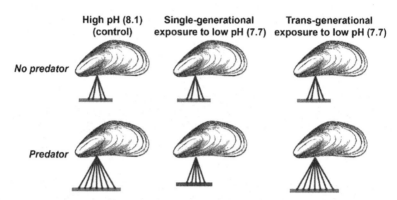

Figure 3.3 Experimental study of transgenerational effects, pH, and predation risk on byssal production. *From Zhao, L. Q., Liang, J., Liang, J. P., Liu, B. Z., Deng, Y. W., Sun, X., Li, H., Lu, Y. N., & Yang, F. (2020). Experimental study of transgenerational effects, pH and predation risk on byssus production in a swiftly spreading invasive fouling Asian mussel, Musculista senhousia (Benson). Environmental Pollution, 260, 114111. https://doi.org/10.1016/j.envpol.2020.114111.*

senhousia was evaluated, both with the presence of a predator and without, under OA (Fig. 3.3) (Zhao et al., 2020). The results showed that in the presence of the predator, *M. senhousia* significantly increased its byssal production. Moreover, the descendants of the mussels accumulated in low pH produced more byssal threads than those under normal pH, indicating positive transgenerational effects. Considering that mussel byssus has proteinic properties, seawater chemistry might affect the byssal performance through byssal protein synthesis. Facing predation risk, mussels might redistribute their energy for synthesizing proteins to resist stress.

Marine bivalves need to balance the energy for growth, reproduction, and defense, and protein synthesis is a highly energy-consuming process; such a process can profoundly influence other necessary maintenance activities. It is rational to hypothesize that *M. senhousia* can adapt to the test acidified seawater conditions because the observed results may not be just transgenerational plasticity but a kind of adaptation. In natural environments, *M. senhousia* is daily exposed to a wide range of pH variability. However, the exposure time in a laboratory experiment might be too short for observing drastic changes in byssal production. On the other hand, the predator might be less active under acidified conditions and it might be inadequate to alter the byssal production of mussels under OA. The experiment was performed at a constant pH state; it may be vital to maintain byssal production under acidification (because the predator can also adapt), and this can be completed fast via transgenerational acclimation or adaptation. This explains the tremendous potential of *M. senhousia* to inhabit various areas of the environment. Additionally, the way in which OA affects the ability of other predators, such as predatory fish, sea stars, and gastropods, to prey upon mussels is still unknown. Considering that the interactions between predator and prey are likely to scale up the effects of OA from effects in individual organisms to population dynamics and ecosystem changes (Kroeker et al., 2014), further investigations are necessary to gain a better, comprehensive understanding of how mussels interact with their predators in a rapidly acidifying ocean.

As a kind of defense mechanism, bivalves can close their valve in response to predation. Clements et al. used high-frequency noninvasive biosensors to test whether 3 weeks of exposure to acidification could negatively impact the behavioral responses of adult Mediterranean mussels (*M. galloprovincialis*) in the presence of predators (Clements et al., 2020). The results showed that when pCO_2 was elevated, although mussels did initiate adaptive valve closures to react to predator cues, they did not completely close the valve gapes. Compared with chemical cues, physical cues particularly lead to the full valve closures of marine mussels. The complete valve gape closure prevents as far as the threat from predators, but such a defense behavior could result in ingestion reduction and disordered breathing, which then leads to some adverse impacts such as reduced growth and reproductive capacity in the long term. In this case, *M. galloprovincialis* individuals chose to partially close the valves; such responses allow constant ingestion and respiration to some extent while offering only partial protection from predators. Thus it appears that bivalves are capable of evaluating predation risk and make behavioral decisions accordingly to adaptively achieve a tradeoff between predator avoidance and physiological functioning.

This case concluded that a 3-week exposure to elevated CO_2 actually did not affect the behavioral responses of mussels to predation risk, which suggests that the effects of acidification on marine animal behaviors may be much weaker than we currently believe. However, it should be noted that this does not mean that mussels are not vulnerable to climate change, as the increased temperature caused by ocean warming has potential impacts on the valve gaping behaviors of marine bivalves (Clements et al., 2020). This case also demonstrates that predation could have adverse impacts on the ecosystem services that bivalves provide. Bivalves have incredible filtering capacity and can improve water quality by helping to remove nutrients in nearshore coastal areas. Considering that the valve gaping and filtration rates of mussels could be affected due to the presence of a predator, such adverse impacts could influence the ecosystem services provided by mussels.

In bivalves' predator avoidance, clustering can also serve as a strategy to resist predators, since predators normally tend to capture them on their aggregations' edge. Individual mussels can form clusters within 24 hours, and they prefer to gather together when exposed to predator's chemical signals. Kong et al. (2019) conducted a laboratory experiment to test the anti-predator defenses of two sympatric *Mytilus* species from China, *M. coruscus* and *M. edulis*, in the presence and absence of predator cues under acidification exposure. In this case, the number of individuals was higher in the acidic treatment, and the presence of a predator significantly reduced that number when compared with the absent predator group. In general, the percentage of clustering in *M. coruscus* was higher than that in *M. edulis*. These results suggest that OA could have adverse effects on the ecological functions of the clustering behavior in mussels. Additionally, such effects could not only impair the antipredator defense mechanisms but also have vital implications for other ecological functions It has been noted that the clustering behavior plays an important role in material stability, and the bed structure of bivalves might get changed under the predicted OA. To a certain extent, these results demonstrate the important effects of acidification on bivalves' defense response; its effects on other behavioral functions still need further investigation. Notwithstanding its limitation, this study only quantified the effect of oceanic climate change on anti-predator defenses without testing relevant responses of predators, and this can also influence the ecological outcome of predator-prey interactions. Considering the impacts of acidification on sensory systems, alternative approaches are necessary. In this case, we also have noticed that even upon exposure to the same conditions, sympatric mussels could have species-specific differences in antipredatory responses. The two species observed are dominant bivalves that populate in coastal seawater in the East China Sea, and such differential responses might have effects on the local marine ecosystem structure.

Case study 3—habitat selection

Habitat identification and selection refer to the behavior of fish to find, locate, and select a suitable habitat for settlement under the guidance of

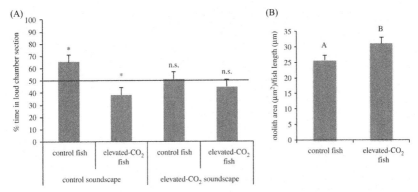

Figure 3.4 Effect of oceanacidification on the auditory preferences and otolith (ear bone) size of settlement-stage larval fish. (A) Mean (± s.e.) percentage of time spent by mulloway larvae (N = 64 per treatment) in the half of the choice chamber closest to the broadcasting speaker playing elevated-CO_2 or control temperate reef sounds. Stars indicate distributions significantly different ($p < 0.05$) from a random distribution of 50%. (B) Mean otolith surface area of mulloway (N = 22 per treatment) standardized to larval body length. Different letters indicate statistically significant ($p < 0.05$) differences. n.s., not significant. *From Rossi, T., Pistevos, J. C. A., Connell, S. D., & Nagelkerken, I. (2018). On the wrong track: Ocean acidification attracts larval fish to irrelevant environmental cues.* Scientific Reports, 8(1). https://doi.org/10.1038/s41598-018-24026-6.

chemical (smell), sound (hearing), and light (visual) signals. Auditory cues are vital for habitat selection during settlement, and OA can interfere with the ability of fish larvae to detect the cues. Rossi et al. (2018) demonstrated that auditory responses of the barramundi larvae could be affected by the acidic condition, which could hinder their survival. The authors found that OA might influence larval settlement by altering behaviors toward natural settlement-habitat cues and by altering the quality of biological auditory cues (Fig. 3.4; Rossi et al., 2018). Three recording types were used for the behavioral trials, namely estuarine soundscapes acting as a natural auditory cue and two irrelevant cues: temperate reefs and white noise. The results showed that under the control pH 8.0, the larvae showed a significant attraction to its natural habitat soundscape. While under the low pH 7.6, the larvae showed not only avoidance toward its natural auditory cue but also attraction toward ecologically irrelevant sounds. Such behavioral alterations might mislead the larvae to settle in inappropriate habitats and thus delay their settlement as they are incapable of distinguishing between natural habitat cues and irrelevant cues (Rossi et al., 2018). It is reported that OA can enlarge the size of fish otoliths, which are used for audition, location, and balance. Theoretical models predict that enlarged otoliths under future OA conditions might extend the hearing range of larval fish. However, this effect has not been empirically proven. In this study, the barramundi exposed to acidified conditions throughout their larval development stage showed an increase of otoliths. However, they still exhibited a failure to respond to ecologically relevant habitat sounds, suggesting that although acidified condition leads to enlarged otoliths and possibly increases the hearing ability, the fish could not compensate for the changed auditory preferences caused by elevated pCO_2. If biological acoustic cues have little to offer in location sensing, then the larvae will have to compensate by using other chemical cues. But OA can impair the olfaction and larval traits such as swimming speed and development. Due to the delayed settlement, a prolonged oceanic life phase is likely to raise predation risk, delay the occupancy of food-rich benthic habitats, and undermine population replenishment. The settlement, the transition from the larval stage to benthic life, is a crucial life process for most marine benthic organisms. Sea urchins, which display various developmental processes, have been widely considered as model organisms in OA research. Echinoids take an essential position in many coastal ecosystems, such as kelp forests and coral reefs. The spatiotemporal difference in the abundances of sea urchin population is in connection with differences in each developmental stage, such as settlement, metamorphosis, and recruitment. Thus any changing environmental factor (e.g., OA) that significantly influences the process of settlement in such keystone taxa might bring about ecological shifts in coastal ecosystems, where they play a key role.

Espinel-Velascoa et al. (2020) investigated the effects of OA on the settlement of larval sea urchin *Evechinus chloroticus*, a critical ecological coastal species in New Zealand. They aimed to test whether reduced seawater pH would alter the settlement and metamorphosis of the larvae. The results showed that settlement success did not decrease significantly when the larvae were exposed to low pH treatments. Thus in this study, the larval sea urchin *E. chloroticus* was capable of resisting acidic seawater during settlement. To fully understand the impact of OA on the life cycle of marine invertebrates and future coastal marine ecosystems, a series of taxa with different settlement selectivities and behaviors need to be further studied. A successful settlement includes the pre-and postsettlement processes; both processes can be affected by OA, individually or interactively with other factors. By now, little is known about how pH acts on early postsettlement. However, previous studies have demonstrated that decreased or fluctuating pH in the boundary layers could hinder the growth and survival of some newly settled organisms. Additionally, the degree of settlement alternations responding to predicted future acidification may be species-specific and linked with the degree of settlement selectivity. Taxa with higher degrees of settlement selectivity (such as corals) are considered to be more severely influenced by OA due to changed substrate characteristics or loss of suitable settlement substrate.

If dispersing larvae find their suitable habitat and survive until the reproductive stage, the replenishment and continuity of marine species can be achieved. Due to the environmental cues, pelagic larvae can make behavioral decisions to respond to environmental changes. For example, chemical cues play an important role in habitat selection at the settlement stage. Lecchini et al. (2017) experimented to investigate how acidic seawater impacts the chemical cues of fish and crustacean larvae. Under current seawater conditions (400 ppm), both the larval fish *C. viridis* and the crustacean *Stenopus hispidus* showed a significant preference to the conspecific chemical cues. When exposed to acidification (1000 ppm CO_2), the larvae only spent 70% of the time using the conspecific cues. For both species, those previously exposed to acidic treatment showed a preference for conspecific chemical cues again when treated with the current pH condition. This case suggests that fish and crustacean larvae show a reduced preference for vital conspecific chemical cues. Such decreased attraction could bring about reduced recruitment to reefs due to oceanic climate changes.

As marine larvae displayed a reduced preference to conspecific chemical cues, those larvae might locate unsuitable settlement habitats, leading to higher predation and less recruitment. Understanding how oceanic climate change impacts coral reef ecosystems is necessary. Larvae dispersal plays an important role in the population dynamics of reef ecosystems. Thus if the recruitment process of marine larvae is interrupted, those larvae might ignore or misread the substantial chemical cues, which results in less recruitment and impaired reefs. Clarifying the relationship between reef structure and recruitment possibility could help governments take measures for coral maintenance and reef conservation.

Volcanic CO_2 seeps bring dissolved CO_2 and pH gradients, and although it is not a perfect parallel for OA, it indeed could provide practical insights into coastal ecosystems and community changes. In an attempt to estimate the effects of OA on the dynamic changes of fish communities in Japan, (Cattano et al., 2020) used fish and benthic community assessments to evaluate how these properties changed spatiotemporally along the CO_2 seep gradient. They found that fish assemblages had significant regional disparities along the CO_2 gradient, and between elevated CO_2 and reference sites. Elevated CO_2 can affect marine fishes in direct and indirect ways and then affect the communities' structure. OA could be detrimental to the physiological and behavioral performance via inducing impairments in acid–base balance regulation and sensory message handling. Moreover, elevated CO_2 may also indirectly affect the composition and complexity of habitats by promoting biogenic habitat changes, and habitat and food web simplification. In this case, the complexity of the biogenic habitat was reduced with elevated pCO_2. When the habitat complexity decreases, such transition may result in reduced available resources such as food and living space, which might impact vital behavioral processes, such as foraging, larval settlement, and antipredation. In terms of the richness of fish species, the average value was found to be significantly decreased in acidification conditions. The abundance of carnivorous species was decreased with the increment in pCO_2. It normally appears that herbivores tend to increase while carnivores tend to decrease in acidification systems. Thus piscivorous species are more attracted to complex habitats with higher prey abundance. Overall, this case documented shifts in biogenic habitat and reduced benthic complexity under acidification conditions. Such changes in fish communities would have great impacts on the marine food web, influencing the trophic transfer as well as fisheries, eventually. This case concluded that elevated pCO_2 would affect coastal ecosystems indirectly and have a far-reaching influence on reef fish communities. It is supposed that the predicted OA soon will hinder the coral reef–associated fish expansion due to global warming.

Case study 4—social hierarchy

Oceanic climate change poses a considerable threat to coral reef ecosystems. Although previous studies have shown that OA has negative impacts on the fitness of reef corals, the long-term effects of competition for space on coral growth rates remain unknown. OA can trigger competition among species because it affects both the supply and demand of resources, thus influencing the social hierarchy to a certain extent. Horwitz et al. (2017) found a competitive hierarchy change and overall coral cover

Present day   Acidified conditions

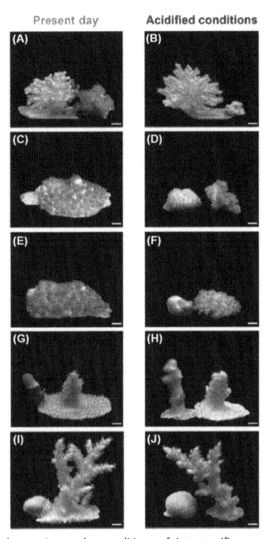

Figure 3.5 Corals growing under conditions of interspecific competition. Images show representative coral fragments from pair-wise interactions under present-day (pH 8.1) and acidified (pH 7.6) conditions for: *Pocillopora damicornis* (A) vs. *Galaxea fascicularis* (B); *Cyphastrea chalcidicum* (C) vs. *G. fascicularis* (D); *Porites lutea* (E) vs. *G. fascicularis* (F); *Stylophora pistillata* (G) vs. *Acropora variabilis* (H); *P. lutea* (I) vs. *A. variabilis* (J). Scale bar length is 1 cm. *From Horwitz, R., Hoogenboom, M. O., & Fine, M. (2017). Spatial competition dynamics between reef corals under ocean acidification. Scientific Reports, 7, 40288. https://doi.org/10.1038/srep40288.*

decrease under low pH using a spatial competition model (Fig. 3.5). In their study, six common Indo-Pacific coral species, namely *Galaxea fascicularis, Pocillopora damicornis, A. variabilis, Cyphastrea chalcidicum, Stylophora pistillata*, and *Porites lutea*, were examined under normal (pH 8.1) and OA (pH 7.6) conditions for 1 year. The corals were accumulated on prelabeled glass slides with a distance of 1 cm between two rivals. The authors aimed to describe the direct rivalry within and between reef-building coral species.

They found that OA significantly obstructed coral growth under conditions of intraspecific competition and that it might alter the spatial competition dynamics between reef corals. Compared with environmental stressors, some competitors can more severely inhibit the growth of all species that interact with them under low pH. These decreases in growth indicated that the spatial competition models might be a prediction of the percentage cover decrease of corals on reefs under OA in comparison with standard conditions. Apart from the decreased coral cover, the reduced growth rates can lead to decreased lifetime reproductive outputs of colonies, because the coral fecundity has a positive relationship with its size. Previous studies, by growing tested coral fragments separately from each other or in short-term tests, have observed changes in coral growth responding to OA but potentially ignored the long-term effects on coral populations. This study focused on more extreme acidification effects on coral growth in a population with intraspecific competition. These results highlight the importance of accounting for competition-dependent alternations in coral growth when branching out experiments to clarify the ecosystem functions in this field. Overall, the present experimental and modeling results expressed that the interactions between coral colonies on reefs, whether within or between species, may alter with pH declines. Such changes may result in a shift in species composition and biodiversity of coral assemblages on reefs and possibly undermine reef functions, such as lower structural complexity, and the quality and quantity of reef ecological products and services may also be reduced eventually.

At the predicted level of OA, changes in energy demand and food supply may interact with each other, thereby affecting the growth and development of marine organisms. Then, changes in individual growth rates may subsequently affect the characteristics of social groups, especially in the formation of species based on size classes. McMahon et al. investigated the interactive effects of pH and food availability on the growth of a juvenile reef fish and if such effects can flow on to influence their social groups (McMahon et al., 2019). Compared with the control treatment, the low pH conditions reduced the standard length and weight of the juveniles by 9% and 11%, respectively. The body length and weight of the juveniles were reduced by 7% and 15%, respectively, due to food deficiency. Parents exposed to low pH were able to reclaim the juvenile length but not the weight, indicating there was an energetic tradeoff that favored linear growth at the cost of poorer body condition, with lower mass for a given length in the juveniles produced by parents exposed to low pH conditions. The body size proportion of the juvenile group (relative size of superior fish) was not affected by any treatment, indicating that even if the individual size and condition change, the relative robustness of the group-level structure remains. This study shows that food supply and OA may affect the physical properties of juveniles, but such changes may not disrupt the emerging group structure of the social species, at least in juveniles. Parental exposure to OA led to increased linear growth. It alleviated some of the adverse effects on development in acidified conditions. However, these results were found to be at the expense of reduced physical conditions. Maintaining a particular length in a social hierarchy can likely generate this tradeoff. Comprehending the effects of acidification and food availability on marine animals within their natural social systems plays an essential role in improving our ability to predict the effect of these environmental stressors on marine ecosystems (McMahon et al., 2019). In an ideal situation, further studies should involve investigations about the interactive impacts of other environmental factors such as ocean warming.

The ability of marine species to adapt to oceanic change and the way in which the changed characteristics influence vital physiological processes reflect how marine communities are affected by OA. McCormick et al. researched how the projected OA affects interspecific competition between two damselfish species and the effects of decreasing habitat quality on these interactions (McCormick et al., 2013). This study showed a reversal of competitive dominance with mortal results, suggesting that OA would lead to more complicated results than previously predicted. For the fish exposed to current or future projected CO_2 levels, such competitive interactions could indirectly generate adverse outcomes, because the aggressive behaviors would expel the subordinate from the shelter and make them at high risk. The alternations demonstrated in this case suggest that OA could lead to changes in the way marine species interact with each other, and this will also change the processes shaping communities and ecosystems.

Under natural conditions, competitive reversals between marine species are very common, as organisms with strong competitive asymmetry could coexist. The interference competition is especially important in influencing the small-scale distribution of some physiological characteristics, such as growth, body size, and distribution concerning predation risk. While previous studies have demonstrated differences in sensitivity to CO_2 among damselfish species, how these differences may affect behavioral interactions under natural conditions of predation risk is still unknown. This case is the first experiment for understanding how the community process changes under elevated CO_2. Besides, more studies should be conducted to reveal the effects on predators. As shown in this case, competition can impair survival by altering individuals' exposure to predators, either directly due to aggression pushing subordinates into riskier habitats, or indirectly by influencing the balance between foraging and vigilance. As both intra- and interspecific competitions critically affect prey vulnerability to predators, there remains a need for understanding communities' regulation in the coming acidic ocean environment.

OA, the other problem caused by CO_2, actively poses significant threats to marine ecosystems. This brief review demonstrates that OA would reduce the fitness of marine organisms by affecting several types of critical behaviors. OA and the subsequent ecological impacts on marine ecosystems are inevitable, and future research should take steps to seek potential solutions rather than record the impacts via laboratory experiments. Understanding the underlying physiological mechanisms behind the effects of OA on marine organisms will play an essential role in these efforts.

References

Allan, B. J. M., Domenici, P., McCormick, M. I., Watson, S. A., & Munday, P. L. (2013). Elevated CO_2 affects predator-prey interactions through altered performance. *PLoS One*, *8*(3). Available from https://doi.org/10.1371/journal.pone.0058520.

Alma, L., Kram, K. E., Holtgrieve, G. W., Barbarino, A., Fiamengo, C. J., & Padilla-Gamino, J. L. (2020). Ocean acidification and warming effects on the physiology, skeletal properties, and microbiome of the purple-hinge rock scallop. *Comparative Biochemistry and Physiology Part A: Molecular and Integrative Physiology*, *240*, 110579. Available from https://doi.org/10.1073/pnas.1805186115.

Andrade, J. F., Hurst, T. P., & Miller, J. A. (2018). Behavioral responses of a coastal flatfish to predation-associated cues and elevated CO_2. *Journal of Sea Research*, *140*, 11–21. Available from https://doi.org/10.1016/j.seares.2018.06.013.

Ashur, M. M., Johnston, N. K., & Dixson, D. L. (2017). Impacts of ocean acidification on sensory function in marine organisms. *Integrative and Comparative Biology*, *57*(1)), 63–80. Available from https://doi.org/10.1093/icb/icx010.

Atema, J., Kingsford, M. J., & Gerlach, G. (2002). Larval reef fish could use odour for detection, retention and orientation to reefs. *Marine Ecology Progress Series*, *241*, 151–160. Available from https://doi.org/10.3354/meps241151.

Babarro, J. M. F., Abad, M. J., Gestoso, I., Silva, E., & Olabarria, C. (2018). Susceptibility of two co-existing mytilid species to simulated predation under projected climate change conditions. *Hydrobiologia*, *807*(1), 247−261. Available from https://doi.org/10.1007/s10750-017-3397-7.

Bamber, T. O., Jackson, A. C., & Mansfield, R. P. (2018). The effects of ocean acidification on feeding and contest behaviour by the beadlet anemone actinia equina. *Ocean Science Journal*, *53*(2), 215−224. Available from https://doi.org/10.1007/s12601-018-0023-1.

Barclay, K. M., Gaylord, B., Jellison, B. M., Shukla, P., Sanford, E., & Leighton, L. R. (2019). Variation in the effects of ocean acidification on shell growth and strength in two intertidal gastropods. *Marine Ecology Progress Series*, *626*, 109−121. Available from https://doi.org/10.3354/meps13056.

Barry, J. P., Lovera, C., Buck, K. R., Peltzer, E. T., Taylor, J. R., Walz, P., Whaling, P. J., & Brewer, P. G. (2014). Use of a free ocean CO2 enrichment (FOCE) system to evaluate the effects of ocean acidification on the foraging behavior of a deep-sea urchin. *Environmental Science and Technology*, *48*(16), 9890−9897. Available from https://doi.org/10.1021/es501603r.

Bednaršek, N., Feely, R. A., Reum, J. C. P., Peterson, B., Menkel, J., Alin, S. R., & Hales, B. (2014). Limacina helicina shell dissolution as an indicator of declining habitat suitability owing to ocean acidification in the California Current Ecosystem. *Proceedings of the Royal Society B: Biological Sciences*, *281*(1785). Available from https://doi.org/10.1098/rspb.2014.0123.

Borges, F. O., Santos, C. P., Sampaio, E., Figueiredo, C., Paula, J. R., Antunes, C., Rosa, R., & Grilo, T. F. (2019). Ocean warming and acidification may challenge the riverward migration of glass eels. *Biology Letters*, *15*(1). Available from https://doi.org/10.1098/rsbl.2018.0627.

Briffa, M., de la Haye, K., & Munday, P. L. (2012). High CO_2 and marine animal behaviour: Potential mechanisms and ecological consequences. *Marine Pollution Bulletin*, *64*(8), 1519−1528. Available from https://doi.org/10.1016/j.marpolbul.2012.05.032.

Brose, U., Archambault, P., Barnes, A. D., Bersier, L. F., Boy, T., Canning-Clode, J., Conti, E., Dias, M., Digel, C., Dissanayake, A., Flores, A. A. V., Fussmann, K., Gauzens, B., Gray, C., Häussler, J., Hirt, M. R., Jacob, U., Jochum, M., Kéfi, S., & Iles, A. C. (2019). Predator traits determine food-web architecture across ecosystems. *Nature Ecology and Evolution*, *3*(6), 919−927. Available from https://doi.org/10.1038/s41559-019-0899-x.

Byers, J. E. (2002). Physical habitat attribute mediates biotic resistance to non-indigenous species invasion. *Oecologia*, *130*(1), 146−156. Available from https://doi.org/10.1007/s004420100777.

Byrne, M., & Przeslawski, R. (2013). Multistressor impacts of warming and acidification of the ocean on marine invertebrates' life histories. *Integrative and Comparative Biology*, *53*(4), 582−596. Available from https://doi.org/10.1093/icb/ict049.

Castro, J. M., Amorim, M. C. P., Oliveira, A. P., Gonçalves, E. J., Munday, P. L., Simpson, S. D., & Faria, A. M. (2017). Painted goby larvae under high-CO_2 fail to recognize reef sounds. *PLoS One*, *12*(1). Available from https://doi.org/10.1371/journal.pone.0170838.

Cattano, C., Agostin, S., Harvey, B. P., Wada, S., Quattrocchi, F., & Turco, G. (2020). Changes in fish communities due to benthic habitat shifts under ocean acidification conditions. *Science of the Total Environment*, *725*, 138501. Available from https://doi.org/10.1016/j.scitotenv.2020.138501.

Cattano, C., Calò, A., Di Franco, A., Firmamento, R., Quattrocchi, F., Sdiri, K., Guidetti, P., & Milazzo, M. (2017). Ocean acidification does not impair predator recognition but increases juvenile growth in a temperate wrasse off CO_2 seeps. *Marine Environmental Research*, *132*, 33−40. Available from https://doi.org/10.1016/j.marenvres.2017.10.013.

Cattano, C., Claudet, J., Domenici, P., & Milazzo, M. (2018). Living in a high CO_2 world: A global *meta*-analysis shows multiple trait-mediated fish responses to ocean acidification. *Ecological Monographs, 88*(3), 320−335. Available from https://doi.org/10.1002/ecm.1297.

Cattano, C., Fine, M., Quattrocchi, F., Holzman, R., & Milazzo, M. (2019). Behavioural responses of fish groups exposed to a predatory threat under elevated CO_2. *Marine Environmental Research, 147*, 179−184. Available from https://doi.org/10.1016/j.marenvres.2019.04.011.

Cigliano, M., Gambi, M. C., Rodolfo-Metalpa, R., Patti, F. P., & Hall-Spencer, J. M. (2010). Effects of ocean acidification on invertebrate settlement at volcanic CO_2 vents. *Marine Biology, 157*(11), 2489−2502. Available from https://doi.org/10.1007/s00227-010-1513-6.

Clark, T. D., Raby, G. D., Roche, D. G., Binning, S. A., Speers-Roesch, B., Jutfelt, F., & Sundin, J. (2020). Ocean acidification does not impair the behaviour of coral reef fishes. *Nature, 577*(7790), 370−375. Available from https://doi.org/10.1038/s41586-019-1903-y.

Clements, J. C., & Hunt, H. L. (2015). Marine animal behaviour in a high CO_2 ocean. *Marine Ecology Progress Series, 536*, 259−279. Available from https://doi.org/10.3354/meps11426.

Clements, J. C., Poirier, L. A., Pérez, F. F., Comeau, L. A., & Babarro, J. M. F. (2020). Behavioural responses to predators in Mediterranean mussels (*Mytilus galloprovincialis*) are unaffected by elevated pCO_2. *Marine Environmental Research, 161*. Available from https://doi.org/10.1016/j.marenvres.2020.105148.

Contolini, G. M., Reid, K., & Palkovacs, E. P. (2020). Climate shapes population variation in dogwhelk predation on foundational mussels. *Oecologia, 192*(2), 553−564. Available from https://doi.org/10.1007/s00442-019-04591-x.

Davis, B. E., Komoroske, L. M., Hansen, M. J., Poletto, J. B., Perry, E. N., Miller, N. A., ... Fangue, N. A. (2018). Juvenile rockfish show resilience to CO_2-acidification and hypoxia across multiple biological scales. *Conservation Physiology, 6*, coy038. Available from https://doi.org/10.1093/conphys/coy038.

Devine, B. M., & Munday, P. L. (2013). Habitat preferences of coral-associated fishes are altered by short-term exposure to elevated CO_2. *Marine Biology, 160*(8), 1955−1962. Available from https://doi.org/10.1007/s00227-012-2051-1.

Devine, B. M., Munday, P. L., & Jones, G. P. (2012a). Homing ability of adult cardinalfish is affected by elevated carbon dioxide. *Oecologia, 169*, 269−276.

Devine, B. M., Munday, P. L., & Jones, G. P. (2012b). Rising CO_2 concentrations affect settlement behaviour of larval damselfishes. *Coral Reefs, 31*(1), 229−238. Available from https://doi.org/10.1007/s00338-011-0837-0.

Dixson, D. L., Jennings, A. R., Atema, J., & Munday, P. L. (2015). Odor tracking in sharks is reduced under future ocean acidification conditions. *Global Change Biology, 21*(4), 1454−1462. Available from https://doi.org/10.1111/gcb.12678.

Dixson, D. L., Munday, P. L., & Jones, G. P. (2010). Ocean acidification disrupts the innate ability of fish to detect predator olfactory cues. *Ecology Letters, 13*(1), 68−75. Available from https://doi.org/10.1111/j.1461-0248.2009.01400.x.

Dodd, L. F., Grabowski, J. H., Piehler, M. F., Westfield, I., & Ries, J. B. (2015). Ocean acidification impairs crab foraging behavior. *Proceedings of the Royal Society B: Biological Sciences, 282*(1810). Available from https://doi.org/10.1098/rspb.2015.0333.

Doropoulos, C., & Diaz-Pulido, G. (2013). High CO_2 reduces the settlement of a spawning coral on three common species of crustose coralline algae. *Marine Ecology Progress Series, 475*, 93−99. Available from https://doi.org/10.3354/meps10096.

Doropoulos, C., Ward, S., Diaz-Pulido, G., Hoegh-Guldberg, O., & Mumby, P. J. (2012). Ocean acidification reduces coral recruitment by disrupting intimate larval-algal settlement interactions. *Ecology Letters, 15*, 338−346. Available from https://doi.org/10.1111/j.1461-0248.2012.01743.x.

Espinel-Velasco, N., Agüera, A., & Lamare, M. (2020). Sea urchin larvae show resilience to ocean acidification at the time of settlement and metamorphosis. *Marine Environmental Research, 159*, 104977. Available from https://doi.org/10.1016/j.marenvres.2020.104977.

Fabricius, K. E., De'ath, G., Noonan, S., & Uthicke, S. (2014). Ecological effects of ocean acidification and habitat complexity on reef-associated macroinvertebrate communities. *Proceedings of the Royal Society B: Biological Sciences, 281*(1775). Available from https://doi.org/10.1098/rspb.2013.2479.

Fabry, V. J., Seibel, B. A., Feely, R. A., & Orr, J. C. (2008). Impacts of ocean acidification on marine fauna and ecosystem processes. *ICES Journal of Marine Science, 65*(3)), 414−432. Available from https://doi.org/10.1093/icesjms/fsn048.

Faleiro, F., Baptista, M., Santos, C., Aurélio, M. L., Pimentel, M., Pegado, M. R., Paula, J. R., Calado, R., Repolho, T., & Rosa, R. (2015). Seahorses under a changing ocean: The impact of warming and acidification on the behaviour and physiology of a poor-swimming bony-armoured fish. *Conservation Physiology, 3*(1). Available from https://doi.org/10.1093/conphys/cov009.

Ferrari, M. C. O., Mccormick, M. I., Munday, P. L., Meekan, M. G., Dixson, D. L., Lönnstedt, O., & Chivers, D. P. (2012). Effects of ocean acidification on visual risk assessment in coral reef fishes. *Functional Ecology, 26*(3), 553−558. Available from https://doi.org/10.1111/j.1365-2435.2011.01951.x.

Ferrari, M. C. O., Munday, P. L., Rummer, J. L., Mccormick, M. I., Corkill, K., Watson, S. A., Allan, B. J. M., Meekan, M. G., & Chivers, D. P. (2015). Interactive effects of ocean acidification and rising sea temperatures alter predation rate and predator selectivity in reef fish communities. *Global Change Biology, 21*(5), 1848−1855. Available from https://doi.org/10.1111/gcb.12818.

Froehlich, K. R., & Lord, J. P. (2020). Can ocean acidification interfere with the ability of mud snails (*Tritia obsoleta*) to sense predators? *Journal of Experimental Marine Biology and Ecology, 526*, 151355. Available from https://doi.org/10.1016/j.jembe.2020.151355.

García, E., Clemente, S., & Hernández, J. C. (2018). Effects of natural current pH variability on the sea urchin Paracentrotus lividus larvae development and settlement. *Marine Environmental Research, 139*, 11−18. Available from https://doi.org/10.1016/j.marenvres.2018.04.012.

Gazeau, F., Parker, L. M., Comeau, S., Gattuso, J. P., O'Connor, W. A., Martin, S., Pörtner, H. O., & Ross, P. M. (2013). Impacts of ocean acidification on marine shelled molluscs. *Marine Biology, 160*(8), 2207−2245. Available from https://doi.org/10.1007/s00227-013-2219-3.

Glaspie, C. N., Longmire, K., & Seitz, R. D. (2017). Acidification alters predator-prey interactions of blue crab Callinectes sapidus and soft-shell clam *Mya arenaria*. *Journal of Experimental Marine Biology and Ecology, 489*, 58−65. Available from https://doi.org/10.1016/j.jembe.2016.11.010.

Gosselin, L. A., & Qian, P. Y. (1997). Juvenile mortality in benthic marine invertebrates. *Marine Ecology Progress Series, 146*(1−3), 265−282. Available from https://doi.org/10.3354/meps146265.

Hadfield, M. G. (2011). Biofilms and marine invertebrate larvae: What bacteria produce that larvae use to choose settlement sites. *Annual Review of Marine Science, 3*, 453−470. Available from https://doi.org/10.1146/annurev-marine-120709-142753.

Hale, R., Calosi, P., Mcneill, L., Mieszkowska, N., & Widdicombe, S. (2011). Predicted levels of future ocean acidification and temperature rise could alter community structure and biodiversity in marine benthic communities. *Oikos, 120*(5), 661−674. Available from https://doi.org/10.1111/j.1600-0706.2010.19469.x.

Heinrich, D. D. U., Watson, S.-A., Rummer, J. L., Brandl, S. J., Simpfendorfer, C. A., Heupel, M. R., & Munday, P. L. (2016). Foraging behaviour of the epaulette shark *Hemiscyllium ocellatum* is not affected by elevated CO_2. *ICES Journal of Marine Science, 73*(3), 633−640. Available from https://doi.org/10.1093/icesjms/fsv085.

Heuer, R. M., & Grosell, M. (2014). Physiological impacts of elevated carbon dioxide and ocean acidification on fish. *American Journal of Physiology — Regulatory Integrative and Comparative Physiology*, *307*(9), R1061—R1084. Available from https://doi.org/10.1152/ajpregu.00064.2014.

Heuer, R. M., Welch, M. J., Rummer, J. L., Munday, P. L., & Grosell, M. (2016). Altered brain ion gradients following compensation for elevated CO_2 are linked to behavioural alterations in a coral reef fish. *Scientific Reports*, *6*, 33216. Available from https://doi.org/10.1038/srep33216.

Hoegh-Guldberg, O., Poloczanska, E. S., Skirving, W., & Dove, S. (2017). Coral reef ecosystems under climate change and ocean acidification. *Frontiers in Marine Science*, *4* (158), 158. Available from https://doi.org/10.3389/fmars.2017.00158.

Horwitz, R., Hoogenboom, M. O., & Fine, M. (2017). Spatial competition dynamics between reef corals under ocean acidification. *Scientific Reports*, *7*, 40288. Available from https://doi.org/10.1038/srep40288.

Horwitz, R., Norin, T., Watson, S. A., Pistevos, J. C. A., Beldade, R., Hacquart, S., Gattuso, J. P., Rodolfo-Metalpa, R., Vidal-Dupiol, J., Killen, S. S., & Mills, S. C. (2020). Near-future ocean warming and acidification alter foraging behaviour, locomotion, and metabolic rate in a keystone marine mollusc. *Scientific Reports*, *10*(1), 5461. Available from https://doi.org/10.1038/s41598-020-62304-4.

Huijbers, C. M., Nagelkerken, I., Lössbroek, P. A. C., Schulten, I. E., Siegenthaler, A., Holderied, M. W., & Simpson, S. D. (2012). A test of the senses: Fish select novel habitats by responding to multiple cues. *Ecology*, *93*(1), 46—55. Available from https://doi.org/10.1890/10-2236.1.

Hunt, H. L., & Scheibling, R. E. (1997). Role of early post-settlement mortality in recruitment of benthic marine invertebrates. *Marine Ecology Progress Series*, *155*, 269—301. Available from https://doi.org/10.3354/meps155269.

Ishimatsu, A., Hayashi, M., & Kikkawa, T. (2008). Fishes in high-CO_2, acidified oceans. *Marine Ecology Progress Series*, *373*, 295—302. Available from https://doi.org/10.3354/meps07823.

Jarrold, M. D., Humphrey, C., McCormick, M. I., & Munday, P. L. (2017). Diel CO_2 cycles reduce severity of behavioural abnormalities in coral reef fish under ocean acidification. *Scientific Reports*, *7*(1), 10153. Available from https://doi.org/10.1038/s41598-017-10378-y.

Jellison, B. M., & Gaylord, B. (2019). Shifts in seawater chemistry disrupt trophic links within a simple shoreline food web. *Oecologia*, *190*(4), 955—967. Available from https://doi.org/10.1007/s00442-019-04459-0.

Jellison, B. M., Ninokawa, A. T., Hill, T. M., Sanford, E., & Gaylord, B. (2016). Ocean acidification alters the response of intertidal snails to a key sea star predator. *Proceedings of the Royal Society B: Biological Sciences*, *283*(1833). Available from https://doi.org/10.1098/rspb.2016.0890.

Jutfelt, F., & Hedgärde, M. (2015). Juvenile Atlantic cod behavior appears robust to near-future CO_2 levels. *Frontiers in Zoology*, *12*(1), 11. Available from https://doi.org/10.1186/s12983-015-0104-2.

Kendall, M. S., Poti, M., Wynne, T. T., Kinlan, B. P., & Bauer, L. B. (2013). Consequences of the life history traits of pelagic larvae on interisland connectivity during a changing climate. *Marine Ecology Progress Series*, *489*, 43—59. Available from https://doi.org/10.3354/meps10432.

Kim, T. W., Taylor, J., Lovera, C., & Barry, J. P. (2016). CO_2-driven decrease in pH disrupts olfactory behaviour and increases individual variation in deep-sea hermit crabs. *ICES Journal of Marine Science*, *73*(3), 613—619. Available from https://doi.org/10.1093/icesjms/fsv019.

Kong, H., Clements, J. C., Dupont, S., Wang, T., Huang, X., Shang, Y., Huang, W., Chen, J., Hu, M., & Wang, Y. (2019). Seawater acidification and temperature modulate anti-predator defenses in two co-existing Mytilus species. *Marine Pollution Bulletin*, *145*, 118−125. Available from https://doi.org/10.1016/j.marpolbul.2019.05.040.

Kroeker, K. J., Kordas, R. L., Crim, R., Hendriks, I. E., Ramajo, L., Singh, G. S., Duarte, C. M., & Gattuso, J. P. (2013). Impacts of ocean acidification on marine organisms: Quantifying sensitivities and interaction with warming. *Global Change Biology*, *19*(6), 1884−1896. Available from https://doi.org/10.1111/gcb.12179.

Kroeker, K. J., Micheli, F., & Gambi, M. C. (2013). Ocean acidification causes ecosystem shifts via altered competitive interactions. *Nature Climate Change*, *3*(2), 156−159. Available from https://doi.org/10.1038/nclimate1680.

Kroeker, K. J., Sanford, E., Jellison, B. M., & Gaylord, B. (2014). Predicting the effects of ocean acidification on predator-prey interactions: A conceptual framework based on coastal molluscs. *Biological Bulletin*, *226*(3), 211−222. Available from https://doi.org/ 10.1086/BBLv226n3p211.

Lassoued, J., Babarro, J. M. F., Padín, X. A., Comeau, L. A., Bejaoui, N., & Pérez, F. F. (2019). Behavioural and eco-physiological responses of the mussel *Mytilus galloprovincialis* to acidification and distinct feeding regimes. *Marine Ecology Progress Series*, *626*, 97−108. Available from https://doi.org/10.3354/meps13075.

Laubenstein, T. D., Rummer, J. L., McCormick, M. I., & Munday, P. L. (2019). A negative correlation between behavioural and physiological performance under ocean acidification and warming. *Scientific Reports*, *9*(1). Available from https://doi.org/10.1038/ s41598-018-36747-9.

Lecchini, D., Dixson, D. L., Lecellier, G., Roux, N., Frédérich, B., Besson, M., Tanaka, Y., Banaigs, B., & Nakamura, Y. (2017). Habitat selection by marine larvae in changing chemical environments. *Marine Pollution Bulletin*, *114*(1), 210−217. Available from https://doi.org/10.1016/j.marpolbul.2016.08.083.

Lecchini, D., Shima, J., Banaigs, B., & Galzin, R. (2005). Larval sensory abilities and mechanisms of habitat selection of a coral reef fish during settlement. *Oecologia*, *143* (2), 326−334. Available from https://doi.org/10.1007/s00442-004-1805-y.

Leis, J. M., Siebeck, U., & Dixson, D. L. (2011). How nemo finds home: The neuroecology of dispersal and of population connectivity in larvae of marine fishes. *Integrative and Comparative Biology*, *51*(5), 826−843. Available from https://doi.org/10.1093/icb/icr004.

Leung, J. Y. S., Russell, B. D., Connell, S. D., Ng, J. C. Y., & Lo, M. M. Y. (2015). Acid dulls the senses: Impaired locomotion and foraging performance in a marine mollusc. *Animal Behaviour*, *106*, 223−229. Available from https://doi.org/10.1016/j. anbehav.2015.06.004.

Li, F., Mu, F. H., Liu, X. S., Xu, X. Y., & Cheung, S. G. (2020). Predator prey interactions between predatory gastropod *Reishia clavigera*, barnacle *Amphibalanus amphitrite amphitrite* and mussel *Brachidontes variabilis* under ocean acidification. *Marine Pollution Bulletin*, *152*, 110895. Available from https://doi.org/10.1016/j.marpolbul.2020.110895.

Lord, J. P., Harper, E. M., & Barry, J. P. (2019). Ocean acidification may alter predator−prey relationships and weaken nonlethal interactions between gastropods and crabs. *Marine Ecology Progress Series*, *616*, 83−94. Available from https://doi.org/10.3354/meps12921.

Maboloc, E. A., & Chan, K. Y. K. (2017). Resilience of the larval slipper limpet *Crepidula onyx* to direct and indirect-diet effects of ocean acidification. *Scientific Reports*, *7*(1), 12062. Available from https://doi.org/10.1038/s41598-017-12253-2.

Maneja, R. H., Frommel, A. Y., Browman, H. I., Geffen, A. J., Folkvord, A., Piatkowski, U., Durif, C. M. F., Bjelland, R., Skiftesvik, A. B., & Clemmesen, C. (2015). The swimming kinematics and foraging behavior of larval Atlantic herring (*Clupea harengus* L.) are unaffected by elevated pCO_2. *Journal of Experimental Marine Biology and Ecology*, *466*, 42−48. Available from https://doi.org/10.1016/j.jembe.2015.02.008.

McClintock, J. B., Angus, R. A., McDonald, M. R., Amsler, C. D., Catledge, S. A., & Vohra, Y. K. (2009). Rapid dissolution of shells of weakly calcified Antarctic benthic macroorganisms indicates high vulnerability to ocean acidification. *Antarctic Science, 21* (5), 449–456. Available from https://doi.org/10.1017/S0954102009990198.

McCormick, M. I., Watson, S. A., & Munday, P. L. (2013). Ocean acidification reverses competition for space as habitats degrade. *Scientific Reports, 3.* Available from https://doi.org/10.1038/srep03280.

McCormick, M. I., Watson, S. A., Simpson, S. D., & Allan, B. J. M. (2018). Effect of elevated CO_2 and small boat noise on the kinematics of predator − Prey interactions. *Proceedings of the Royal Society B: Biological Sciences, 285*(1875). Available from https://doi.org/10.1098/rspb.2017.2650.

McMahon, S. J., Munday, P. L., Wong, M. Y. L., & Donelson, J. M. (2019). Elevated CO_2 and food ration affect growth but not the size-based hierarchy of a reef fish. *Scientific Reports, 9*(1). Available from https://doi.org/10.1038/s41598-019-56002-z.

Melzner, F., Mark, F. C., Seibel, B. A., & Tomanek, L. (2020). Ocean acidification and coastal marine invertebrates: Tracking CO_2 effects from seawater to the cell. *Annual Review of Marine Science, 12,* 499–523. Available from https://doi.org/10.1146/annurev-marine-010419-010658.

Menu-Courey, K., Noisette, F., Piedalue, S., Daoud, D., Blair, T., Blier, P. U., ... Calosi, P. (2019). Energy metabolism and survival of the juvenile recruits of the American lobster (*Homarus americanus*) exposed to a gradient of elevated seawater pCO_2. *Marine environmental research, 143,* 111–123. Available from https://doi.org/10.1016/j.marenvres.2018.10.002.

Meseck, S. L., Sennefelder, G., Krisak, M., & Wikfors, G. H. (2020). Physiological feeding rates and cilia suppression in blue mussels (*Mytilus edulis*) with increased levels of dissolved carbon dioxide. *Ecological Indicators, 117,* 106675. Available from https://doi.org/10.1016/j.ecolind.2020.106675.

Moura, É., Pimentel, M., Santos, C. P., Sampaio, E., Pegado, M. R., Lopes, V. M., & Rosa, R. (2019). Cuttlefish early development and behavior under future high CO_2 conditions. *Frontiers in Physiology, 10.* Available from https://doi.org/10.3389/fphys.2019.00975.

Moy, A. D., Howard, W. R., Bray, S. G., & Trull, T. W. (2009). Reduced calcification in modern Southern Ocean planktonic foraminifera. *Nature Geoscience, 2*(4), 276–280. Available from https://doi.org/10.1038/ngeo460.

Munday, P. L., Dixson, D. L., McCormick, M. I., Meekan, M., Ferrari, M. C. O., & Chivers, D. P. (2010). Replenishment of fish populations is threatened by ocean acidification. *Proceedings of the National Academy of Sciences of the United States of America, 107* (29), 12930–12934. Available from https://doi.org/10.1073/pnas.1004519107.

Nadermann, N., Seward, R. K., & Volkoff, H. (2019). Effects of potential climate change -induced environmental modifications on food intake and the expression of appetite regulators in goldfish. *Comparative Biochemistry and Physiology - Part A: Molecular and Integrative Physiology, 235,* 138–147. Available from https://doi.org/10.1016/j.cbpa.2019.06.001.

Nagelkerken, I., & Munday, P. L. (2016). Animal behaviour shapes the ecological effects of ocean acidification and warming: Moving from individual to community-level responses. *Global Change Biology, 22*(3), 974–989. Available from https://doi.org/10.1111/gcb.13167.

Nakamura, M., Ohki, S., Suzuki, A., & Sakai, K. (2011). Coral larvae under ocean acidification: Survival, metabolism, and metamorphosis. *PLoS One, 6*(1). Available from https://doi.org/10.1371/journal.pone.0014521.

Nasuchon, N., Yagi, M., Kawabata, Y., Gao, K., & Ishimatsu, A. (2016). Escape responses of the Japanese anchovy *Engraulis japonicus* under elevated temperature and CO_2 conditions. *Fisheries Science, 82*(3), 435–444. Available from https://doi.org/10.1007/s12562-016-0974-z.

Nelson, K. S., Baltar, F., Lamare, M. D., & Morales, S. E. (2020). Ocean acidification affects microbial community and invertebrate settlement on biofilms. *Scientific Reports*, *10*(1), 3274. Available from https://doi.org/10.1038/s41598-020-60023-4.

Nilsson, G. E., Dixson, D. L., Domenici, P., McCormick, M. I., Sørensen, C., Watson, S. A., & Munday, P. L. (2012). Near-future carbon dioxide levels alter fish behaviour by interfering with neurotransmitter function. *Nature Climate Change*, *2*(3), 201−204. Available from https://doi.org/10.1038/nclimate1352.

Nowicki, J. P., Miller, G. M., & Munday, P. L. (2012). Interactive effects of elevated temperature and CO_2 on foraging behavior of juvenile coral reef fish. *Journal of Experimental Marine Biology and Ecology*, *412*, 46−51. Available from https://doi.org/10.1016/j.jembe.2011.10.020.

Pankhurst, N. W., & Munday, P. L. (2011). Effects of climate change on fish reproduction and early life history stages. *Marine and Freshwater Research*, *62*(9), 1015−1026. Available from https://doi.org/10.1071/MF10269.

Park, S., Ahn, I. Y., Sin, E., Shim, J. H., & Kim, T. (2020). Ocean freshening and acidification differentially influence mortality and behavior of the Antarctic amphipod *Gondogeneia antarctica*. *Marine Environmental Research*, *154*, 104847. Available from https://doi.org/10.1016/j.marenvres.2019.104847.

Pecquet, A., Dorey, N., & Chan, K. Y. K. (2017). Ocean acidification increases larval swimming speed and has limited effects on spawning and settlement of a robust fouling bryozoan, *Bugula neritina*. *Marine Pollution Bulletin*, *124*(2), 903−910. Available from https://doi.org/10.1016/j.marpolbul.2017.02.057.

Pimentel, M. S., Faleiro, F., Marques, T., Bispo, R., Dionísio, G., Faria, A. M., Machado, J., Peck, M. A., Pörtner, H., Pousão-Ferreira, P., Gonçalves, E. J., & Rosa, R. (2016). Foraging behaviour, swimming performance and malformations of early stages of commercially important fishes under ocean acidification and warming. *Climatic Change*, *137*(3−4), 495−509. Available from https://doi.org/10.1007/s10584-016-1682-5.

Pistevos, J. C. A., Nagelkerken, I., Rossi, T., Olmos, M., & Connell, S. D. (2015). Ocean acidification and global warming impair shark hunting behaviour and growth. *Scientific Reports*, *5*, 16293. Available from https://doi.org/10.1038/srep16293.

Przeslawski, R., Ahyong, S., Byrne, M., Wörheide, G., & Hutchings, P. (2008). Beyond corals and fish: The effects of climate change on noncoral benthic invertebrates of tropical reefs. *Global Change Biology*, *14*(12), 2773−2795. Available from https://doi.org/10.1111/j.1365-2486.2008.01693.x.

Raby, G. D., Sundin, J., Jutfelt, F., Cooke, S. J., & Clark, T. D. (2018). Exposure to elevated carbon dioxide does not impair short-term swimming behaviour or shelter-seeking in a predatory coral-reef fish. *Journal of Fish Biology*, *93*(1), 138−142. Available from https://doi.org/10.1111/jfb.13728.

Regan, M. D., Turko, A. J., Heras, J., Andersen, M. K., Lefevre, S., Wang, T., Bayley, M., Brauner, C. J., Huong, D. T. T., Phuong, N. T., & Nilsson, G. E. (2016). Ambient CO_2, fish behaviour and altered GABAergic neurotransmission: Exploring the mechanism of CO_2-altered behaviour by taking a hypercapnia dweller down to low CO_2 levels. *Journal of Experimental Biology*, *219*(1), 109−118. Available from https://doi.org/10.1242/jeb.131375.

Ridgwell, A., & Schmidt, D. N. (2010). Past constraints on the vulnerability of marine calcifiers to massive carbon dioxide release. *Nature Geoscience*, *3*(3), 196−200. Available from https://doi.org/10.1038/ngeo755.

Roberts, D., Howard, W. R., Moy, A. D., Roberts, J. L., Trull, T. W., Bray, S. G., & Hopcroft, R. R. (2011). Interannual pteropod variability in sediment traps deployed above and below the aragonite saturation horizon in the Sub-Antarctic Southern Ocean. *Polar Biology*, *34*(11), 1739−1750. Available from https://doi.org/10.1007/s00300-011-1024-z.

Rodríguez, A., Hernández, J. C., Brito, A., & Clemente, S. (2017). Effects of ocean acidification on juveniles sea urchins: Predator-prey interactions. *Journal of Experimental Marine Biology and Ecology, 493*, 31−40. Available from https://doi.org/10.1016/j.jembe.2017.04.005.

Rodriguez, S. R., Ojeda, F. P., & Inestrosa, N. C. (1993). Settlement of benthic marine invertebrates. *Marine Ecology Progress Series, 97*(2), 193−207. Available from https://doi.org/10.3354/meps097193.

Rong, J. H., Su, W. H., Guan, X. F., Shi, W., Zha, S. J., He, M. L., ... Liu, G. X. (2018). Ocean acidification impairs foraging behavior by interfering with olfactory neural signal transduction in black sea bream, *Acanthopagrus schlegelii. Frontiers in physiology, 9*, 1592. Available from https://doi.org/10.3389/fphys.2018.01592.

Ross, E., & Behringer, D. (2019). Changes in temperature, pH, and salinity affect the sheltering responses of Caribbean spiny lobsters to chemosensory cues. *Scientific Reports, 9* (1). Available from https://doi.org/10.1038/s41598-019-40832-y.

Rossi, T., Pistevos, J. C. A., Connell, S. D., & Nagelkerken, I. (2018). On the wrong track: Ocean acidification attracts larval fish to irrelevant environmental cues. *Scientific Reports, 8*(1). Available from https://doi.org/10.1038/s41598-018-24026-6.

Roy, A. S., Gibbons, S. M., Schunck, H., Owens, S., Caporaso, J. G., Sperling, M., Nissimov, J. I., Romac, S., Bittner, L., Mühling, M., Riebesell, U., LaRoche, J., & Gilbert, J. A. (2013). Ocean acidification shows negligible impacts on high-latitude bacterial community structure in coastal pelagic mesocosms. *Biogeosciences, 10*(1), 555−566. Available from https://doi.org/10.5194/bg-10-555-2013.

Saavedra, L. M., Parra, D., San Martin, V., Lagos, N., Espinel-Velascoa, A., & Vargas, C. A. (2018). Local habitat influences on feeding and respiration of the intertidal mussels *Perumytilus purpuratus* exposed to increased pCO_2 levels. *Estuaries and Coasts, 41*(4), 1118−1129, Estuaries and Coasts.

Sadler, D. E., Lemasson, A. J., & Knights, A. M. (2018). The effects of elevated CO_2 on shell properties and susceptibility to predation in mussels Mytilus edulis. *Marine Environmental Research, 139*, 162−168. Available from https://doi.org/10.1016/j.marenvres.2018.05.017.

Sampaio, E., Maulvault, A. L., Lopes, V. M., Paula, J. R., Barbosa, V., Alves, R., Pousão-Ferreira, P., Repolho, T., Marques, A., & Rosa, R. (2016). Habitat selection disruption and lateralization impairment of cryptic flatfish in a warm, acid, and contaminated ocean. *Marine Biology, 163*(10), 217. Available from https://doi.org/10.1007/s00227-016-2994-8.

Sanford, E., Gaylord, B., Hettinger, A., Lenz, E. A., Meyer, K., & Hill, T. M. (2014). Ocean acidification increases the vulnerability of native oysters to predation by invasive snails. *Proceedings of the Royal Society B: Biological Sciences, 281*(1778). Available from https://doi.org/10.1098/rspb.2013.2681.

Schram, J. B., Schoenrock, K. M., McClintock, J. B., Amsler, C. D., & Angus, R. A. (2014). Multiple stressor effects of near-future elevated seawater temperature and decreased pH on righting and escape behaviors of two common Antarctic gastropods. *Journal of Experimental Marine Biology and Ecology, 90*−96. Available from https://doi.org/10.1016/j.jembe.2014.04.005.

Shang, Y., Wang, X., Kong, H., Huang, W., Hu, M., & Wang, Y. (2019). Nano-ZnO impairs anti-predation capacity of marine mussels under seawater acidification. *Journal of Hazardous Materials, 371*, 521−528. Available from https://doi.org/10.1016/j.jhazmat.2019.02.072.

Sih, A., Bell, A., & Johnson, J. C. (2004). Behavioral syndromes: An ecological and evolutionary overview. *Trends in Ecology and Evolution, 19*(7), 372−378. Available from https://doi.org/10.1016/j.tree.2004.04.009.

Simpson, S. D., Munday, P. L., Wittenrich, M. L., Manassa, R., Dixson, D. L., Gagliano, M., & Yan, H. Y. (2011). Ocean acidification erodes crucial auditory behaviour in a marine fish. *Biology Letters*, 7(6), 917−920. Available from https://doi.org/10.1098/rsbl.2011.0293.

Spady, B. L., Munday, P. L., & Watson, S. A. (2018). Predatory strategies and behaviours in cephalopods are altered by elevated CO_2. *Global Change Biology*, 24(6), 2585−2596. Available from https://doi.org/10.1111/gcb.14098.

Steckbauer, A., Díaz-Gil, C., Alós, J., Catalán, I. A., & Duarte, C. M. (2018). Predator avoidance in the European Seabass after recovery from short-term hypoxia and different CO_2 conditions. *Frontiers in Marine Science*, 5, 350. Available from https://doi.org/10.3389/fmars.2018.00350.

Stella, J. S., Pratchett, M. S., Hutchings, P. A., & Jones, G. P. (2011). Diversity, importance and vulnerability of coral-associated invertebrates. *Oceanography and Marine Biology: An Annual Review*, 49, 43−116.

Sui, Y., Hu, M., Huang, X., Wang, Y., & Lu, W. (2015). Anti-predatory responses of the thick shell mussel *Mytilus coruscus* exposed to seawater acidification and hypoxia. *Marine Environmental Research*, 109, 159−167. Available from https://doi.org/10.1016/j.marenvres.2015.07.008.

Sui, Y., Liu, Y., Zhao, X., Dupont, S., Hu, M., Wu, F., Huang, X., Li, J., Lu, W., & Wang, Y. (2017). Defense responses to short-term hypoxia and seawater acidification in the thick shell mussel *Mytilus coruscus*. *Frontiers in Physiology*, 8, 145. Available from https://doi.org/10.3389/fphys.2017.00145.

Sunday, J. M., Fabricius, K. E., Kroeker, K. J., Anderson, K. M., Brown, N. E., Barry, J. P., Connell, S. D., Dupont, S., Gaylord, B., Hall-Spencer, J. M., Klinger, T., Milazzo, M., Munday, P. L., Russell, B. D., Sanford, E., Thiyagarajan, V., Vaughan, M. L. H., Widdicombe, S., & Harley, C. D. G. (2017). Ocean acidification can mediate biodiversity shifts by changing biogenic habitat. *Nature Climate Change*, 7(1), 81−85. Available from https://doi.org/10.1038/nclimate3161.

Thiyagarajan, V. (2010). A review on the role of chemical cues in habitat selection by barnacles: New insights from larval proteomics. *Journal of Experimental Marine Biology and Ecology*, 392(1−2), 22−36. Available from https://doi.org/10.1016/j.jembe.2010.04.030.

Tresguerres, M., & Hamilton, T. J. (2017). Acid-base physiology, neurobiology and behaviour in relation to CO_2-induced ocean acidification. *Journal of Experimental Biology*, 220(12), 2136−2148. Available from https://doi.org/10.1242/jeb.144113.

Tylianakis, J. M., Didham, R. K., Bascompte, J., & Wardle, D. A. (2008). Global change and species interactions in terrestrial ecosystems. *Ecology Letters*, 11(12), 1351−1363. Available from https://doi.org/10.1111/j.1461-0248.2008.01250.x.

Uthicke, S., Pecorino, D., Albright, R., Negri, A. P., Cantin, N., Liddy, M., Dworjanyn, S., Kamya, P., Byrne, M., & Lamare, M. (2013). Impacts of ocean acidification on early life-history stages and settlement of the coral-eating sea star *Acanthaster planci*. *PLoS One*, 8(12). Available from https://doi.org/10.1371/journal.pone.0082938.

Venn, A. A., Tambutté, E., Holcomb, M., Laurent, J., Allemand, D., & Tambutté, S. (2013). Impact of seawater acidification on pH at the tissue-skeleton interface and calcification in reef corals. *Proceedings of the National Academy of Sciences of the United States of America*, 110(5), 1634−1639. Available from https://doi.org/10.1073/pnas.1216153110.

Viyakarn, V., Lalitpattarakit, W., Chinfak, N., Jandang, S., Kuanui, P., Khokiattiwong, S., & Chavanich, S. (2015). Effect of lower pH on settlement and development of coral, *Pocillopora damicornis* (Linnaeus, 1758). *Ocean Science Journal*, 50, 475−480. Available from https://doi.org/10.1007/s12601-015-0043-z.

Wang, T., & Wang, Y. (2020). Behavioral responses to ocean acidification in marine invertebrates: New insights and future directions. *Journal of Oceanology and Limnology*, *38*(3), 759−772. Available from https://doi.org/10.1007/s00343-019-9118-5.

Wang, X., Song, L., Chen, Y., Ran, H., & Song, J. (2017). Impact of ocean acidification on the early development and escape behavior of marine medaka (*Oryzias melastigma*). *Marine Environmental Research*, *131*, 10−18. Available from https://doi.org/10.1016/j.marenvres.2017.09.001.

Watson, S. A., Fields, J. B., & Munday, P. L. (2017). Ocean acidification alters predator behaviour and reduces predation rate. *Biology Letters*, *13*(2). Available from https://doi.org/10.1098/rsbl.2016.0797.

Webster, N. S., Uthicke, S., Botte, E. S., Flores, F., & Negri, A. P. (2013). Ocean acidification reduces induction of coral settlement by crustose coralline algae. *Global Chang Biology*, *19*, 303−315. Available from https://doi.org/10.1111/gcb.12008.

Wessel, N., Martin, S., Badou, A., Dubois, P., Huchette, S., Julia, V., Nunes, F., Harney, E., Paillard, C., & Auzoux-Bordenave, S. (2018). Effect of CO$_2$−induced ocean acidification on the early development and shell mineralization of the European abalone (*Haliotis tuberculata*). *Journal of Experimental Marine Biology and Ecology*, *508*, 52−63. Available from https://doi.org/10.1016/j.jembe.2018.08.005.

Widdicombe, S., McNeill, C. L., Stahl, H., Taylor, P., Queirós, A. M., Nunes, J., & Tait, K. (2015). Impact of sub-seabed CO$_2$ leakage on macrobenthic community structure and diversity. *International Journal of Greenhouse Gas Control*, *38*, 182−192. Available from https://doi.org/10.1016/j.ijggc.2015.01.003.

Wittmann, A. C., & Pörtner, H. O. (2013). Sensitivities of extant animal taxa to ocean acidification. *Nature Climate Change*, *3*(11), 995−1001. Available from https://doi.org/10.1038/nclimate1982.

Wright, J. M., Parker, L. M., O'Connor, W. A., Scanes, E., & Ross, P. M. (2018). Ocean acidification affects both the predator and prey to alter interactions between the oyster *Crassostrea gigas* (Thunberg, 1793) and the whelk *Tenguella marginalba* (Blainville, 1832). *Marine Biology*, *165*(3), 46. Available from https://doi.org/10.1007/s00227-018-3302-6.

Wu, F., Wang, T., Cui, S., Xie, Z., Dupont, S., Zeng, J., Gu, H., Kong, H., Hu, M., Lu, W., & Wang, Y. (2017). Effects of seawater pH and temperature on foraging behavior of the Japanese stone crab *Charybdis japonica*. *Marine Pollution Bulletin*, *120*(1−2), 99−108. Available from https://doi.org/10.1016/j.marpolbul.2017.04.053.

Xia, X., Wang, Y., Yang, Y., Luo, T., Van Nostrand, J. D., Zhou, J., Jiao, N., & Zhang, R. (2019). Ocean acidification regulates the activity, community structure, and functional potential of heterotrophic bacterioplankton in an oligotrophic gyre. *Journal of Geophysical Research: Biogeosciences*, *124*(4), 1001−1017. Available from https://doi.org/10.1029/2018JG004707.

Yuan, H., Xu, X., Yang, F., Zhao, L., & Yan, X. (2020). Impact of seawater acidification on shell property of the Manila clam Ruditapes philippinarum grown within and without sediment. *Journal of Oceanology and Limnology*, *38*(1), 236−248. Available from https://doi.org/10.1007/s00343-019-8281-z.

Zhao, L. Q., Liang, J., Liang, J. P., Liu, B. Z., Deng, Y. W., Sun, X., Li, H., Lu, Y. N., & Yang, F. (2020). Experimental study of transgenerational effects, pH and predation risk on byssus production in a swiftly spreading invasive fouling Asian mussel, Musculista senhousia (Benson). *Environmental Pollution*, *260*, 114111. Available from https://doi.org/10.1016/j.envpol.2020.114111.

Zhao, X., Guo, C., Han, Y., Che, Z., Wang, Y., Wang, X., Chai, X., Wu, H., & Liu, G. (2017). Ocean acidification decreases mussel byssal attachment strength and induces molecular byssal responses. *Marine Ecology Progress Series*, *565*, 67−77. Available from https://doi.org/10.3354/meps11992.

Zhao, X., Shi, W., Han, Y., Liu, S., Guo, C., Fu, W., Chai, X., & Liu, G. (2017). Ocean acidification adversely influences metabolism, extracellular pH and calcification of an economically important marine bivalve, Tegillarca granosa. *Marine Environmental Research*, *125*, 82−89. Available from https://doi.org/10.1016/j.marenvres.2017.01.007.

Zunino, S., Canu, D. M., Bandelj, V., & Solidoro, C. (2017). Effects of ocean acidification on benthic organisms in the Mediterranean Sea under realistic climatic scenarios: A meta-analysis. *Regional Studies in Marine Science*, *10*, 86−96. Available from https://doi.org/10.1016/j.rsma.2016.12.011.

CHAPTER FOUR

Potential mechanisms underpinning the impacts of ocean acidification on marine animals

Wei Shi and Guangxu Liu
College of Animal Sciences, Zhejiang University, Hangzhou, P.R. China

Introduction

Due to anthropogenic activities (e.g., fossil fuel burning, land use changes, and industrial activities), large quantities of carbon dioxide (CO_2) have been emitted and the atmospheric CO_2 has increased at an alarming rate. On the basis of the geological record, the predicted change in global oceanic pH is higher and increasing faster than any occurrence experienced over the past 300 million years of Earth's history. Approximately, 30% of the emitted CO_2 is eventually absorbed by the world's oceans, resulting in (1) a continuous reduction in oceanic pH and (2) carbonate chemistry shifts (Caldeira & Wickett, 2003; Doney et al., 2009)—a process known as ocean acidification (OA). According to the Representative Concentration Pathway (RCP) 8.5 scenario of the Intergovernmental Panel on Climate Change, the average surface seawater pH has declined by 0.1 units (a 26% increase in hydrogen ion concentration) since the beginning of the Industrial Age and is expected to further decrease by 0.3−0.4 units before the end of the 21st century (Doney et al., 2009). In addition to lowering seawater pH, oceanic uptake of excessive CO_2 also brings about a decrease in $[CO_3^{2-}]$ and thereby the calcium carbonate saturation state (Ω), which is determined by $[CO_3^{2-}][Ca^{2+}]/K_{sp}$ (K_{sp} is the stoichiometric solubility product of $CaCO_3$) (Caldeira & Wickett, 2003; Feely et al., 2004). Since various physiological functions of marine organisms are dependent on intracellular pH homeostasis, especially those of the organisms that actively control and build $CaCO_3$ structures such as

Ocean Acidification and Marine Wildlife.
DOI: https://doi.org/10.1016/B978-0-12-822330-7.00005-8

shells, spines, and ossicles, alterations in these seawater parameters will undoubtedly pose a threat to marine organisms (Evans & Watson-Wynn, 2014; Feely et al., 2004; Fitzer et al., 2014). A substantial portion of the literature has documented that OA can affect a broad range of biological processes and physiological functions of marine organisms, including fertilization, larval development, calcification, metabolism, immune responses, growth, and behavior (Collard et al., 2014; Fitzer et al., 2014; Hernroth et al., 2011; Hiebenthal et al., 2013; Hüning et al., 2013; Kurihara, 2008). Therefore these changes will threaten not only the survival of marine organisms but also the sustainability of fish and shellfish aquaculture.

In order to better understand the impacts of elevated CO_2 levels on marine animals, the underlying affecting mechanisms involved in these OA-induced effects on marine organisms must be studied and are a key area in future research. However, there have been fewer detailed investigations of elevated atmospheric CO_2 inhibition on the physiological processes of marine organisms at the biochemical and cellular levels; therefore the mechanisms responsible for these effects remain largely unknown. The present chapter reviews the potential mechanisms underpinning the impacts of OA on marine animals, which include the following: (1) disturbance of acid−base homeostasis and energy reallocation by OA; (2) alteration in the normal function of neurotransmitters due to OA; (3) interference with the transduction of neural signals by OA; and (4) OA's influence on the expression patterns of genes and proteins involved in key biological processes.

Acid−base homeostasis and energy reallocation

The uptake of anthropogenic CO_2 by the oceans results in the elevated concentration of CO_2 in seawater (Caldeira & Wickett, 2003; Doney et al., 2009). Due to its high diffusion capability, CO_2 in seawater can diffuse easily across the epithelial surfaces of marine organisms and enter their intracellular compartments (Collard et al., 2014; Fitzer et al., 2014; Hernroth et al., 2011; Hiebenthal et al., 2013; Hüning et al., 2013; Kurihara, 2008). Excessive CO_2 inside of the cell would react with internal body fluids and form HCO_3^- and H^+ ions, which subsequently alter intracellular acid−base balance (Evans & Watson-Wynn, 2014). According to previous studies (Ern & Esbaugh, 2016; Hüning et al., 2013;

Lannig et al., 2010), the OA-induced extracellular acid—base disturbances may exert negative effects on marine organisms directly by causing extracellular acidosis and indirectly by altering energy allocation.

OA-induced hypercapnia is recognized as the main cause for its toxicity on many marine organisms. For vertebrates like fish, exposure to elevated CO_2 in the environment is known to result in a proportional increase in the partial pressure of CO_2 (pCO_2) in their blood (Tresguerres & Hamilton, 2017). The blood plasma pCO_2 levels of the Atlantic cod *Gadus morhua* were detected to increase significantly from approximately 3000 ppm to 3600 ppm and 4600 ppm as the seawater pCO_2 levels elevated from 400 ppm (pH 8.0) to 1000 ppm (pH 7.8) and 2000 ppm (pH 7.6), respectively (Larsen et al., 1997). After 24 hours of exposure to hypercapnic conditions at 1000 ppm, the blood pCO_2 levels and total blood CO_2 of the red drum *Sciaenops ocellatus* increased from 3.1 mm Hg and 7.5 mM in normocapnic water (567 ppm) to 4.1 mm Hg and 9.8 mM, respectively (Ern & Esbaugh, 2016). Compared to vertebrates, the internal pH homeostasis in marine invertebrates like bivalves and echinoderms, with relatively weaker acid—base regulators, are reported to be more susceptible to OA (Pörtner et al., 2004). Lannig et al. (2010) found that 1 month of incubation under OA conditions (pH 7.09) decreases the hemolymph pH (7.1 under OA vs 7.6 under control) and increased the hemolymph pCO_2 levels (0.5 kPa under OA vs 0.2 kPa under control) of the oyster *Crassostrea gigas*, indicating a weak regulation of the extracellular acid—base status. Furthermore, it was found that even a short-term (6—12 days) exposure to OA (pH 7.7 and 7.4) significantly reduced the coelomic fluid pH of the two sea cucumbers (*Holothuria scabra* and *Holothuria parva*) and thus resulted in extracellular acidosis (Collard et al., 2014). Moderately hypercapnic seawater (3.8 mm Hg; pH 7.3) caused a three- to fourfold elevation in the hemolymph pCO_2 level of the mussel *Mytilus galloprovincialis*, while the same was 1.2 mm Hg under the ambient seawater conditions (Michaelidis et al., 2005a, 2005b). Zhao et al. (2017a) reported that the hemolymph pH values of the blood clam *Tegillarca granosa* decreased from 7.5 to 7.2 as the seawater pH declined from 8.1 (554 ppm) to 7.4 (3120 ppm), respectively. Significant reductions in extracellular pH were observed in the green sea urchin *Strongylocentrotus droebachiensis* and the sea star *Leptasterias polaris* after 1 day of exposure to elevated pCO_2 levels (Dupont & Thorndyke, 2012). Similarly, OA-induced disturbances in extracellular acid—base homeostasis were also reported in other marine invertebrates such as the sea urchin *Psammechinus miliaris* (Spicer et al., 2007), the blue mussel *Mytilus edulis* (Thomsen & Melzner, 2010), and the sea star *Asterias rubens* (Hernroth et al., 2011). The extracellular acidosis due to OA exposure may exert various negative effects such as suppressed aerobic metabolic rate, slower growth rate, shell dissolution, and higher mortality of marine organisms, especially for invertebrates and newly hatched vertebrates that lack sufficient capacity for internal pH regulation (Larsen et al., 1997; Pörtner et al., 2004). Zhao et al. (2017b) suggested that the OA-induced extracellular acidosis (hemolymph pH 7.41, 7.26, and 7.21 at seawater pH 7.8, 7.6, and 7.4, respectively) of the blood clam *T. granosa* would affect the activities of metabolic enzymes and thus result in metabolic depression (indicated by the decreased clearance, reduced respiration rates, and increased ammonium excretion rates). Pimentel et al. (2014) reported that oxygen consumption rates and swimming capabilities of recently hatched larvae (3 days posthatching) of the tropical dolphinfish *Coryphaena hippurus* were significantly reduced by extracellular acidosis that resulted from future environmental hypercapnia (ΔpH 0.5; 1600 ppm), which may therefore greatly influence larval growth, feeding, predation rate, and survival. Moreover, since the extracellular fluids are in direct contact with the inner shell or mineral surfaces of marine calcifiers, OA-induced extracellular acidosis would result in shell dissolution (Avignon et al., 2020; Zhao et al., 2017a, 2020). For example, the extracellular acidosis caused by OA was shown to impair the calcification process and the inner shell surface integrity of the clam *T. granosa* and the mussel *Mytilus coruscus* (Zhao et al., 2017a, 2020) (Fig. 4.1).

Similarly, shell properties, namely the shell growth, calcification, and microstructure, of an adult European abalone *Haliotis tuberculata* were all negatively impacted throughout a 2-month exposure period to OA (pH 7.7), possibly due to a reduction of pH in their internal fluid, which may have resulted in a reduction in their shell strength and, subsequently, protection from environmental disturbances and predators (Avignon et al., 2020).

To maintain normal physiological functions, marine organisms have developed various mechanisms to cope with extracellular acidosis (Larsen et al., 1997), which can partly or even fully compensate for extracellular acidosis under moderate OA conditions. Under elevated seawater pCO$_2$ levels (1000 and 2000 ppm), the cod *G. morhua* would compensate for the extracellular acidosis and sustain the stability of blood pH by CO$_2$ secretion (Larsen et al., 1997). By the same token, 21 days of exposure to increased

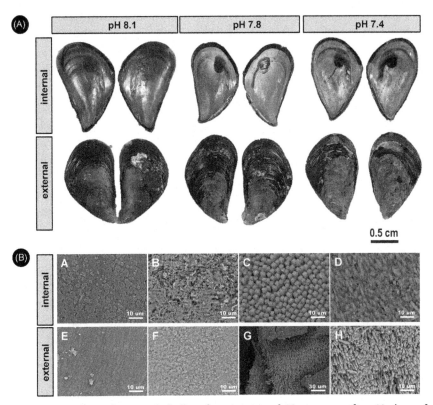

Figure 4.1 Inner and external shell surface images of *M. coruscus* after 40 days of treatment. (A) Representative stereomicroscopic images; (B) representative scanning electron microscope (SEM) images. Capital letters indicate where the SEM images were taken. A: the normal nacreous layer on the internal shell surface; B: the corroded nacreous layer on the internal shell surface; C: the normal prismatic layer on the internal shell surface; D: the corroded prismatic layer on the internal shell surface; E: the normal region of periostracum on the external shell surface; F: the discoloration region of periostracum on the external shell surface; G: the breakage and lifting region of periostracum on the external shell surface; H: the periostracum loss and prismatic layer dissolution region on the external shell surface. *From Zhao, X., Han, Y., Chen, B., Xia, B., Qu, K., & Liu, G. (2020). CO₂-driven ocean acidification weakens mussel shell defense capacity and induces global molecular compensatory responses. Chemosphere, 243, 125415. https://doi.org/10.1016/j.chemosphere.2019.125415.*

seawater pCO_2 conditions (1000 ppm) had no significant effect on the blood pH level of the European sea bass *Dicentrarchus labrax* juveniles (Shrivastava et al., 2019). According to previous studies (Larsen et al., 2014; Perry, 2011; Perry & Gilmour, 2006), the two main mechanisms, given as follows, are adopted by marine organisms to regulate acid—base status.

Active ion transport and the production of protons contribute substantially to the regulation of the acid—base balance in marine organisms under OA (Whittamore, 2012). It has been demonstrated that excess CO_2 in the interstitial fluids of many species can be hydrated into HCO_3^- and H^+ by carbonic anhydrase (Perry & Gilmour, 2006). When the blood passes through the gills of marine organisms, the hydration of CO_2 is exported against the steep concentration gradients by pumps, such as $V(H^+)$-ATPase, Na^+/K^+-ATPase, and Na^+-H^+ exchangers located in the gills (Larsen et al., 2014). In addition to gills, the intestines of some marine organisms may also be responsible for their whole-body acid—base balance, due to their role in regulating plasma HCO_3^- levels and compensating for the elevation of blood HCO_3^- (Heuer et al., 2012; Perry & Gilmour, 2006). For example, the Gulf toadfish *Opsanus beta* exposed to higher pCO_2 levels (5000, 10,000, 15,000, and 20,000 ppm) would activate intestinal apical anion exchange, which thus results in intestinal HCO_3^- loss, causing the retention of HCO_3^- in its body to prevent increases in blood plasma pH levels (Heuer et al., 2012). In addition to active ion transport, the buffering of intra- and extracellular compartments through increasing hemolymph bicarbonate levels was also found to be essential for the compensation of acid—base disturbances in marine calcifiers (Genz et al., 2008). Previous studies have suggested that the hemolymph bicarbonates used for the buffering of hypercapnic acidosis are mainly derived from the dissolution of exoskeletons (Michaelidis et al., 2005a, 2005b; Spicer et al., 2007). The mussel *M. galloprovincialis* was observed to produce bicarbonates to compensate for the respiratory acidosis in extracellular fluids when the seawater pH decreased from pH 8.05 to 7.3 (Michaelidis et al., 2005a, 2005b). The extracellular acid—base balance of the velvet swimming crab *Necora puber* under OA conditions (pH 7.31, 6.74, and 6.05) was also maintained by the dissolution of its exoskeleton (Spicer et al., 2007).

However, extracellular pH regulation in marine organisms through ion transport and base secretion are energy-dependent processes (Stapp et al., 2015). For example, under OA conditions (pH 7.0), the hepatic energy expenditure of the Atlantic cod *G. morhua* for both protein synthesis and Na^+/K^+-ATPase was elevated, indicated increased costs for ion regulation and cellular reorganization (Stapp et al., 2015). As compared to that reared in ambient seawater (pH 8.3; 338 ppm), the energy of a juvenile oyster *Crassostrea virginica* under high CO_2 levels (pH 7.5; 3500 ppm) was found to be diverted away from being used for the growth of the shell and soft body to being used for acid−base regulation (Beniash et al., 2010). Similarly, the rates of protein synthesis and ion transport in the sea urchin *Strongylocentrotus purpuratus* increased approximately 50% under OA, which accounted for more than 80% of the available ATP; meanwhile, only 40% of the ATP were allocated for these two processes under ambient conditions (Pan et al., 2015). Besides, if the disturbances in extracellular pCO_2 levels exceed compensatory capacities, the organisms would possibly adopt the strategy of passive tolerance, which involves suppressing metabolism to sustain survival (Avignon et al., 2020; Hu et al., 2011). For example, the activity of the branchial acid−base transporter Na^+/K^+-ATPase in the gill of the cephalopod *Sepia officinalis* was increased significantly by approximately 15% upon short-term (2−11 days) exposure to elevated pCO_2 levels (pH 7.63 and 7.28; 0.14 and 0.4 kPa) and this disturbance was not compromised after long-term (42 days) exposure (Hu et al., 2011). In addition, the ability to adopt passive tolerance requires several days to months (Avignon et al., 2020); marine organisms in their early larval stage are unlikely to have the time to compensate for acidosis and this extracellular acidosis may remain largely uncompensated.

Consequently, although these compensatory mechanisms may temporarily compensate for the OA-induced acid−base disturbance, they may influence the normal functions of various physiological processes due to energetic repercussions in the long term. It has been confirmed that the energy allocated for these maintenance activities of acid−base balance is traded off and used for other biological functions such as growth, shell formation, behavior, immune responses, and toxicant metabolism (Deigweiher et al., 2010; Roberts et al., 2013; Shi et al., 2016; Stumpp et al., 2011). Deigweiher et al. (2010) found that exposure to elevated pCO_2 levels (10,000 ppm) resulted in increased energetic demands for protein synthesis and Na^+/K^+-ATPase activity in the gills of the Antarctic notothenioids *Gobionotothen gibberifrons* and *Notothenia coriiceps*, while the overall metabolic rates of the gill tissues remained relatively constant. Stumpp et al. (2011) reported that the sea urchin *S. droebachiensis* larvae fully compensated for the OA-induced (pH 7.8 and 7.4) intracellular acidosis by using a bicarbonate buffer mechanism that enables calcification to proceed despite decreased extracellular pH; however, this resulted in enhanced costs for calcification or cellular homeostasis and therefore led to modifications in energy partitioning, which would then impact the growth and the survival rate of echinoid larvae. Liu et al. (2016) suggested

that OA-induced immunity depression such as inhibited hemocyte phagocytosis in the blood clam *T. granosa* may be an indirect consequence of
energy redistribution within the animal. Similarly, due to the shift in the
allocation of energy, from being used for oxidative stress defense to being
used for pH compensation, which has high physiological costs, significantly increased DNA damage was detected in the amphipod *Corophium
volutator* after 9 days of exposure to sediments contaminated by toxic
metals under OA conditions (750 ppm) (Roberts et al., 2013). Calculating
the scope for the growth of the sea urchin *S. purpuratus* revealed that only
39%−45% of the available energy was used for somatic growth under OA
conditions (1271 ppm; pH 8.2), while approximately 80% was allocated
for the growth processes in the urchins under ambient seawater conditions
(380 ppm; pH 7.7), which explains the delayed growth and development
after OA exposure (Stumpp et al., 2011). For marine bivalves such as
M. edulis, T. granosa, and *Meretrix meretrix,* exposure to OA would reduce
their energy availability for cadmium (Cd) exportation and thus lead to
significant Cd accumulation in these species (Shi et al., 2016).

   In conclusion, although the impacts of OA-induced acidosis on physiological processes have been studied in many marine organisms, their
responses are species-specific and vary within phyla (Ross et al., 2011).
Therefore future studies are needed to identify the potential mechanisms
underpinning the differences among different species.

Neurotransmitter disturbance and signaling transduction

   In addition to exerting negative effects on fundamental biological
processes including the metabolism, growth, calcification, and reproduction of marine organisms, the impacts of OA on their olfactory discrimination, auditory preferences, swimming activity, and learning ability have
also been confirmed (Jutfelt et al., 2013; Lai et al., 2015; Peng et al.,
2017). For example, the time for the shark *Heterodontus portusjacksoni*
reared under elevated pCO_2 levels (1000 ppm) to locate its prey in mesocosms through olfaction was found to be four times longer than that for
the sharks reared under ambient seawater (Pistevos et al., 2015). Peng
et al. (Jutfelt et al., 2013; Lai et al., 2015; Peng et al., 2017) reported that
future OA conditions (pH 8.1, 7.8, and 7.4) would lead to a significant

reduction in the burrowing behavior (as shown by the digging depth) of the razor clam *Sinonovacula constricta* following 1 week of exposure. Exposure to acidified seawater (pH 8.1, 7.8, and 7.4) for 15 days resulted in significant reductions in the gustation-mediated feeding behavior of the black sea bream *Acanthopagrus schlegelii* as indicated by the consumption rate and swallowing rate of feed-containing agar pellets (Rong et al., 2020). Jutfelt et al. (2013) suggested that future OA conditions would severely affect several behaviors of the three-spined stickleback (*Gasterosteus aculeatus*), including boldness, exploratory behavior, lateralization, and learning. According to previous studies (Lai et al., 2015; Rong et al., 2020), these OA-induced behavioral impairments on marine organisms are mainly caused by the interference of OA in their neurotransmitter function and signaling transduction.

The neurotransmitter system is a modulatory system that regulates various functions of animals (Borowsky et al., 1995; Dvoryanchikov et al., 2011; Kim et al., 2009). In the neurotransmitter system, gamma-aminobutyric acid (GABA) is a primary inhibitory neurotransmitter that is found to be involved in the neural mechanisms causing behavioral abnormalities (Dvoryanchikov et al., 2011). Since the receptor ($GABA_A$) of GABA is an ion channel that allows the movement of Cl^- and HCO_3^- across the neural membrane, the inflow of negatively charged ions into the postsynaptic neuron will prevent the depolarization of the neuronal membrane and thus reduce neural activity (Dvoryanchikov et al., 2011). It is reported that many marine organisms would maintain their acid—base balance to avoid acidosis by accumulating HCO_3 to buffer intracellular pH under OA conditions (Kim et al., 2009). However, this intracellular pH regulatory mechanism may result in altered transmembrane Cl^- and HCO_3^- gradients in the body of marine organisms, and thus in turn potentiate the function or reverse the action of $GABA_A$, thereby causing dramatic shifts in sensory preferences and behavioral changes (Nilsson et al., 2012; Rong et al., 2020).

The neurotransmitter interference—induced behavioral or sensory alterations under OA conditions have been widely reported in a variety of marine vertebrate species (Jutfelt et al., 2013; Lai et al., 2015; Nilsson et al., 2012). Nilsson et al. (2012) found that 11 days of OA exposure (450 and 900 ppm) would significantly affect the olfactory preference (indicated by their olfactory responses to the predator odor) of the clownfish *Amphiprion percula* larvae. In their research, the clownfish larvae reared under ambient seawater strongly repelled from the predator odor, while

those larvae that had undergone OA treatment (900 ppm) were attracted
(>90% of the time) to the odor. However, these alternations in the
olfactory ability of those OA-treated larvae were reversed by the addition
of gabazine (a specific $GABA_A$ receptor antagonist) into seawater, with
the larvae spending less than 12% of their time in the water stream containing the predator odor. Similarly, they also found that 4 days of exposure to OA (450 and 900 ppm) would impair the behavioral lateralization
of the larvae of the damselfish *Neopomacentrus azysron*, as indicated by
them turning at random in a nonlateralized manner, which can also be
reversed by treatment with gabazine. In addition, a study conducted on
the three-spined stickleback (*G. aculeatus*) also suggested that 50 days of
OA exposure (pH 8.02 and pH 7.69) would alter their behavioral lateralization (Lai et al., 2015). This research further analyzed the recovery effect
of gabazine for the lateralization of *G. aculeatus*. Their results demonstrated
that the lateralization for the samples in the OA group was fully restored
upon gabazine treatment, while the same showed no difference in the
control group. OA-induced impairment of the ability to learn to recognize predators was also reported for the damselfish *Pomacentrus amboinensis*
(Chivers et al., 2013). In the control group, the fish displayed an antipredator response including reduced activity and feeding when exposed to the
odor of moon wrasse, while the fish exposed to OA did not respond differently to the cues. These OA-induced behavioral changes can be
reversed by 5 days of gabazine treatment, indicating that the presence of
gabazine had allowed learning to occur in those fish during the learning
session. Hamilton et al. (2014) tested the anxiety of the juvenile rockfish
Sebastes diploproa reared in OA conditions (483 and 1125 ppm) by using
behavioral analysis, which measures light/dark preference (scototaxis) and
proximity to an object. After 1 week of OA exposure, increased anxiety,
shown by the significant preference of the fish for the dark zone, was
observed, while no significant preference for either the light or the dark
zone was observed for the control fish. This study also found that the
administration of gabazine caused a significant preference for the dark in
the control group but did not alter the preference of the OA group.
However, the administration of muscimol, a normal anxiolytic, produced
an opposing and significant shift in the location preference of the control
and OA-exposed rockfish, indicating that OA causes shifts in the action of
$GABA_A$ receptors, from shunting (inhibitory) to depolarizing (excitatory)
(Hamilton et al., 2014). While relatively well described in fishes, the
behavioral impacts of OA on marine invertebrates are poorly understood

and a mechanistic understanding of these impacts is lacking (Jutfelt et al., 2013; Lai et al., 2015; Peng et al., 2017). To date, only limited studies reported that the $GABA_A$ neurotransmitter that function in marine invertebrates may modulate their behavioral responses under OA (Clements et al., 2016). For example, Clements et al. found that the burrowing behavior (measured as the proportion of clams burrowed into sediment) of the juvenile clams *Mya arenaria* was significantly inhibited after exposure to elevated CO_2, while gabazine-treated clams showed similar burrowing responses under both control and low pH conditions. All of these studies indicated that altered $GABA_A$ receptor function is the underlying reason for the behavioral abnormalities displayed by various marine organisms under OA conditions (Clements et al., 2016). In addition to overexcitation of the GABA receptors, the direct impact of OA on the in vivo contents of neurotransmitters and key molecules from the signal transduction cascade pathway is also shown to induce behavioral impairments in marine organisms (Rong et al., 2018, 2020) examined the effects of elevated pCO_2 (pH 8.1, 7.8, and 7.4) on the foraging behavior of the larvae of the black sea bream *A. schlegelii*. The results revealed that 15 days of OA exposure exerted a significant negative impact on the foraging behavior of black sea breams by reducing curvilinear velocity and the linearity of the swimming path while also increasing the latency time (the time it takes for individuals to leave the acclimated area), the response time (the time it takes for individuals to reach the food source), and the wobble of the swimming path. To explore the underlying physiological mechanisms causing these impacts, the in vivo contents of important neurotransmitters and the expression of genes encoding key modulatory enzymes from the olfactory transduction pathway were investigated. The results showed that the in vivo contents of neurotransmitters [GABA and acetylcholine (Ach)] were significantly decreased under OA and the expressions of genes encoding positive regulators in the olfactory transduction pathway were significantly altered. On the basis of the data obtained in this research, the impaired foraging behavior may have resulted from reduced sensitivity to olfactory cues due to the interference in the transduction of olfactory neural signals.

Although the effects of OA-induced neurotransmitter disruption on the behaviors of marine organisms have been widely reported, its impacts on other physiological functions remain largely unknown. According to previous studies, neurotransmitters are one of the most essential messengers that mediate the communication between the nervous system and the immune system; any disturbances in neurotransmitters would result in immune disorder (Holzmann, 2012). Once an invasive pathogen is recognized by the host, the nervous system would temporally release neurotransmitters such as GABA, ACh, and norepinephrine to modulate the immune response (Li et al., 2016; Liu et al., 2017). For example, Li et al. (2016) detected that the phagocytosis rate and apoptosis rate of the hemolymph of the Pacific oyster (*C. gigas*) increased obviously after lipopolysaccharide stimulation, whereas this increase can be repressed with the addition of GABA, suggesting that GABA plays a crucial role in avoiding excess immune reactions and maintaining the immune homeostasis. In addition, neurotransmitters also play important roles in regulating many other physiological functions in marine organisms such as larval development, metabolism, and growth (Morse et al., 1979). Although these physiological functions of marine organisms have been reported to be affected by OA, their relationship with OA-induced neurotransmitter disturbances is still unclear. Therefore further research is needed to determine whether OA-induced neurotransmitter disturbances and signaling transduction are the causes for these impacts.

Potential molecular mechanisms revealed by omics technologies

Rapid progress in the development of biological sciences such as sequencing technologies has provided many valuable insights into solving unanswered questions across various aspects in recent years (Mardis, 2008). According to previous studies, genes, mRNA, proteins, and metabolites are frequently modified in response to environmental disturbances, such as pH, temperature, salinity, and toxicant (Gleason, 2019). In this regard, the knowledge of the molecular-level response of an organism to environmental changes is essential for elucidating the molecular functions and affecting mechanisms. Nowadays, many omics technologies, including transcriptomics, proteomics, and metabolomics, have been applied for the study of the potential biological responses of organisms to OA and for understanding the action mechanisms of OA on marine organisms (Strader et al., 2020). Comparative transcriptomics has proven to be an effective method for examining organism−environment interactions in various species (Li et al., 2017). The advantages of comparative transcriptomics are particularly valuable in marine systems, where differential gene expression analysis has been employed to investigate how marine organisms respond to environmental alternations. A number of studies in recent years have investigated the transcriptomic responses of marine organisms to future OA conditions, which revealed the underlying molecular mechanisms behind the OA-induced alternations in traits such as ion transport, metabolism, growth, reproduction, calcification, and survival (Li et al., 2017; Zhao et al., 2020). For example, to explore the molecular responses of polar organisms, living in the places that are predicted to be "first in time" to experience the impacts of OA, Johnson and Hofmann (2017) assessed the gene expression profiles of juvenile *L. h. Antarctica* after 21 days of exposure to different seawater pH conditions (pH 7.71, 7.9, and 8.13). The results showed that a total of 241 transcripts were upregulated by 21 days of OA exposure (pH 7.71), while 872 transcripts were downregulated. These upregulated transcripts were found to be mainly involved in translation (24 transcripts), being structural components of ribosomes (17 transcripts), formation of membrane-bound organelles (24 transcripts), catalytic activity (16 transcripts), metabolic processes (20 transcripts), and ion binding (6 transcripts). Enrichment analysis suggested that these upregulated transcripts were associated with membrane biogenesis and assembly. Among the downregulated transcripts, 77 transcripts were associated with catalytic processes, 57 with metabolic processes, 59 with ion binding, 40 with calcium ion binding, and 16 transcripts with the cytoskeleton, which were enriched to 9 gene ontologies associated with ATP binding, ATPase activity, microtubule motor activity, and transport. In conclusion, changes in the transcriptome of *L. h. antarctica* revealed that key processes, such as shell formation, cellular stress response, metabolism, and neural function, which are physiologically significant to calcifying marine organisms, would be affected by future OA conditions; this may challenge biogenic calcification and energy allocation in a juvenile marine calcifier. In addition, transcriptomic profiling has also been applied for exploring the intracellular acclimation or adaptation of marine organisms to OA (Evans & Watson-Wynn, 2014; Evans et al., 2013). It has been reported that Sydney rock oysters from the B2 breeding line (a selected breeding line with fast growth and disease resistance) exhibit resilience to OA at the physiological level (Parker et al., 2012). In order to investigate the molecular basis of this physiological resilience, Goncalves et al. (2017) analyzed the gill transcriptome of B2 oysters that had been exposed to near-future OA conditions over two consecutive generations. The results showed that the transgenerational exposure of B2 oysters to seawater with declined pH (pH 7.77) led to changes in the expression levels of 2909 gene clusters as compared to those reared under ambient conditions (pH 8.13). Further analysis found that most of these differential genes were predicted to be associated with various cellular processes such as the cell cycle (cell division, cell growth and differentiation, cell migration, cell death and maintenance of cellular homeostasis), metabolism (metabolic and catabolic processes, oxidative and reductive reactions, glycolysis and lipid metabolism), response to stress, and the immune system, which was reflected in the inducible responses in B2 oysters to control the cell cycle and maintain cellular homeostasis under OA conditions. Overall, the data obtained in this research revealed that B2 oysters might undergo rapid acclimation to decreasing pH through complex and dynamic modifications at the molecular level, primarily due to the mitigation of apoptosis by regulation of the cell cycle.

Microbiota composition and functional analysis can help to better understand the adaptation, dynamics, and evolution of microbial communities in various ecosystems. According to previous studies, alternations in the seawater chemical characteristics may favor distinct bacterial groups and lead to changes in the bacterial community (Joint et al., 2011). Therefore OA may have the potential to affect the bacterial community in seawater, and its impacts on the microbial community composition in seawater have been investigated through 16S rRNA high-throughput sequencing in various studies (Meron et al., 2011). To date, OA-induced alterations in the composition of the microbial community have been reported in the Great Barrier Reef, the Arctic Ocean, and the Antarctica Ocean (Witt et al., 2011). As shifts in bacterial communities in seawater can impact the health statuses of marine organisms, the analysis of biomass and the community structure of seawater microbiota in response to the changing environment may help explain some OA-induced effects on marine organisms. For example, Zha et al. (2017) examined the effects of OA on the bacterial community in seawater by using 16S rRNA high-throughput sequencing and analyzed the host—pathogen interaction between the blood clam *T. granosa* and bacterial groups. The microbial community analysis showed that the microbial community of the water column was significantly altered after 7 days of OA treatment. The microbes of the microbial community in the control seawater (pH 8.1) were mainly ($>90\%$) composed of *Proteobacteria* (69.6%), *Bacteroidetes* (12.4%), and *Firmicutes* (11.1%), while the abundance of *Firmicutes* was decreased in the acidified seawater (pH 7.4). Notably, at the genus level, the proportion of classified *Vibrio*, a major group of pathogens in many marine invertebrates, was significantly increased under acidified seawater (Fig. 4.2). Meanwhile, the hemolytic activity of *V. harveyi* was shown to be greater when incubated with hemolymph from blood clams after OA exposure, indicating that OA may also affect the immunity of marine organisms indirectly by altering seawater microbiota biomass, especially pathogenic bacteria. In addition, the settlement success of most marine invertebrates is attributed to biofilm recognition or quorum sensing, which is strongly linked to biofilm bacteria. Many studies have suggested that OA may affect the settlement of various marine invertebrates; however, the underlying molecular mechanisms remain unclear (Cigliano et al., 2010). In order to better understand the effects of reduced pH on the development of microbial biofilm communities and subsequent larvae settlement, Nelson et al. (2020) analyzed the microbial community on the biofilms. The results of 16S rRNA gene amplicon sequencing revealed that the presence of *Nitrospiraea*, *Bacteriodetes*, *Proteobacteria*, and *Actinobacteria* were decreased with the declined seawater pH. This research

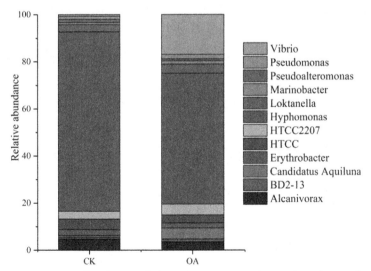

Figure 4.2 Relative abundance of the major genera of microbes detected in the microbial communities of the control trails (CK) and pCO_2-acidified seawater trails (OA). *From Zha, S., Liu, S., Su, W., Shi, W., Xiao, G., Yan, M., & Liu, G. (2017). Laboratory simulation reveals significant impacts of ocean acidification on microbial community composition and host-pathogen interactions between the blood clam and* Vibrio harveyi. Fish & Shellfish Immunology, 71, 393–398. https://doi.org/10.1016/j.fsi.2017.10.034.

supports the hypothesis that alterations in microbial biofilm composition under OA conditions may partly contribute to the OA-induced shifts in the settlement of marine invertebrates. According to previous studies, changes in environmental factors including salinity, pH, and temperature can influence the gut microbiota structure of marine organisms (Benson et al., 2010). As a result, the effects of OA on the gut microbiota and the subsequent physiological statuses of marine organisms have also been studied (Fonseca et al., 2019). Fonseca et al. (2019) have conducted a study to investigate the bacterial community associated with the intestinal fluid of the sea bream *Sparus aurata* under acidified seawater. The analysis of the results from 16S rRNA sequencing showed distinct dysbiosis in the intestinal lumen of the fish in response to OA, with the phylum *Firmicutes* absent from the bacterial communities of the fish exposed to seawater at elevated pCO_2 levels; whereas, the abundance of *Proteobacteria* was increased under OA. These data indicated that the polysaccharide and protein degradation of marine organisms may be affected by OA-induced changes in the bacterial community in the fish gut.

Metabolomics is defined as the systematical qualitative and quantitative analysis of metabolites in organisms, which is then used to describe changes in endogenous metabolites after exposure to different conditions (Bino et al., 2004). Therefore metabolomics is a useful tool for observing the endogenous metabolite perturbations induced by environmental changes such as OA (Sardans et al., 2011). For example, Liu et al. (2020) investigated the underlying mechanisms of the OA-hampered initial shell formation in oysters using a combination of the metabolomic and transcriptomic approaches. The results of the metabolomic analysis indicated that a high level of energy metabolism was required for the formation of calcified shells during the "middle" stage (15 hpf) of the oyster larvae since most of the chemical compounds were overexpressed during this period. However, mRNA expression levels of genes, including those of DNA-directed RNA polymerase II submit spb 7, sparc-related modular calcium-binding protein 1-like, calcium-dependent protein kinase 31, ATP synthase lipid-binding mitochondrial-like, and ATP synthase subunit mitochondrial-like, were significantly decreased in the larvae of the oysters after OA exposure. These data suggested that OA may delay the larval shell formation of marine calcifiers by suppressing amino acid metabolism, which results in the lack of ATP synthesis.

Although omics technologies have been frequently applied to investigate the potential underlying molecular mechanisms of the OA-induced negative effects on marine organisms, their applications in evaluating the impacts still have great challenges. For example, it is hard to establish relationships between the results from metabolomic studies and other experimental results. In the future, the combination of different omics technologies can be employed for assessing and predicting the effects of OA on marine organisms.

Examples of case studies

Case study 1—Fertilization success

Fertilization success is crucial for the recruitment of marine organisms and the stability of marine ecosystems. Most marine invertebrates are broadcast spawners, which release their gametes directly into the water column for external fertilization (Lotterhos & Levitan, 2010). During the fertilization process, these gametes are in direct contact with the surrounding seawater; therefore they are exposed to multiple potential environmental stressors (Sewell et al., 2014). Consequently, the process of fertilization of marine organisms, especially that of broadcast spawners, is believed to be the most susceptible period in their life cycles. Many studies have confirmed that OA might affect the fertilization success of various marine organisms (Collard et al., 2014; Fitzer et al., 2014; Hernroth et al., 2011; Hiebenthal et al., 2013; Hüning et al., 2013; Kurihara, 2008). However, the underlying mechanisms of these effects are not completely understood.

A study conducted by Shi et al. (2017a, 2017b) investigated the affecting mechanisms of OA-induced reductions in fertilization success. The blood clam *T. granosa* was chosen as the model organism due to its wide distribution and representation as a typical broadcast-spawning invertebrate (Shao et al., 2016). One setting with ambient pH representing current conditions (pH at 8.1) and two settings with lower pH levels (pH 7.8 and pH 7.4, representing the OA scenarios in 2100 and 2300, respectively) were used in this research. The gametes were obtained by artificial spawning, and pair mating was performed (the sperm of one male was crossed with the eggs of one female) after 1 hour of exposure to different OA conditions. The obtained results showed that the fertilization success of *T. granosa* decreased significantly, 85.8% and 71.6% of that of the control group (pH 8.1), respectively, after exposure to acidified seawater at pH 7.8 and 7.4. In order to explore the affecting mechanisms underpinning the OA-induced reductions in fertilization success, the impacts of OA on the essential steps during the fertilization processes, including successful sperm–egg collisions, gamete fusion, and standard generation of Ca^{2+} oscillations, were then analyzed.

The gamete collision frequency is one of the most important factors determining the fertilization success rate of broadcast-spawning marine invertebrates (Styan & Butler, 2000). Numerous studies have shown that sperm with high swimming velocities should collide with eggs more frequently and are more effective at fertilizing eggs in the three-dimensional water column (Styan & Butler, 2000). Therefore the velocity of sperm was determined after 1 hour of exposure to acidified seawater. The velocity average path, curvilinear velocity, and velocity straight line of the sperm were all decreased upon OA exposure. The pCO_2-induced declining sperm velocity observed in this study may reduce the probability of gamete collisions and was predicted to result in about 7.2% and 20.8% drop in fertilization success at pH 7.8 and pH 7.4, respectively, when compared to the control.

As not every sperm–egg collision can result in successful fertilization, the scoring of the probability of gamete fusion per collision plays an important role in determining fertilization success thereafter (Marshall et al., 2000). On the basis of previous studies (Vogel et al., 1982), the probability of gamete fusion per sperm–egg collision can be estimated by Vogel's fertilization kinetics model. After being exposed to OA for 1 hour, the probability of gamete fusion per collision decreased sharply to approximately 69.4% and 46.8% of that of the control for the experimental groups at pH 7.8 and 7.4, respectively. These results suggest that OA would hamper the fusions of the sperm and eggs during the fertilization process.

A successful gamete fusion will trigger a series of intracellular Ca^{2+} oscillations in eggs, which are reported to be an integral part of egg activation and a prerequisite for subsequent fertilization events (Wagner et al., 1998). The Ca^{2+} oscillations of each egg of T. granosa during the fertilization event were determined from the measured fluorescence intensities indicated by Fluo-3/AM calcium indicators. As shown in (Fig. 4.3), the intracellular Ca^{2+} fluorescence intensities in the eggs decreased with the reduced seawater pH. In addition, the amplitude of egg Ca^{2+} oscillations during fertilization also significantly declined after treatment with pH 7.4 seawater. These results confirmed that OA would hinder the increase in

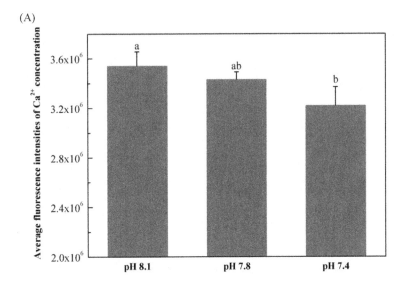

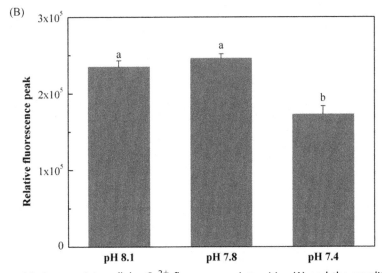

Figure 4.3 Average intracellular Ca^{2+} fluorescence intensities (A) and the amplitude (maximum fluorescence intensity change) (B) during fertilization in the eggs of *T. granosa* under different pCO$_2$ levels. Data were analyzed by a one-way ANOVA, followed by the Tukey *post-hoc* test. Mean values that do not share the same superscript were significantly different. *From Shi, W., Han, Y., Guo, C., Zhao, X., Liu, S., Su, W., Wang, Y., Zha, S., Chai, X., & Liu, G. (2017a). Ocean acidification hampers sperm-egg collisions, gamete fusion, and generation of Ca^{2+} oscillations of a broadcast spawning bivalve,* Tegillarca granosa. Marine Environmental Research, 130, 106−112. https://doi.org/10.1016/j.marenvres.2017.07.016.

intracellular Ca^{2+} and the generation of Ca^{2+} oscillations during the fertilization events.

Overall, the results obtained in the present study suggest that OA could narrow the window for successful fertilization of *T. granosa*, which may be determined by the decreased sperm velocity, thus leading to a reduced probability of sperm—egg collisions, lowered probability of gamete fusion for each gamete collision event, and disrupted intracellular Ca^{2+} oscillations.

Case study 2—Feeding behavior

Feeding, which provides matter and energy for an animal, is one of the most important behaviors for the development, growth, reproduction, and survival of aquatic organisms (Kasumyan & Døving, 2003). This behavior can be divided into several processes, namely detection, capture, ingestion, digestion, and assimilation, which rely on the generation and transduction of gustatory signals (Hisatsune et al., 2007). Recently, several studies have suggested that OA may hinder the feeding behavior of marine organisms by disrupting the olfactory perception (Clements, 2016; Rong et al., 2020).

A study performed by Rong et al. (2020) demonstrated that the gustation-mediated feeding behavior of the black sea bream *A. schlegelii* was significantly impaired by OA. The species *A. schlegelii* is frequently posed with challenges such as food shortage and is one of the major commercial fish species, which makes it an excellent fish species for studying feeding behavior (Nip et al., 2003). The pH values of 8.1, 7.8, and 7.4 were adopted to simulate the pH levels at present and in the years 2100 and 2300, respectively. The feeding behavior of *A. schlegelii* in this research was assessed by evaluating the consumption of agar pellets containing commercial feed and blank agar pellets. During this experiment, the consumption rates (CRs, percentage of agar pellets consumed relative to the total number of pellets provided) and swallowing rates (SRs, the percentage of agar pellets swallowed relative to the total number of pellets grasped) were calculated for the pellets containing feed and for the blank agar pellets. After 15 days of exposure to OA, the CR and the SR of feed-containing and blank agar pellets were significantly reduced. Compared to the control, the CRs for the feed-containing agar pellets declined by approximately 11.2% and 21.5% in the sea bream at pH 7.8 and 7.4, respectively. Similarly, the SRs also dropped to approximately

93.4% and 82.7% of the control for OA treatment groups at pH 7.8 and pH 7.4, respectively. To explore the potential mechanisms underpinning the decreased feeding behavior due to OA, the influence of OA on the crucial events during the feeding behavior, including the generation and transduction of olfactory neural signals, were investigated.

According to previous studies, the generation of olfactory neural signals is dependent on a cascade of cellular signaling events triggered by the binding of odor cues to corresponding receptors (Hisatsune et al., 2007; Morais, 2016). In brief, taste-stimulating molecules from food activate the taste receptors such as taste 1 receptor member 3 (*T1R3*) on the membrane of type II taste receptor cells (*TRCs*) and induce an increase in inositol triphosphate (*IP3*). Then, *IP3* will bind to inositol 1,4,5-trisphosphate receptor type 3 (*IP3R*) and open ion channels on the cell membrane such as transient receptor potential cation channel subfamily M member 5 (*TRPM5*), resulting in membrane depolarization (Morais, 2016). Subsequently, the generated gustatory electrical signal will be transmitted to the brain with assistance from several receptors such as the pyrimidinergic receptor (*P2Y4*) (Huang et al., 2007). Therefore the expression levels of key molecules from the olfactory transduction cascade pathway such as *T1R3*, *IP3R*, *TRPM5*, and *P2Y4* were explored to reveal the underlying molecular mechanism behind the OA-induced feeding behavior. The results from qRT-PCR (real-time quantitative reverse transcription PCR) showed that expression levels of all the four genes were significantly suppressed in the brain of black sea bream reared in acidified seawater, suggesting that OA may disrupt the generation and transmission of the gustatory signal in black sea bream.

Following olfactory neural signal generation, the transduction of olfactory neural signals is mediated by various neurotransmitters including GABA, 5-hydroxytryptamine (5-HT), and ACh; through this transmission, the fish decide whether to ingest or reject the food. Therefore the impacts of OA on the in vivo contents of key neurotransmitters (GABA, ACh, and 5-HT) in black sea breams were investigated. The determined results found that the GABA contents in the brains of black sea breams declined by approximately 21.6% at pH 7.4 and 13.1% at pH 7.8 as compared to the control after 15 days of OA exposure. Similarly, the ACh contents in the brains of black sea breams of the treatment groups at pH 7.4 and 7.8 were only approximately 40.5% and 20.3% of those of the control, respectively. Under these circumstances, TRCs would be less efficient at transmitting the gustatory signal to the sensory afferent fibers, therefore leading to the observed reduction in the gustation-mediated feeding behavior of black sea breams under OA.

In conclusion, future OA conditions would disturb neural signal transduction by altering in vivo neurotransmitter contents and expression levels of genes in the signal transduction pathway, thus hampering the foraging behavior of *A. schlegelii*.

Case study 3—Defense capacity

Defense capacity is critical for the survival of marine organisms. For most marine bivalves with limited movement, inducible morphological defenses play important roles in defending against predation and environmental fluctuations (Zhao et al., 2017b, 2020). However, some studies have revealed that future OA conditions would weaken the defense capacity of various marine organisms including marine bivalves, which will render marine organisms more susceptible to predators, parasites, and pathogens (Liu et al., 2016; Zhao et al., 2020).

Marine mussels, mainly *Mytilus* spp., living on rocky shores in the intertidal zone are ecologically and economically important oceanic organisms. As they live in a complex environment, mussels produce thick shells and byssal thread as defense strategies to protect themselves against predation and current shearing forces (Shi et al., 2020). Zhao et al. (2017a, 2017b, 2020) suggested that future OA would reduce the shell defense capacity and decrease the byssal attachment strength of the mussel *M. coruscus*. In their studies, declined pH levels of 7.8 and 7.4 were set to mimic the projected oceanic surface pH conditions in the years 2100 and 2300, respectively (Caldeira & Wickett, 2003). For assessing the shell defense capacity, the integrity, the structure, the strength, and the closure strength of the shells of mussels under different OA conditions were examined. In brief, the severity of corrosion on the mussel shells was quantified by checking the percentages of corroded areas for both the internal and external shell surfaces. The shell strength was determined as the force required to break the valve of mussels with a material-testing machine. Due to the positive linear correlation between the adductor muscle size and the shell closure strength (Christensen et al., 2012), relative shell closure strength was measured as the relative size of the posterior adductor muscle (calculated as the ratio of the adductor muscle diameter to the shell length). The results of the calculation of the shell integrity showed that both the external and inner shell surfaces of mussels were significantly corroded due to future OA conditions. The shell strength of mussels was also significantly weakened to 56 and 50 N in the pH 7.8 and pH 7.4 groups, respectively, while it was found to be 86 N in the control group. In addition, the shell closure strength was significantly reduced due to future OA conditions as the relative size of the adductor muscle decreased from 0.154 at pH 8.1 to 0.147 and 0.133 at pH 7.8 and 7.4, respectively. Overall, the obtained results revealed that OA would impair the shell defense capacity of mussels by reducing their shell strength and shell closure.

To identify the underlying physiological mechanisms behind the alterations in the shell defense capacity, the hemolymph pH and Ca^{2+} concentration of mussels were determined. After 40 days of exposure at pH 7.8 and 7.4, the hemolymph Ca^{2+} concentration of mussels also reduced from 297 mg/L (control group) to 283 and 278 mg/L, respectively. According to previous studies (Lannig et al., 2010), this reduction in hemolymph Ca^{2+} concentration may be ascribed to the suppressed filter-feeding behavior of organisms under OA conditions, which thus leads to decreased Ca^{2+} uptake from food and seawater. Similarly, the obtained results showed that the hemolymph pH level significantly decreased from 7.52 in ambient seawater to approximately 7.38 and 7.24 in the acidified seawater at pH 7.8 and 7.4, respectively. Therefore the observed corrosion on the inner shell surfaces may have resulted from the OA-induced extracellular acidosis. Since the mantle of mussels plays a critical role in shell formation and growth, transcriptome sequencing of the mantle tissues of *M. coruscus* was conducted to uncover the underlying molecular responses. After 40 days of exposure, the whole mantle tissues of the mussel from the pH 8.1 and pH 7.4 groups were isolated for the investigation of gene expression profiles. The results from the transcriptome sequencing showed that 2448 unigenes were differentially expressed (1624 upregulated and 824 downregulated) in acidified seawater as compared to the control. Gene Ontology (GO) enrichment analyses revealed that five terms in the "Biological Process" category were significantly enriched among the upregulated differentially expressed unigenes (DEGs), while four were significantly enriched among the downregulated DEGs. Kyoto Encyclopedia of Genes and Genomes (KEGG) enrichment analyses revealed that seven pathways were significantly enriched among the upregulated DEGs. The mantle transcriptome analyses indicated that OA would not only induce potential mantle tissue injury but also trigger a series of molecular compensatory responses. On one hand, genes from the "apoptosis" KEGG pathway were significantly upregulated after OA exposure, suggesting potential mantle cell death in mussels. According to GO enrichment analyses, the cell death may be due to the increased protein degradation (indicated by the upregulation of "cysteine-type endopeptidase activity" and "protease binding"), which would inhibit the formation of protein quaternary structures (indicated by the downregulation of "protein homooligomerization") and suppress the repair of DNA damage (indicated by the downregulation of "DNA recombination"). On the other hand, there were also compensatory responses related to shell formation and maintenance. For example, OA exposure would induce the promotion of chitin synthesis and the inhibition of chitin degradation, as shown by the significantly enriched "chitin binding" and "adult chitin-containing cuticle pigmentation" among the upregulated DEGs.

For assessing the byssal attachment strength of mussels, the mechanical properties (such as strength and extensibility) and the number of byssal threads produced by the mussel *M. coruscus* were analyzed. After 7 days of exposure to OA, the byssal threads were excised and the quantitative values (e.g., length, diameter, and number) of these newly produced byssal threads were examined. Tensile tests were then performed to detect the breaking force, breaking stress, thread extensibility, and thread toughness of the byssal threads in *M. coruscus* under different OA conditions. The results showed that *M. coruscus* reared under acidified seawater (pH 7.4) produced significantly fewer threads as compared to that reared in ambient seawater (pH 8.1). The breaking force, breaking stress, and toughness of the byssal threads produced by *M. coruscus* under OA conditions was significantly decreased compared to the control. These reductions in the mechanical properties and the number of byssal threads would greatly hamper the ability of the mussels to attach to a substratum (the potential maximal byssal attachment strength for mussels exposed to acidified seawater at pH 7.4 was 35% of the control).

Since the byssal thread is mainly composed of different types of 3,4-dihydroxyphenylalanine (DOPA)-rich proteins, such as mussel foot proteins (mfps), precursor collagen proteins (preCOLs), and proximal thread matrix proteins (PTMPs), the secretion and the assembly of these proteins are crucial for the mechanical properties of the byssal thread (Silverman & Roberto, 2007). Therefore to explore the molecular byssal responses of this mussel to OA, the expression levels of genes encoding important byssal proteins were analyzed. Compared to the ambient pH group, the expressions of the genes *preCOL-NG* and *mfp-4* were downregulated after OA exposure. Since *preCOL-NG* and *mfp-4* are closely related to the mechanical properties of the inner core and the distal region-adhesive plaque junction of the mussel byssal thread (Silverman & Roberto, 2010), their decreased expression may lead to declined thread strength and extensibility under OA conditions. Furthermore, the expression levels of *mfp-2, -3, -5,* and *-6* were all declined in the seawater at pH 7.4, which may explain the structural failure at the adhesive plaque that frequently occurred under the OA (pH 7.40) scenario. Interestingly, the genes *preCOL-P, PTMP,* and *preCOL-D*, which are essential for the elasticity and extensibility to the proximal region of the byssal thread, were upregulated in response to OA, indicating that mussels may compensate for the weakened byssal threads under OA by elevating the elasticity of the proximal region and the stiffness of the distal region of thread.

In conclusion, OA may limit both the shell defense capacity and the byssal attachment strength of marine mussels. Although mussels have adopted molecular compensatory responses to mitigate these adverse effects, the fragile shell structure, reduced mechanical strength, weak shell closure strength, and declined byssal attachment strength under OA conditions would tremendously elevate mussels' vulnerability to pathogens, parasites, shell-crushing predators, and shell-drilling and shell-entering predators.

Case study 4—Chemical communication

Chemical cues, based on every form of biological molecule, from amino acids to nucleic acids and carbohydrates, are omnipresent in marine systems and regulate critical aspects of the behaviors of various marine organisms (Hay, 2009). As these molecules are often produced unintentionally and mostly evoke highly specific responses, they are defined as signaling cues. Most of these signaling cues are water soluble due to their chemical properties; any alternations in seawater physicochemical properties may have the possibility to influence the structure and thus the biological function of these signaling cues. Although the potential vulnerability of biological molecules to pH has already been hypothesized (Kim et al., 2009), the impacts of OA on the functions of the signaling cues still remain unclear.

A study conducted by Roggatz et al.(2016) investigated the influence of future oceanic pH conditions (pH 7.8 and 7.4) on the signaling cues—mediated egg ventilation behavior of the shore crabs *Carcinus maenas*. This behavior is a naturally occurring stereotyped behavior of female decapods carrying an egg clutch that allows the chemical communication between the female and her brood (Forward et al., 1987). Two peptides, namely cues glycyl-L-histidyl-L-lysine (GHK) and glycyl-glycyl-L-arginine (GGR) were chosen in this research due to their identified function in inducing egg ventilation in crustaceans (Forward et al., 1987). A bowl assay was performed to determine the egg ventilation frequency of ovigerous crabs before and after the addition of signaling cues (GHK and GGR) under different OA conditions. In brief, after a 1 -minute habituation phase and a 5 -minute interval of counting the abdominal pumps, the peptide cue was added close to the crab's abdomen and the egg ventilation frequency was counted for 5 minutes. A higher ventilation frequency after the addition of the cue was counted as a positive response. The obtained results showed that future oceanic pH conditions negatively inhibited the behavioral response of shore crabs to both cues. Crabs under the seawater at pH 7.7 required at least tenfold higher concentrations of the cues (GHK and GGR) to induce the egg ventilation behavior.

To explore the underlying affecting mechanisms of the OA disturbed ventilation behavior, the protonation states of peptide signaling molecules under different pH conditions, the NMR spectroscopy with quantum chemical calculations was conducted to obtain a more complete picture of the direct molecular impacts of OA. The determined pK_a values of GHK and GGR indicated that future OA would influence the N-terminal glycine and L-leucine residues with pK_a values within an oceanic pH range Table 4.1. On the basis of the determined pK_a values and the Henderson–Hasselbalch equation, the GHK and GGR protonation states were reduced by 23% and 22%, respectively. In other words, there will be an approximately 20% reduction in the current bioactive peptide forms

Table 4.1 pK_a values (mean ± SD) of the ionizable groups of the signaling peptides glycyl-L-histidyl-L-lysine (GHK) and glycyl-glycyl-L-arginine (GGR).

Peptide	Ionizable group	pK_a
GHK	Gly NH_2	7.98 ± 0.04
	His side chain	6.45 ± 0.05
	Lys COOH	2.8 ± 0.4
	Lys side chain	11.44 ± 0.06
GGR	Gly NH_2	8.00 ± 0.05
	Arg COOH	2.89 ± 0.08
	Arg side chain	15 ± 9

available under OA conditions, and therefore these organisms would require a 1.3-times higher concentration of the molecule for chemical communication. The potential structural changes of these cues were further studied using quantum chemical calculations. The calculated conformers of peptides suggested that the conformations of the protonation states of each peptide differ significantly under OA conditions. In addition, the calculation of the molecular electrostatic potential found that the different protonation states of peptides under different seawater pH also differed considerably when represented on their electron density isosurfaces. According to previous studies (Hamilton et al., 2013), the receptor–ligand interaction of a signaling molecule with its receptor depends on the molecule's functional groups, charge, shape, hydrophobicity, and flexibility; therefore OA-induced alternations in some of these characteristics would result in decreased successful interactions of the protonated peptide cues and the proposed receptor.

Case study 5—Immune responses

Exposed to a complex environment, marine organisms are often challenged by various pathogenic microorganisms. Therefore maintaining a robust immune response is crucial for the survivorship of bivalve species in the complex marine environment. Benthic filter feeders, such as bivalve mollusks, are some of the most effective sinks of pollutants and thus regarded as the model organism for studying immunotoxicity (Asplund et al., 2013). For bivalves, the immune strategy mainly depends on innate defense from hemocytes and humoral factors. The hemocytes are regarded as their main immune cells that remove invading exogenous particles such as pathogens through the process of phagocytosis. Many studies have reported that OA can exert negative effects on the immune responses of marine bivalves, including alterations in both total hemocyte count and cell-type composition, and phagocytosis. For example, Liu et al. (2016) reported that OA would reduce the total number of hemocyte cells and the phagocytosis frequency of marine bivalves. However, the underlying mechanisms behind the immunomodulatory functions of marine animals against OA still remain largely unclear, especially in marine invertebrates.

Su et al. (2018) performed a study to explore the impact and mechanism of OA on the process of phagocytosis in the marine clam *T. granosa*. After 14 days of OA treatment (pH 7.8 and 7.4), some clams were randomly selected from each treatment group for the phagocytosis assays of hemocytes. The results showed that OA exposure exerted negative impacts on the rate of phagocytosis of hemocytes. Compared to the control, the phagocytic rate of hemocytes reduced to only approximately 69.2% after exposure to acidified seawater at pH 7.4. In order to examine the affecting mechanisms, the effects of OA on the cellular processes of phagocytosis were investigated.

The process of phagocytosis consists of a series of sequential steps, including the attachment of hemocytes to the targeted particles, the recognition of pathogenic particles, the engulfment of the pathogenic particles, and the degradation of the engulfed target (Song et al., 2010). During this process, the engulfment of the pathogenic particles is facilitated by the cytoskeleton, mainly via the actin-myosin contractile system; any alternations in the cytoskeleton of hemocytes would influence the phagocytosis of hemocytes. In this study, the cytoskeleton of the hemocytes in *T. granosa* after 14 days of exposure to OA was determined by fluorescence staining of F-actin (one of the essential components of microfilaments). The fluorescence results showed that the abundance of cytoskeletons of the hemocytes, in terms of the microfilaments, was significantly reduced upon OA exposure. This reduction in the abundance of the cytoskeleton component would hamper the phagocytic activity of hemocytes and thus suppress the immune responses of marine bivalves. It is worth noting that the expression levels of genes related to the actin cytoskeleton regulation pathway (Insall et al., 2001), such as actin-related protein 2 (*ARPC2*), actin-related protein 3 (*ARPC3*), GTPase Kras (*KRAS*), GTPase Mras (*MRAS*), and Ras-related C3 botulinum toxin substrate 1 (*Rac1*), which function as a complex to assemble the monomers into polymers in the process of producing F-actin from G-actin, were shown to be significantly induced after OA exposure. According to previous studies (Tomanek et al., 2011), the upregulation of these genes could be a feedback to compensate for the reduction in cytoskeleton components caused by OA exposure.

Following the engulfment of pathogenic particles, the engulfed target will be destroyed within phagosomes, mainly by lysozymes (LZM) (Buggé et al., 2007). Therefore the concentration and enzymatic activity of LZM in the hemocytes of *T. granosa* under the present (pH 8.1) and future OA scenarios (pH 7.8 and 7.4) were examined. The concentration of LZM in the hemocytes of *T. granosa* significantly reduced to only approximately 56.6% of that of the control after OA treatments (pH 7.4). Similarly, the enzymatic activity of LZM in the hemocytes was also reduced by about 10% following OA exposure. These reductions suggested a hampered degradation of the engulfed particles through the oxygen-independent pathway, which may also account for the reduction in phagocytosis under OA scenarios.

It is reported that nitric oxide (NO) plays versatile roles in immune responses such as phagocytosis, in which NO not only functions as an active antimicrobial molecule in vivo but also negatively regulates immune responses (Bogdan, 2001). Since reduced pH could promote the generation of NO in mammals or plants, OA-induced decreases in the pH of the hemolymph may influence the production of in vivo NO

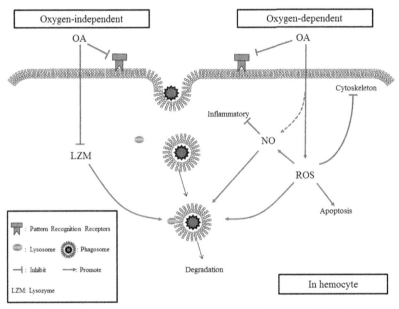

Figure 4.4 The summarized affecting mechanisms of OA-induced reduction in phagocytosis according to results obtained in the present study and those published previously. (1) OA hampers pattern recognition receptors (PRRs); (2) OA induces the buildup of reactive oxygen species, which increases the apoptosis of hemocytes, hampers cytoskeleton-mediated engulfment, and alters oxygen-dependent degradation; (3) OA induces the production of NO, which may directly enhance the oxygen-dependent degradation but negatively regulate immune responses; (4) OA reduces the concentration and activity of lysozymes, which hampers the oxygen-independent degradation of the engulfed particles. *From Wenhao, S., Rong, J., Zha, S., Yan, M., Fang, J., & Liu, G. (2018). Ocean acidification affects the cytoskeleton, lysozymes, and nitric oxide of hemocytes: A possible explanation for the hampered phagocytosis in blood clams,* Tegillarca granosa. Frontiers in Physiology, 9. https://doi.org/10.3389/fphys.2018.00619.

(Röszer, 2012). Therefore the concentrations of NO in the hemocytes of *T. granosa* after 14 days of OA exposure were determined. However, NO concentrations in the hemocytes were detected to increase with the increase in the pCO_2 level in the seawater the subjects were exposed to, which were approximately 1.86 and 6.13 times that of the control for the pH 7.8 and 7.4 groups, respectively. As an active antimicrobial molecule, this detected increase in NO contents in vivo upon exposure to OA seems to indicate an enhanced ability to degrade engulfed particles. However, the OA-induced NO production would also suppress immune responses such as phagocytosis of *T. granosa* by inhibiting the activation of the inflammasome *NLRP3* (Mao et al., 2013).

This research suggested that OA suppresses the process of phagocytosis in marine bivalves by inhibiting the process of engulfment by reducing the abundance of cytoskeleton components, as well as by altering the oxygen-dependent and oxygen-independent degradation, which occurs through the enhancement in the production of NO while a reduction in the concentration and activity of LZM (Fig. 4.4).

References

Asplund, M., Baden, S., Russ, S., Ellis, R., Gong, N., & Hernroth, B. (2013). Ocean acidification and host-pathogen interactions: Blue mussels, *Mytilus edulis*, encountering *Vibrio tubiashii*. *Environmental Microbiology*, 16. Available from https://doi.org/10.1111/1462-2920.12307.

Avignon, S., Bordenave, S., Martin, S., Dubois, P., Badou, A., Coheleach, M., Richard, N., Di Giglio, S., Malet, L., Servili, A., Gaillard, F., Huchette, S., & Roussel, S. (2020). An integrated investigation of the effects of ocean acidification on adult abalone (*Haliotis tuberculata*). *ICES Journal of Marine Science*, 77. Available from https://doi.org/10.1093/icesjms/fsz257.

Beniash, E., Ivanina, A., Lieb, N. S., Kurochkin, I., & Sokolova, I. M. (2010). Elevated level of carbon dioxide affects metabolism and shell formation in oysters *Crassostrea virginica*. *Marine Ecology Progress Series*, *419*, 95–108. Available from http://www.int-res.com/abstracts/meps/v419/p95-108/.

Benson, A. K., Kelly, S. A., Legge, R., Ma, F., Low, S. J., Kim, J., Zhang, M., Oh, P. L., Nehrenberg, D., Hua, K., Kachman, S. D., Moriyama, E. N., Walter, J., Peterson, D. A., & Pomp, D. (2010). Individuality in gut microbiota composition is a complex polygenic trait shaped by multiple environmental and host genetic factors. *Proceedings of the National Academy of Sciences of the United States of America*, *107*(44), 18933. Available from https://doi.org/10.1073/pnas.1007028107.

Bino, R. J., Hall, R. D., Fiehn, O., Kopka, J., Saito, K., Draper, J., Nikolau, B. J., Mendes, P., Roessner-Tunali, U., Beale, M. H., Trethewey, R. N., Lange, B. M., Wurtele, E. S., & Sumner, L. W. (2004). Potential of metabolomics as a functional genomics tool. *Trends in Plant Science*, *9*(9), 418–425. Available from https://doi.org/10.1016/j.tplants.2004.07.004.

Bogdan, C. (2001). Nitric oxide and the immune response. *Nature Immunology*, *2*(10), 907–916. Available from https://doi.org/10.1038/ni1001-907.

Borowsky, B., Hoffman, B. J., Bradley, R. J., & Harris, R. A. (1995). *Neurotransmitter transporters: Molecular biology, function, and regulation* (38, pp. 139−199). New York, NY: Academic Press, https://doi.org/10.1016/S0074-7742(08)60526-7.

Buggé, D. M., Hégaret, H., Wikfors, G. H., & Allam, B. (2007). Oxidative burst in hard clam (*Mercenaria mercenaria*) haemocytes. *Fish & Shellfish Immunology, 23*(1), 188−196. Available from https://doi.org/10.1016/j.fsi.2006.10.006.

Caldeira, K., & Wickett, M. E. (2003). Anthropogenic carbon and ocean pH. *Nature, 425* (6956), 365. Available from https://doi.org/10.1038/425365a.

Chivers, D., Mccormick, M., Nilsson, G., Munday, P., Watson, S.-A., Meekan, M., Mitchell, M., Corkill, K., & Ferrari, M. (2013). Impaired learning of predators and lower prey survival under elevated CO_2: A consequence of neurotransmitter interference. *Global Change Biology, 20.* Available from https://doi.org/10.1111/gcb.12291.

Christensen, H. T., Dolmer, P., Petersen, J. K., & Tørring, D. (2012). Comparative study of predatory responses in blue mussels (*Mytilus edulis* L.) produced in suspended long line cultures or collected from natural bottom mussel beds. *Helgoland Marine Research, 66*(1), 1−9. Available from https://doi.org/10.1007/s10152-010-0241-0.

Cigliano, M., Gambi, M. C., Rodolfo-Metalpa, R., Patti, F. P., & Hall-Spencer, J. M. (2010). Effects of ocean acidification on invertebrate settlement at volcanic CO_2 vents. *Marine Biology, 157*(11), 2489−2502. Available from https://doi.org/10.1007/s00227-010-1513-6.

Clements, J. C. (2016). Meta-analysis reveals taxon- and life stage-dependent effects of ocean acidification on marine calcifier feeding performance. *BioRxiv,* 066076. Available from https://doi.org/10.1101/066076.

Clements, J. C., Woodard, K. D., & Hunt, H. L. (2016). Porewater acidification alters the burrowing behavior and post-settlement dispersal of juvenile soft-shell clams (*Mya arenaria*). *Journal of Experimental Marine Biology and Ecology, 477,* 103−111. Available from https://doi.org/10.1016/j.jembe.2016.01.013.

Collard, M., Eeckhaut, I., Dehairs, F., & Dubois, P. (2014). Acid−base physiology response to ocean acidification of two ecologically and economically important holothuroids from contrasting habitats, *Holothuria scabra* and *Holothuria parva*. *Environmental Science and Pollution Research, 21*(23), 13602−13614. Available from https://doi.org/10.1007/s11356-014-3259-z.

Deigweiher, K., Hirse, T., Bock, C., Lucassen, M., & Pörtner, H. O. (2010). Hypercapnia induced shifts in gill energy budgets of Antarctic notothenioids. *Journal of Comparative Physiology B, 180*(3), 347−359. Available from https://doi.org/10.1007/s00360-009-0413-x.

Doney, S. C., Fabry, V. J., Feely, R. A., & Kleypas, J. A. (2009). Ocean acidification: The other CO_2 problem. *Annual Review of Marine Science, 1*(1), 169−192. Available from https://doi.org/10.1146/annurev.marine.010908.163834.

Dupont, S., & Thorndyke, M. (2012). Relationship between CO_2-driven changes in extracellular acid−base balance and cellular immune response in two polar echinoderm species. *Journal of Experimental Marine Biology and Ecology, 424−425,* 32−37. Available from https://doi.org/10.1016/j.jembe.2012.05.007.

Dvoryanchikov, G., Huang, Y. A., Barro-Soria, R., Chaudhari, N., & Roper, S. D. (2011). GABA, its receptors, and GABAergic inhibition in mouse taste buds. *Journal of Neuroscience, 31*(15), 5782. Available from https://doi.org/10.1523/JNEUROSCI.5559-10.2011.

Ern, R., & Esbaugh, A. J. (2016). Hyperventilation and blood acid−base balance in hypercapnia exposed red drum (*Sciaenops ocellatus*). *Journal of Comparative Physiology B, 186* (4), 447−460. Available from https://doi.org/10.1007/s00360-016-0971-7.

Evans, T. G., Chan, F., Menge, B. A., & Hofmann, G. E. (2013). Transcriptomic responses to ocean acidification in larval sea urchins from a naturally variable pH environment. *Molecular Ecology, 22*(6), 1609−1625. Available from https://doi.org/10.1111/mec.12188.

Evans, T. G., & Watson-Wynn, P. (2014). Effects of seawater acidification on gene expression: Resolving broader-scale trends in sea urchins. *The Biological Bulletin, 226* (3), 237−254. Available from https://doi.org/10.1086/bblv226n3p237.

Feely, R. A., Sabine, C. L., Lee, K., Berelson, W., Kleypas, J., Fabry, V. J., & Millero, F. J. (2004). Impact of anthropogenic CO_2 on the $CaCO_3$ system in the oceans. *Science, 305*(5682), 362. Available from https://doi.org/10.1126/science.1097329.

Fitzer, S. C., Cusack, M., Phoenix, V. R., & Kamenos, N. A. (2014). Ocean acidification reduces the crystallographic control in juvenile mussel shells. *Journal of Structural Biology, 188*(1), 39−45. Available from https://doi.org/10.1016/j.jsb.2014.08.007.

Fonseca, F., Cerqueira, R., & Fuentes, J. (2019). Impact of ocean acidification on the intestinal microbiota of the marine sea bream (*Sparus aurata* L.). *Frontiers in Physiology, 10*, 1446. Available from https://doi.org/10.3389/fphys.2019.01446.

Forward, R., Rittschof, D., & Vries, M. (1987). Peptide pheromones synchronize crustacean egg hatching and larval release. *Chemical Senses, 12*, 491−498. Available from https://doi.org/10.1093/chemse/12.3.491.

Genz, J., Taylor, J., & Grosell, M. (2008). Effects of salinity on intestinal bicarbonate secretion and compensatory regulation of acid-base balance in *Opsanus beta*. *The Journal of Experimental Biology, 211*, 2327−2335. Available from https://doi.org/10.1242/jeb.016832.

Gleason, L. U. (2019). Applications and future directions for population transcriptomics in marine invertebrates. *Current Molecular Biology Reports, 5*(3), 116−127. Available from https://doi.org/10.1007/s40610-019-00121-z.

Goncalves, P., Jones, D. B., Thompson, E. L., Parker, L. M., Ross, P. M., & Raftos, D. A. (2017). Transcriptomic profiling of adaptive responses to ocean acidification. *Molecular Ecology, 26*(21), 5974−5988. Available from https://doi.org/10.1111/mec.14333.

Hamilton, T., Holcombe, A., & Tresguerres, M. (2013). CO_2-induced ocean acidification increases anxiety in Rockfish via alteration of GABAA receptor functioning, supplement to: Hamilton, Trevor James; Holcombe, Adam; Tresguerres, Martin (2013): CO_2-induced ocean acidification increases anxiety in Rockfish via alteration of GABAA receptor functioning. *Proceedings of the Royal Society B-Biological Sciences, 281* (1775), 20132509. Available from https://doi.org/10.1594/PANGAEA.834309.

Hamilton, T. J., Holcombe, A., & Tresguerres, M. (2014). CO_2-induced ocean acidification increases anxiety in Rockfish via alteration of GABAA receptor functioning. *Proceedings of the Royal Society B: Biological Sciences, 281*(1775), 20132509. Available from https://doi.org/10.1098/rspb.2013.2509.

Hay, M. E. (2009). Marine chemical ecology: Chemical signals and cues structure marine populations, communities, and ecosystems. *Annual Review of Marine Science, 1*(1), 193−212. Available from https://doi.org/10.1146/annurev.marine.010908.163708.

Hernroth, B., Baden, S., Thorndyke, M., & Dupont, S. (2011). Immune suppression of the echinoderm *Asterias rubens* (L.) following long-term ocean acidification. *Aquatic Toxicology, 103*(3), 222−224. Available from https://doi.org/10.1016/j.aquatox.2011.03.001.

Heuer, R. M., Esbaugh, A. J., & Grosell, M. (2012). Ocean acidification leads to counterproductive intestinal base loss in the Gulf toadfish (*Opsanus beta*). *Physiological and Biochemical Zoology, 85*(5), 450−459. Available from https://doi.org/10.1086/667617.

Hiebenthal, C., Philipp, E. E. R., Eisenhauer, A., & Wahl, M. (2013). Effects of seawater pCO_2 and temperature on shell growth, shell stability, condition and cellular stress of Western Baltic Sea *Mytilus edulis* (L.) and *Arctica islandica* (L.). *Marine Biology, 160*(8), 2073−2087. Available from https://doi.org/10.1007/s00227-012-2080-9.

Hisatsune, C., Yasumatsu, K., Takahashi-Iwanaga, H., Ogawa, N., Kuroda, Y., Yoshida, R., Ninomiya, Y., & Mikoshiba, K. (2007). Abnormal taste perception in mice lacking the type 3 inositol 1,4,5-trisphosphate receptor. *Journal of Biological Chemistry, 282* (51), 37225−37231. Available from http://www.jbc.org/content/282/51/37225. abstract.

Holzmann, B. (2012). Nerve-driven immunity: The effects of neurotransmitters on immune cells, *functions and disease*. Vienna: Springer, https://doi.org/10.1007/978-3-7091-0888-8_10.

Hu, M., Tseng, Y.-C., Stumpp, M., Gutowska, M., Kiko, R., Lucassen, M., & Melzner, F. (2011). Elevated seawater pCO_2 differentially affects branchial acid-base transporters over the course of development in the cephalopod *Sepia officinalis*. *American Journal of Physiology. Regulatory, Integrative and Comparative Physiology, 300*, R1100−R1114. Available from https://doi.org/10.1152/ajpregu.00653.2010.

Huang, Y.-J., Maruyama, Y., Dvoryanchikov, G., Pereira, E., Chaudhari, N., & Roper, S. D. (2007). The role of pannexin 1 hemichannels in ATP release and cell-cell communication in mouse taste buds. *Proceedings of the National Academy of Sciences of the United States of America, 104*(15), 6436−6441. Available from https://doi.org/10.1073/pnas.0611280104.

Hüning, A. K., Melzner, F., Thomsen, J., Gutowska, M. A., Krämer, L., Frickenhaus, S., Rosenstiel, P., Pörtner, H.-O., Philipp, E. E. R., & Lucassen, M. (2013). Impacts of seawater acidification on mantle gene expression patterns of the Baltic Sea blue mussel: Implications for shell formation and energy metabolism. *Marine Biology, 160*(8), 1845−1861. Available from https://doi.org/10.1007/s00227-012-1930-9.

Insall, R., Müller-Taubenberger, A., Machesky, L., Köhler, J., Simmeth, E., Atkinson, S., Weber, I., & Gerisch, G. (2001). Dynamics of the Dictyostelium Arp2/3 complex in endocytosis, cytokinesis, and chemotaxis. *Cell Motility and the Cytoskeleton, 50*(3), 115−128. Available from https://doi.org/10.1002/cm.10005.

Johnson, K. M., & Hofmann, G. E. (2017). Transcriptomic response of the antarctic pteropod limacina helicina antarctica to ocean acidification. *BMC Genomics, 18*(1), 812. Available from https://doi.org/10.1186/s12864-017-4161-0.

Joint, I., Doney, S. C., & Karl, D. M. (2011). Will ocean acidification affect marine microbes? *The ISME Journal, 5*(1), 1−7. Available from https://doi.org/10.1038/ismej.2010.79.

Jutfelt, F., Bresolin de Souza, K., Vuylsteke, A., & Sturve, J. (2013). Behavioural disturbances in a temperate fish exposed to sustained high-CO_2 levels. *PLOS ONE, 8*(6), e65825. Available from https://doi.org/10.1371/journal.pone.0065825.

Kasumyan, A., & Døving, K. (2003). Taste preferences in fish. *Fish and Fisheries, 4*, 289−347. Available from https://doi.org/10.1046/j.1467-2979.2003.00121.x.

Kim, D.-Y., Fenoglio, K. A., Kerrigan, J. F., & Rho, J. M. (2009). Bicarbonate contributes to GABAA receptor-mediated neuronal excitation in surgically resected human hypothalamic hamartomas. *Epilepsy Research, 83*(1), 89−93. Available from https://doi.org/10.1016/j.eplepsyres.2008.09.008.

Kurihara, H. (2008). Effects of CO_2-driven ocean acidification on the early developmental stages of invertebrates. *Marine Ecology Progress Series, 373*, 275−284. Available from http://www.int-res.com/abstracts/meps/v373/p275-284/.

Lai, F., Jutfelt, F., & Nilsson, G. E. (2015). Altered neurotransmitter function in CO_2-exposed stickleback (*Gasterosteus aculeatus*): A temperate model species for ocean acidification research. *Conservation Physiology, 3*(1), cov018. Available from https://doi.org/10.1093/conphys/cov018.

Lannig, G., Eilers, S., Pörtner, H. O., Sokolova, I. M., & Bock, C. (2010). Impact of ocean acidification on energy metabolism of oyster, *Crassostrea gigas*—changes in metabolic pathways and thermal response. *Marine Drugs*, *8*(8), 2318—2339. Available from https://doi.org/10.3390/md8082318.

Larsen, B. K., Pörtner, H.-O., & Jensen, F. B. (1997). Extra- and intracellular acid-base balance and ionic regulation in cod (*Gadus morhua*) during combined and isolated exposures to hypercapnia and copper. *Marine Biology*, *128*(2), 337—346. Available from https://doi.org/10.1007/s002270050099.

Larsen, E. H., Deaton, L., Onken, H., O'Donnell, M., Grosell, M., Dantzler, W., & Weihrauch, D. (2014). Osmoregulation and excretion. *Comprehensive Physiology*, *4*, 405—573. Available from https://doi.org/10.1002/cphy.c130004.

Li, M., Qiu, L., Wang, L., Wang, W., Xin, L., Li, Y., Liu, Z., & Song, L. (2016). The inhibitory role of γ-aminobutyric acid (GABA) on immunomodulation of pacific oyster *Crassostrea gigas*. *Fish & Shellfish Immunology*, *52*, 16—22. Available from https://doi.org/10.1016/j.fsi.2016.03.015.

Li, S., Liu, C., Zhan, A., Xie, L., & Zhang, R. (2017). Influencing mechanism of ocean acidification on byssus performance in the pearl oyster *Pinctada fucata*. *Environmental Science and Technology*, *51*(13), 7696——7706. Available from https://doi.org/10.1021/acs.est.7b02132.

Liu, S., Shi, W., Guo, C., Zhao, X., Han, Y., Peng, C., Chai, X., & Liu, G. (2016). Ocean acidification weakens the immune response of blood clam through hampering the NF-kappa β and toll-like receptor pathways. *Fish & Shellfish Immunology*, *54*, 322—327. Available from https://doi.org/10.1016/j.fsi.2016.04.030.

Liu, Z., Zhang, Y., Zhou, Z., Zong, Y., Zheng, Y., Liu, C., Kong, N., Gao, Q., Wang, L., & Song, L. (2020). Metabolomic and transcriptomic profiling reveals the alteration of energy metabolism in oyster larvae during initial shell formation and under experimental ocean acidification. *Scientific Reports*, *10*(1), 6111. Available from https://doi.org/10.1038/s41598-020-62963-3.

Liu, Z., Zhou, Z., Jiang, Q., Wang, L., Yi, Q., Qiu, L., & Song, L. (2017). The neuroendocrine immunomodulatory axis-like pathway mediated by circulating haemocytes in pacific oyster *Crassostrea gigas*. *Open Biology*, *7*(1), 160289. Available from https://doi.org/10.1098/rsob.160289.

Lotterhos, K. E., & Levitan, D. (2010). Gamete release and spawning behavior in broadcast spawning marine invertebrates. *The Evolution of Primary Sexual Characters*, 99—120.

Mao, K., Chen, S., Chen, M., Ma, Y., Wang, Y., Huang, B., He, Z., Zeng, Y., Hu, Y., Sun, S., Li, J., Wu, X., Wang, X., Strober, W., Chen, C., Meng, G., & Sun, B. (2013). Nitric oxide suppresses NLRP3 inflammasome activation and protects against LPS-induced septic shock. *Cell Research*, *23*(2), 201—212. Available from https://doi.org/10.1038/cr.2013.6.

Mardis, E. R. (2008). The impact of next-generation sequencing technology on genetics. *Trends in Genetics*, *24*(3), 133—141. Available from https://doi.org/10.1016/j.tig.2007.12.007.

Marshall, D. J., Styan, C. A., & Keough, M. J. (2000). Intraspecific co-variation between egg and body size affects fertilisation kinetics of free-spawning marine invertebrates. *Marine Ecology Progress Series*, *195*, 305—309. Available from http://www.int-res.com/abstracts/meps/v195/p305-309/.

Meron, D., Atias, E., Iasur Kruh, L., Elifantz, H., Minz, D., Fine, M., & Banin, E. (2011). The impact of reduced pH on the microbial community of the coral *Acropora eurystoma*. *The ISME Journal*, *5*(1), 51—60. Available from https://doi.org/10.1038/ismej.2010.102.

Michaelidis, B., Ouzounis, C., Paleras, A., & Pãf Ârtner, H. O. (2005a). Effects of long-term moderate hypercapnia on acid—base balance and growth rate in marine mussels *Mytilus galloprovincialis*. *Marine Ecology Progress Series*, *293*, 109—118. Available from http://www.int-res.com/abstracts/meps/v293/p109-118.

Michaelidis, B., Ouzounis, C., Paleras, A., & Pörtner, H.-O. (2005b). Effects of long-term moderate hypercapnia on acid-base balance and growth rate in marine mussels *Mytilus galloprovincialis*. *Marine Ecology-Progress Series, 293*, 109—118. Available from https://doi.org/10.3354/meps293109.

Morais, S. (2016). The physiology of taste in fish: Potential implications for feeding stimulation and gut chemical sensing. *Reviews in Fisheries Science & Aquaculture, 25*, 1—17. Available from https://doi.org/10.1080/23308249.2016.1249279.

Morse, D. E., Hooker, N., Duncan, H., & Jensen, L. (1979). γ-aminobutyric acid, a neurotransmitter, induces planktonic abalone larvae to settle and begin metamorphosis. *Science, 204*(4391), 407. Available from https://doi.org/10.1126/science.204.4391.407.

Nelson, K. S., Baltar, F., Lamare, M. D., & Morales, S. E. (2020). Ocean acidification affects microbial community and invertebrate settlement on biofilms. *Scientific Reports, 10*(1), 3274. Available from https://doi.org/10.1038/s41598-020-60023-4.

Nilsson, G. E., Dixson, D. L., Domenici, P., McCormick, M. I., Sørensen, C., Watson, S.-A., & Munday, P. L. (2012). Near-future carbon dioxide levels alter fish behaviour by interfering with neurotransmitter function. *Nature Climate Change, 2*(3), 201—204. Available from https://doi.org/10.1038/nclimate1352.

Nip, T. H. M., Ho, W.-Y., & Kim Wong, C. (2003). Feeding ecology of larval and juvenile black seabream (*Acanthopagrus schlegeli*) and Japanese seaperch (*Lateolabrax japonicus*) in Tolo Harbour, Hong Kong. *Environmental Biology of Fishes, 66*(2), 197—209. Available from https://doi.org/10.1023/A:1023611207492.

Pan, T.-C. F., Applebaum, S. L., & Manahan, D. T. (2015). Experimental ocean acidification alters the allocation of metabolic energy. *Proceedings of the National Academy of Sciences of the United States of America, 112*(15), 4696. Available from https://doi.org/10.1073/pnas.1416967112.

Parker, L. M., Ross, P. M., O'Connor, W. A., Borysko, L., Raftos, D. A., & Pörtner, H.-O. (2012). Adult exposure influences offspring response to ocean acidification in oysters. *Global Change Biology, 18*(1), 82—92. Available from https://doi.org/10.1111/j.1365-2486.2011.02520.x.

Peng, C., Zhao, X., Liu, S., Shi, W., Han, Y., Guo, C., Peng, X., Chai, X., & Liu, G. (2017). Ocean acidification alters the burrowing behaviour, Ca^{2+}/Mg^{2+}-ATPase activity, metabolism, and gene expression of a bivalve species, *Sinonovacula constricta*. *Marine Ecology Progress Series, 575*, 107—117. Available from http://www.int-res.com/abstracts/meps/v575/p107-117/.

Perry, S. F. (2011). Carbon dioxide excretion in fishes. *Canadian Journal of Zoology, 64*, 565—572. Available from https://doi.org/10.1139/z86-083.

Perry, S. F., & Gilmour, K. M. (2006). Acid—base balance and CO_2 excretion in fish: Unanswered questions and emerging models. *Frontiers in Comparative Physiology II: Respiratory Rhythm, Pattern and Responses to Environmental Change, 154*(1), 199—215. Available from https://doi.org/10.1016/j.resp.2006.04.010.

Pimentel, M., Pegado, M., Repolho, T., & Rosa, R. (2014). Impact of ocean acidification in the metabolism and swimming behavior of the dolphinfish (*Coryphaena hippurus*) early larvae. *Marine Biology, 161*(3), 725—729. Available from https://doi.org/10.1007/s00227-013-2365-7.

Pistevos, J. C. A., Nagelkerken, I., Rossi, T., Olmos, M., & Connell, S. D. (2015). Ocean acidification and global warming impair shark hunting behaviour and growth. *Scientific Reports, 5*(1), 16293. Available from https://doi.org/10.1038/srep16293.

Pörtner, H. O., Langenbuch, M., & Reipschläger, A. (2004). Biological impact of elevated ocean CO_2 concentrations: Lessons from animal physiology and earth history. *Journal of Oceanography, 60*(4), 705—718. Available from https://doi.org/10.1007/s10872-004-5763-0.

Roberts, D., Birchenough, S., Lewis, C., Sanders, M., Bolam, T., & Sheahan, D. (2013). Ocean acidification increases the toxicity of contaminated sediments. *Global Change Biology*, *19*, 340–351. Available from https://doi.org/10.1111/gcb.12048.

Roggatz, C., Lorch, M., Hardege, J., & Benoit, D. (2016). Ocean acidification affects marine chemical communication by changing structure and function of peptide signalling molecules. *Global Change Biology*, *22*. Available from https://doi.org/10.1111/gcb.13354.

Rong, J., Su, W., Guan, X., Shi, W., Zha, S., He, M., Wang, H., & Liu, G. (2018). Ocean acidification impairs foraging behavior by interfering with olfactory neural signal transduction in black sea bream, *Acanthopagrus schlegelii*. *Frontiers in Physiology*, *9*, 1592.

Rong, J., Tang, Y., Zha, S., Han, Y., Shi, W., & Liu, G. (2020). Ocean acidification impedes gustation-mediated feeding behavior by disrupting gustatory signal transduction in the black sea bream, *Acanthopagrus schlegelii*. *Marine Environmental Research*, *162*, 105182. Available from https://doi.org/10.1016/j.marenvres.2020.105182.

Ross, P., Parker, L., O'Connor, W., & Bailey, E. (2011). The impact of ocean acidification on reproduction, early development and settlement of marine organisms. *Water*, *3*, 1005–1030. Available from https://doi.org/10.3390/w3041005.

Röszer, T. (2012). *The biology of subcellular nitric oxide* (pp. 179–185). https://doi.org/10.1007/978-94-007-2819-6_11.

Sardans, J., Peñuelas, J., & Rivas-Ubach, A. (2011). Ecological metabolomics: Overview of current developments and future challenges. *Chemoecology*, *21*(4), 191–225. Available from https://doi.org/10.1007/s00049-011-0083-5.

Sewell, M. A., Millar, R. B., Yu, P. C., Kapsenberg, L., & Hofmann, G. E. (2014). Ocean acidification and fertilization in the antarctic sea urchin *Sterechinus neumayeri*: The importance of polyspermy. *Environmental Science and Technology*, *48*(1), 713–722. Available from https://doi.org/10.1021/es402815s.

Shao, Y., Chai, X., Xiao, G., Zhang, J., Lin, Z., & Liu, G. (2016). Population genetic structure of the blood clam, *Tegillarca granosa*, along the pacific coast of Asia: Isolation by distance in the sea. *Malacologia*, *59*(2), 303–312. Available from https://doi.org/10.4002/040.059.0208.

Shi, W., Guan, X., Sun, S., Han, Y., Du, X., Tang, Y., Zhou, W., & Liu, G. (2020). Nanoparticles decrease the byssal attachment strength of the thick shell mussel *Mytilus coruscus*. *Chemosphere*, *257*, 127200. Available from https://doi.org/10.1016/j.chemosphere.2020.127200.

Shi, W., Zhao, X., Han, Y., Che, Z., Chai, X., & Liu, G. (2016). Ocean acidification increases cadmium accumulation in marine bivalves: A potential threat to seafood safety. *Scientific Reports*, *6*(1), 20197. Available from https://doi.org/10.1038/srep20197.

Shi, W., Han, Y., Guo, C., Zhao, X., Liu, S., Su, W., Wang, Y., Zha, S., Chai, X., & Liu, G. (2017a). Ocean acidification hampers sperm-egg collisions, gamete fusion, and generation of Ca^{2+} oscillations of a broadcast spawning bivalve, *Tegillarca granosa*. *Marine Environmental Research*, *130*, 106–112. Available from https://doi.org/10.1016/j.marenvres.2017.07.016.

Shi, W., Zhao, X., Han, Y., Guo, C., Liu, S., Su, W., Wang, Y., Zha, S., Chai, X., Fu, W., Yang, H., & Liu, G. (2017b). Effects of reduced pH and elevated pCO_2 on sperm motility and fertilisation success in blood clam, *Tegillarca granosa*. *Null*, *51*(4), 543–554. Available from https://doi.org/10.1080/00288330.2017.1296006.

Shrivastava, J., Ndugwa, M., Caneos, W., & De Boeck, G. (2019). Physiological trade-offs, acid-base balance and ion-osmoregulatory plasticity in European sea bass (*Dicentrarchus labrax*) juveniles under complex scenarios of salinity variation, ocean acidification and high ammonia challenge. *Aquatic Toxicology*, *212*, 54−69. Available from https://doi.org/10.1016/j.aquatox.2019.04.024.

Silverman, H. G., & Roberto, F. F. (2007). Understanding marine mussel adhesion. *Marine Biotechnology*, *9*(6), 661−681. Available from https://doi.org/10.1007/s10126-007-9053-x.

Silverman, H. G., & Roberto, F. F. (2010). *Byssus Formation in Mytilus* (pp. 273−283). Vienna: Springer, https://doi.org/10.1007/978-3-7091-0286-2_18.

Song, L., Wang, L., Qiu, L., & Zhang, H. (2010). *Bivalve immunity* (pp. 44−65). United States: Springer, https://doi.org/10.1007/978-1-4419-8059-5_3.

Spicer, J. I., Raffo, A., & Widdicombe, S. (2007). Influence of CO_2-related seawater acidification on extracellular acid−base balance in the velvet swimming crab *Necora puber*. *Marine Biology*, *151*(3), 1117−1125. Available from https://doi.org/10.1007/s00227-006-0551-6.

Stapp, L., Kreiss, C., Pörtner, H.-O., & Lannig, G. (2015). Differential impacts of elevated CO_2 and acidosis on the energy budget of gill and liver cells from Atlantic cod, *Gadus morhua*. *Comparative Biochemistry and Physiology. Part A, Molecular & Integrative Physiology*, 187. Available from https://doi.org/10.1016/j.cbpa.2015.05.009.

Strader, M., Wong, J., & Hofmann, G. (2020). Ocean acidification promotes broad transcriptomic responses in marine metazoans: A literature survey. *Frontiers in Zoology*, 17. Available from https://doi.org/10.1186/s12983-020-0350-9.

Stumpp, M., Wren, J., Melzner, F., Thorndyke, M. C., & Dupont, S. T. (2011). CO_2 induced seawater acidification impacts sea urchin larval development I: Elevated metabolic rates decrease scope for growth and induce developmental delay. *Comparative Biochemistry and Physiology Part A: Molecular & Integrative Physiology*, *160*(3), 331−340. Available from https://doi.org/10.1016/j.cbpa.2011.06.022.

Styan, C. A., & Butler, A. J. (2000). Fitting fertilisation kinetics models for free-spawning marine invertebrates. *Marine Biology*, *137*(5), 943−951. Available from https://doi.org/10.1007/s002270000401.

Su, W., Rong, J., Zha, S., Yan, M., Fang, J., & Liu, G. (2018). Ocean acidification affects the cytoskeleton, lysozymes, and nitric oxide of hemocytes: A possible explanation for the hampered phagocytosis in blood clams, *Tegillarca granosa*. *Frontiers in Physiology*, *9*. Available from https://doi.org/10.3389/fphys.2018.00619.

Thomsen, J., & Melzner, F. (2010). Moderate seawater acidification does not elicit long-term metabolic depression in the blue mussel *Mytilus edulis*. *Marine Biology*, *157*(12), 2667−2676. Available from https://doi.org/10.1007/s00227-010-1527-0.

Tomanek, L., Zuzow, M. J., Ivanina, A. V., Beniash, E., & Sokolova, I. M. (2011). Proteomic response to elevated pCO_2 level in eastern oysters, *Crassostrea virginica*: Evidence for oxidative stress. *Journal of Experimental Biology*, *214*(11), 1836. Available from https://doi.org/10.1242/jeb.055475.

Tresguerres, M., & Hamilton, T. J. (2017). Acid−base physiology, neurobiology and behaviour in relation to CO_2-induced ocean acidification. *Journal Experimental Biology*, *220*(12), 2136. Available from https://doi.org/10.1242/jeb.144113.

Vogel, H., Czihak, G., Chang, P., & Wolf, W. (1982). Fertilization kinetics of sea urchin eggs. *Mathematical Biosciences*, *58*(2), 189−216. Available from https://doi.org/10.1016/0025-5564(82)90073-6.

Wagner, J., Li, Y.-X., Pearson, J., & Keizer, J. (1998). Simulation of the fertilization Ca^{2+} wave in *Xenopus laevis* eggs. *Biophysical Journal*, *75*(4), 2088−2097. Available from https://doi.org/10.1016/S0006-3495(98)77651-9.

Whittamore, J. M. (2012). Osmoregulation and epithelial water transport: Lessons from the intestine of marine teleost fish. *Journal of Comparative Physiology B*, *182*(1), 1−39. Available from https://doi.org/10.1007/s00360-011-0601-3.

Witt, V., Wild, C., Anthony, K., Diaz-Pulido, G., & Uthicke, S. (2011). Effects of ocean acidification on microbial community composition of, and oxygen fluxes through, biofilms from the Great Barrier Reef. *Environmental Microbiology*, *13*, 2976—2989. Available from https://doi.org/10.1111/j.1462-2920.2011.02571.x.

Zha, S., Liu, S., Su, W., Shi, W., Xiao, G., Yan, M., & Liu, G. (2017). Laboratory simulation reveals significant impacts of ocean acidification on microbial community composition and host-pathogen interactions between the blood clam and *Vibrio harveyi*. *Fish & Shellfish Immunology*, *71*, 393—398. Available from https://doi.org/10.1016/j.fsi.2017.10.034.

Zhao, X., Guo, C., Han, Y., Che, Z., Wang, Y., Wang, X., Chai, X., Wu, H., & Liu, G. (2017a). Ocean acidification decreases mussel byssal attachment strength and induces molecular byssal responses. *Marine Ecology Progress Series*, *565*, 67—77. Available from http://www.int-res.com/abstracts/meps/v565/p67-77/.

Zhao, X., Han, Y., Chen, B., Xia, B., Qu, K., & Liu, G. (2020). CO_2-driven ocean acidification weakens mussel shell defense capacity and induces global molecular compensatory responses. *Chemosphere*, *243*, 125415. Available from https://doi.org/10.1016/j.chemosphere.2019.125415.

Zhao, X., Shi, W., Han, Y., Liu, S., Guo, C., Fu, W., Chai, X., & Liu, G. (2017b). Ocean acidification adversely influences metabolism, extracellular pH and calcification of an economically important marine bivalve, *Tegillarca granosa*. *Marine Environmental Research*, *125*, 82—89. Available from https://doi.org/10.1016/j.marenvres.2017.01.007.

CHAPTER FIVE

Interactive effects of ocean acidification and other environmental factors on marine organisms

Tianyu Zhang[1,2,3], Qianqian Zhang[1,2], Yi Qu[1,2,3], Xin Wang[1,2,3] and Jianmin Zhao[1,2,4]

[1]Muping Coastal Environmental Research Station, Yantai Institute of Coastal Zone Research, Chinese Academy of Sciences, Yantai, Shandong, P.R. China
[2]Key Laboratory of Coastal Biology and Biological Resources Utilization, Yantai Institute of Coastal Zone Research, Chinese Academy of Sciences, Yantai, Shandong, P.R. China
[3]University of Chinese Academy of Sciences, Beijing, P.R. China
[4]Center for Ocean Mega-Science, Chinese Academy of Sciences, Qingdao, Shandong, P.R. China

Introduction

Due to the increasing amount of anthropogenic activities (e.g., fossil fuel combustion and deforestation), the concentration of atmospheric carbon dioxide (CO_2) on the planet has reached nearly 400 ppm, from 280 ppm during preindustrial times (Friedlingstein et al., 2019). Much of the carbon dioxide (CO_2) in the atmosphere is absorbed by oceans, which leads to increased hydrogen ion (H^+) concentration and the associated chemical changes, including elevated bicarbonate ion (HCO_3^-) concentration and reductions in the number of carbonate ions (CO_3^{2-}) in the seawater (Friedlingstein et al., 2019; Raven & Falkowski, 1999; Sabine et al., 2004). These alterations in pH and carbonate chemical balance of the seawater are known as ocean acidification (OA), which has become one of the most serious environmental problems nowadays. Globally, the average pH of ocean surface water has reduced approximately 0.1 units since preindustrial times and is expected to fall a further app. 0.3−0.4 units by 2100 according to the business-as-usual emission scenario projected by the Intergovernmental Panel on Climate Change (IPCC) (Bopp et al., 2013; Caldeira & Wickett, 2005; Gattuso et al., 2015). No doubt, these will bring up serious threats to the biotic interactions, community structures, and

Ocean Acidification and Marine Wildlife.
DOI: https://doi.org/10.1016/B978-0-12-822330-7.00001-0

ecosystem function in the ocean. A range of studies has proven that OA could affect the physiological processes of marine organisms through various mechanisms. Firstly, changes in carbonate chemical balance in would seawater suppress the ability of marine calcifying organisms to absorb Ca^{2+} and CO_3^{2-}, which are important materials in building shells or skeletons of the organisms. OA could significantly inhibit the calcification process of the bivalves *Mytilus coruscus*, *Mytilus galloprovincialis*, and *Tegillarca granosa*, leading to stunted and deformed shell structures (Michaelidis et al., 2005; Zhao et al., 2017, 2020). Attributed to the reduced seawater pH and state of calcium carbonate saturation, it is predicted that OA will accelerate the loss of net calcium carbonate in the coral rubble community in the future (Andersson et al., 2007; Chan & Eggins, 2017; Stubler & Peterson, 2016). Moreover, reductions in seawater pH can inhibit the acid—base regulatory capabilities of marine organisms and thereby change their physiological states, especially in marine invertebrates. For instance, it has been revealed that individuals of the early benthic juvenile stage of the European lobster *Homarus gammarus* consume more energy to maintain acid—base balance, leading to increased death rates during the molting period. In the scallop *Patinopecten yessoensis* the hemolymph pH of the organism decreased due to the low seawater pH, accompanied by oxidative stress via exceeded reactive oxygen species production (Liao et al., 2019).

In nature, a biota has rarely been affected by one single environmental factor. Apart from OA, marine organisms simultaneously face threats from other environmental stressors including ocean warming, hypoxia, salinity fluctuation, elevated UV radiation and pollutants, and anomalous alterations in other environmental factors (Sage, 2020). Although a body of evidence has revealed the impacts of the individual stressors on one single species, it is difficult to predict the combined impact of multiple simultaneous stressors on certain marine organisms. A combination of different environmental stressors usually induces interactive effects on both individual species and the vulnerability of a wide range of ecosystems (Bailey & van der Grient, 2020). The interactive effects vary among additive (total effect equal to the sum of individual effects), synergistic (total effect greater than the sum of individual effects), and antagonistic (total effect less than the sum of individual effects), depending on the species and life stages of the studied organism, and the nature of the stressors themselves (Crain et al., 2008; Darling & Côté, 2008; Harvey et al., 2013; Piggott et al., 2015). Apparently, a full understanding of the impacts on oceanic biota by global change will only be attained when complex interactions between environmental parameters are appreciated (Boyd & Hutchins, 2012). Traditionally, field surveys explore correlations between climatic conditions and biological properties, which have predictive power about the survival of marine organisms in the future. Alternatively, mesocosm experiments are also reliable in predicting the ecological consequences of climate change, which have become crucial in the recent climate change—related studies (Stewart et al., 2013; Taulbee et al., 2009). Here, we have reviewed studies with related experiments in recent years, aiming to discuss interactive effects of OA combined with four other typical environmental stressors (ocean warming, hypoxia, salinity fluctuation, and heavy metal pollution) on marine organisms.

Interaction of ocean acidification with global warming

Ocean warming is another important environmental stressor driven by the increasing levels of atmospheric CO_2. Infrared radiation reflected by the ground is trapped by greenhouse gases (mainly CO_2), leading to excessive heat storage in the atmosphere and an increase in the global temperature, which is known as the greenhouse effect (Schneider, 1989). The ocean has absorbed over 90% of the excessive heat caused by the greenhouse effect, and the global sea surface temperatures have risen by 0.74°C over the last 100 years (IPCC, 2014; Trenberth & Fasullo, 2013). According to the prediction by the IPCC, the sea surface temperatures will rise a further 1°C—4°C by the year 2100 relative to 1990, with consequences in organisms that are adapted to specific temperature ranges both in terms of latitudinal ranges as well as depth ranges (IPCC, 2014; Trenberth & Fasullo, 2013). Ubiquitously, temperature affects both the rates of physiological processes and the integrity of molecular cellular structures, thus playing a crucial role in controlling the distributions of organisms in the ocean (Willmer, 2002). Experimental manipulations simulating predicted future ocean temperatures have suggested that warming will lead to exceeded metabolic costs of plants (O'Connor et al., 2009) and increased consumption rates of animals (Sanford, 1999). Increased temperature also damages the structures of proteins and DNA, which activate cellular stress response to repair the damage. Cellular homeostasis is restored when the repair is successful, while programmed cell death (apoptosis) and necrosis, regulated by complex signaling pathways, may occur if the repair is inadequate (Braby & Somero, 2006; Lockwood et al., 2010; Lockwood & Somero, 2012; Somero, 2012). In addition, the rising temperature may have deleterious effects on the survival and growth of embryos by promoting premature hatching, suppressing metabolic activity, and inducing oxidative stress on the eggs of marine organisms. An investigation on the fertilization and development of the sea urchin *Heliocidaris erythrogramma* showed that a temperature of 4°C–6°C higher than the normal value induced a cleavage reduction of 40%–60%, respectively, and the development was largely inhibited when the temperatures were even higher (Byrne et al., 2009). Collectively, the influence of environmental temperature on the physiology of aquatic ectothermic animals has been extensively studied, which shows consistency with the predicted effects of temperature on biological and chemical processes.

According to the theory of oxygen- and capacity-limited thermal tolerance, each of the aquatic animal species has a specific threshold to seawater temperature (Deutsch et al., 2015). Additional environmental factors, like hypoxia, would reduce the thermal tolerance ability of marine organisms. As marine ecosystems are facing increases in both temperature and CO_2 levels, a rapidly growing body of experimental research is attempting to predict the interactive biological impacts of OA combined with warming on species, communities, and, ultimately, the ecosystems. Despite an increasing volume of literature addressing the cumulative impacts, there is a huge gap in the knowledge on how different pressures may interact, what their interactive effects are, and what conservation and management decisions should be made to protect and restore the marine ecosystems. A search was carried out on the ISI Web of Science to summarize the peer-reviewed papers that explicitly investigated the interactive effects of OA and ocean warming on marine organisms, using the keywords "ocean acidification and warming" or "elevated temperature and acidification." A total of 105 published articles were accessed, referring to 160 various biological and physiological responses of 105 marine species coexposed to OA and warming.

Reviews on these studies show that OA and warming have affected various biological processes on different life-history stages of marine organisms. The subjects studied were those that are mostly considered ecologically and economically important marine species, with more than a half belonging to invertebrates (Fig. 5.1A), including corals, bivalves,

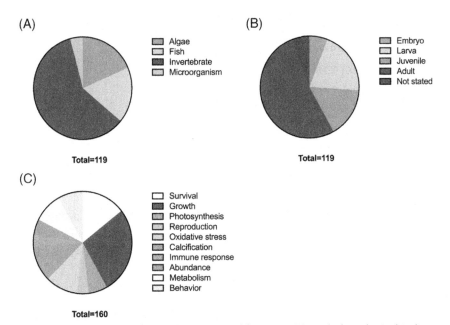

Figure 5.1 Proportions of group species (A), life stages (B), and physiological indices (C) in the studies related to interactions between ocean acidification and warming.

gastropods, and crustaceans. This might be attributed to the fact that invertebrates are more sensitive to changes in environmental factors. A great number of studies investigated the effects of OA and warming on microbial communities, referring to both symbiotic and free-living organisms. Moreover, nearly 50% of the studies have focused on the physiological responses of adult individuals (Fig. 5.1B), concerning key physiological processes including survival, growth, photosynthesis, reproduction, oxidative stress, calcification, immune response, metabolism, and behaviors in marine organisms (Fig. 5.1C).

It has been widely acknowledged that exposure to the combination of OA and warming generally exhibits a stronger effect (either positive or negative) on marine organisms when compared to the effect of exposure to the stressors in isolation. Coexposure studies on the physiological responses of marine organisms to OA and warming showed equal probability for all types of interactive effects (additive, synergistic, and antagonistic) to occur (Fig. 5.2). Due to the variable resilience of the studied species, no specific mechanism could be unified to describe the interactive effect of OA and warming.

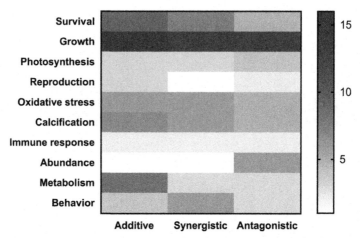

Figure 5.2 Heatmap of different interactions between ocean acidification and warming distributed across different physiological responses.

Our analysis showed that an optimal temperature range is necessary for the survival of each species. Synergistic negative effects were usually observed in the physiological responses of marine organisms to OA−warming coexposure, when the increased temperature exceeded the tolerance range or deviated from the normal physiological tolerance curve of the species. For example, studies have shown a reduction in the fitness of *Diadema africanum* at the warmer limit of its thermal range when combined with low pH (Hernandez et al., 2020). Similarly, OA and warming generated negative synergistic effects on the survival of the sea bream *Sparus aurata*, which suffered mobilization of storage compounds, enhancement in anaerobic pathways, and degradation of the impaired proteases (Araújoa et al., 2018). Synergistic effects of OA and warming were also found on the condition index (CI) of the common cockle *Cerastoderma edule*. The following mechanisms explaining the effects have been proposed: low pH conditions might reduce food intake, meaning it is insufficient to support the higher energy required to compensate for basal maintenance and growth in warmer and more acidic waters (Ong et al., 2017). Also, OA and warming showed significant synergistic effects on the activities of digestive enzymes in the mussels *M. coruscus* (Khan et al., 2020). The swimming speed of the larvae of the neogastropod ubiquitous *Tritia reticulate* was significantly reduced under combined acidic and warmer conditions. Reproduction of the commercially important Northeast Arctic cod (*Gadus morhua*) was synergistically impaired by OA and warming (Hänsel et al., 2020). Warming and elevated CO_2 also showed positively synergistic interactions in stimulating the photosynthesis of temperate phytoplankton communities after 2 weeks of incubation (Cabrerizo et al., 2020). Although elevated temperature and pCO_2 did not affect the carbonic anhydrase activity of the brooded larvae of *Pocillopora damicornis*, synergistically enhanced symbiont carbonic anhydrase activity was revealed, which contributed greatly to the stimulated photosynthetic capacity of the coral (Jiang et al., 2020). Complex and synergistic effects of OA and warming on the swimming activity were also found in the marine trematode parasite *Himasthla* sp. (Franzova et al., 2019). In the marine sea urchin *Loxechinus albus*, OA and warming induced synergistic effects on the transcription levels of heat shock protein 70, which were found to be 75 times higher than those in the control conditions (Manríquez et al., 2019). Warmer rearing temperature combined with elevated pCO_2 induced negative effects on the survival of the northern sand lance *Ammodytes dubius* (Murray et al., 2019). Exposure to high pCO_2 combined with acute and long-term warming synergistically affected the aerobic capacity to synthesize ATP in the European seabass *Dicentrarchus labrax* L., while no effects were found to be caused by the acidification alone (Howald et al., 2019). The short-time effects of temperature and CO_2 on the net photosynthesis, respiration, and calcification of the coralline algae *Phymatolithon lusitanicum* were synergistically positive (Sordo et al., 2019). Synergistic effects on the net calcification rates of two species of corals, namely *Acropora pharaonis* and *Porites lutea*, were observed, after 72 h of exposure to decreased pH combined with increasing temperature (Behbehani et al., 2019). A synergistic effect of temperature and pCO_2 levels was also detected on the swimming performance of the polar cod *Boreogadus saida*, mainly evoked by the low number of bursts detected in the treatment at 8°C/1170 µatm (Kunz et al., 2018). Negative synergistic effects of OA and warming on the survival of the commercial sea bream *S. aurata* were reported, which perhaps could be attributed to an enhancement in anaerobic pathways and impaired proteasomal degradation (Araújoa et al., 2018). The synergistic effect of OA and warming caused increased dehydrogenase activity in *Squalius torgalensis* (Jesus et al., 2018). Sustained increases in the metabolic rate of the Antarctic emerald rockcod (*Trematomus bernacchii)* after 28 days of exposure to OA in combination with ocean warming, indicating synergistic interactions (Davis et al., 2018). A synergistic effect of elevated CO_2 and temperature on the larval weight of the Atlantic herring *Clupea harengus* was found, where no effect was detected in the treatment with elevated CO_2 concentrations at 12°C, but a negative effect was found in the treatment with elevated CO_2 at 10°C (Sswat et al., 2018). Spore germination of the crustose coralline algae *Porolithon cf. onkodes* was negatively impacted by both the synergistic effects of reduced seawater pH and increased seawater temperature and the effects of each stressor independently (Ordoñez et al., 2017). One significant synergistic case is that the warmer temperature aggravated the inhibition induced by OA on growth rates of the diatom *Pseudo-nitzschia subcurvata* (Zhu et al., 2017). Synergetic effects on the mortality of the red king crab *Paralithodes camtschaticus* were observed under the exposure condition of pH 7.8 and ambient temperature +4°C after a long term (183 days) (Swiney et al., 2017). A synergistic adverse effect on the mortality rates (up to 60%) was observed in the solitary zooxanthellate (i.e., symbiotic) Mediterranean coral *Balanophyllia europaea*, suggesting that the elevated seawater temperatures may have increased their metabolic rates, which, in conjunction with decreasing pH, could have led to the rapid deterioration of cellular processes and performance (Prada et al., 2017). Elevated temperature was found to enhance the effects of seawater acidification on the net calcification rate, the calcium content, and the Ca/Mg ratio in the shells of the ecologically and economically important mussel *Mytilus edulis* (Li et al., 2015). Owing to the synergistic effects, OA and warming could have negative effects on the hatching success, embryonic development, and citrate synthase activities of the Antarctic dragonfish *Gymnodraco acuticeps* (Flynn et al., 2015). Synergistic effects of OA and warming on the growth rate and acyl lipid fatty acid content were detected in the Antarctic sea ice diatom *Nitzschia lecointei* (Torstensson et al., 2013). When coexposed to OA and warming, the energy budget of the calcifying gastropods *Austrocochlea concamerata* had a substantial decrease due to the reduced feeding rate and energy assimilation, leading to reductions in shell growth and shell strength (Leung et al., 2020). The temperature and pCO_2 had a synergistic interactive effect on the net calcification of the crustose coralline algae *Hydrolithon onkodes*, which was driven by the increased calcification response to moderately elevated pCO_2 (Johnson & Carpenter, 2012). In the future, it is predicted that CO_2 and temperature will interact synergistically to positively affect the abundance of the algal turfs *Ecklonia radiata*, whereby they will have twice the biomass and occupy over four times more available space than those under current conditions (Connell & Russell, 2010).

A temperature increase within the optimal range might promote the relevant performance, which is in turn resistant to the negative effects induced by low pH. Interestingly, the effects of OA and warming on autotrophs are likely antagonistic (Crain et al., 2008). The dissolved inorganic carbon sources utilized by marine autotrophs are set to increase, and it is predicted that photosynthesizing organisms will likely be more resilient to climate scenarios at the end of the century, as long as the temperature does not exceed their thermotolerance threshold (Hofmann & Todgham, 2009; Tomanek, 2010); there are no limits from other factors such as inorganic nutrient availability (Langdon & Atkinson, 2005; Ries et al., 2009). For example, Yuan et al. (2018) found that the diatoms *Thalassiosira pseudonana* grown under a pCO_2 level of 1000 μatm did not increase their photoinactivation rate like those grown at a high temperature, indicating that the negative effect of warming upon PSII activity was somewhat alleviated by the pCO_2. Similarly, OA and warming have antagonistic effects on the phototrophic species *Carteriospongia foliascens* in terms of mortality, necrosis, and bleaching (Bennett et al., 2017). In addition, it was commonly found that OA and warming have antagonistic effects on the photosynthesis of marine algae. For example, it has been reported that the two typical stressors have antagonistic effects on PSII function in the diatom *Thalassiosira weissflogii* (Gao et al., 2018). Instead of suppression, the combination of the two stressors notably increased PsbA protein synthesis in the oceanic diatom *Skeletonema dohrnii* (7.22-fold), which is required for PSII repair and restoration of photochemical activity (Thangaraj & Sun, 2020). Apart from marine algae, antagonistic effects of OA and warming were also found in marine animals. Higher temperatures appeared to alleviate the impacts of acidification on the embryonic development of the squid *Doryteuthis pealeii* (Zakroff & Mooney, 2020). Warming could mitigate the negative effects of OA on the skeletal stiffness and strength of the glass sponge *Aphrocallistes vastus*, but the reef formation is still curtailed under this condition (Stevenson et al., 2020). OA and warming were observed to have antagonistic effects on the abundance of the bacterioplankton community by affecting nutrient concentration and trophic interactions (Allen et al., 2020). The negative effects of ocean warming on larval survivorship and settlement of the coral *Orbicella faveolata* were mitigated when combined with OA, for which both indices increased by 41% in the combined treatment relative to the treatment with warming alone (Pitts et al., 2020). OA and ocean warming acted antagonistically on the growth rate of sea bass *D. labrax* after 3 days of coexposure (Cominassi et al., 2020). Similarly, it is shown that elevated pCO_2 conditions may adversely impact the early development of the red sea urchin *Mesocentrotus franciscanus*, while moderate warming may improve its growth and thermal tolerance (Wong & Hofmann, 2020). A significant antagonistic effect between temperature and pH on the post-challenge baseline levels of hemocyte data in the spiny lobster *Jasus lalandii* was recorded (Knapp et al., 2019). Effects on the metabolite 5-HIAA in the cleaner wrasses *Labroides dimidiatus* by CO_2 were suppressed under higher temperatures, suggesting an antagonistic interaction between two stressors (Paula et al., 2019). Temperature and high CO_2 were found to have opposite effects on the mineralization in the cartilage of the crura and jaws of the little skate *Leucoraja erinace* (Di Santo, 2019). Interactive effects of temperature and/or pH were also observed on Cd-mediated metallothionein induction, responsiveness of the antioxidant system, and onset of oxidative damages in lipids, with tissue-specific effect (Nardi, Benedetti, d'Errico, et al., 2018). The combination of thermal stress and low pH stimulated the survival rates of juvenile *Argopecten irradians*, which was higher than the survival rate under either individual stressor singly as predicted (Stevens & Gobler, 2018). Exposure to OA would increase the resting oxygen uptake rates of the larval kingfish *Seriola lalandi*; nevertheless, ocean warming suppresses the effects (Laubenstein et al., 2018). In the tropical top shell *Trochus histrio* the lipid peroxidation declined to background levels when the two stressors were combined, indicating that the hypercapnia-deleterious effects and cellular damage induced by acidification were weakened by the elevated temperature (Grilo et al., 2018). Moderate ocean warming may be beneficial to the sea urchins *M. franciscanus* in early development by helping to offset the negative effects induced by OA or acute warming events (Wong & Hofmann, 2020). In the commercially important fish *Argyrosomus regius*, OA also elicited antagonistic interaction to the impact of warming, namely oxidative stress (including heat shock response), although negative effects were also present in the combined treatments (Sampaio et al., 2018). In the Iberian fish *Squalius carolitertii*, acidification and warming showed antagonistic effects on the levels of lipid peroxidation (Jesus et al., 2018). It was shown by Van Colen et al. (2018) that the impairments in hatching success in the marine clam *Limecola balthica* induced by lower pH were moderately alleviated by higher temperature. The acidification-induced molecular responses in the oysters *Saccostrea glomerata*, for example expressions of actin, Rho-GDI, protein spot 7516, ecSOD, CYP1A, kelch 3, and so on, were alleviated by the addition of warming treatment (Goncalves et al., 2017). It has been recorded that the stunting effect on the growth of the sea urchin *Arbacia lixula* induced by OA would be ameliorated by ocean warming (Visconti et al., 2017). The antagonistic interaction of warming and acidification in the phototrophic species *C. foliascens* was reflected by the reduced mortality, necrosis, and bleaching in the most severe treatment (pH 7.6, 31.5°C) (Bennett et al., 2017). The ongoing OA can be expected to boost antioxidant activity (EC50) of the phaeophyceae *Cystoseira tamariscifolia*, but ocean warming have mitigated it due to the antagonistic effect (Celis-Plá et al., 2017). The negative effects of OA on the phagocytic activity of the sea urchin *H. erythrogramma* were alleviated by ocean warming (Brothers et al., 2016). Copepodite abundance (developmental stages $1-5$) and *nauplii* abundance were antagonistically affected by OA and ocean warming (Garzke et al., 2016). Harney et al. (2016) found that concurrent OA and warming mitigated the independent negative effects of pH on the larval size of the Pacific oyster *Crassostrea gigas*. Growth rates of the native scallop *Argopecten purpuratus* in the combined treatment were not significantly different from those in the control treatment, thus providing evidence of an antagonistic interaction between ocean warming and acidification (Lagos et al., 2016). Within the thermotolerance threshold, the negative effects of OA on the larval development and settlement of the sea urchin *Paracentrotus lividus* would be mitigated by ocean warming (García et al., 2015). Although no main effects of OA or warming as independent stressors were detected on the standard metabolic rate of the Pacific sea urchin *Echinometra* sp. *A*, the effects of the interaction between the two stressors were marginally significant (Uthicke et al., 2014). Elevated pCO_2 partially compensated for the decrease in phytoplankton biomass induced by temperature in some phytoplanktonic groups, thus showing a weak antagonistic interaction (Coello-Camba et al., 2014). Kim et al. (2020) reported that ocean warming could offset the negative effects of acidification on the temperate crustose coralline algae *Chamberlainium* sp. by creating more suitable conditions for photosynthesis and growth

The additive results of OA and warming were common in the studies that focused on the survival and development at an early stage of marine organisms. The detrimental effects of OA and warming on net photosynthesis and total antioxidant capacity of the calcareous algae *Halimeda opuntia* were found to be due to the additive interaction (Marques et al., 2020). Moreover, the interaction between pCO_2 and temperature did not show significant effects on the clearance rate of the oyster *Ostrea chilensis* (Navarro et al., 2020). The two stressors were not found to impact the recruitment (total or mean number) of the reef coral *Pocillopora acuta*, both independently and in combination (Bahr et al., 2020). In the purple-hinge rock scallop *Crassadoma gigantean*, simultaneous exposure to high pCO_2 and temperature reduced shell strength and decreased outer shell density, although the interactive effects were not significant (Alma et al., 2020). Similarly, no interactions were found between OA and ocean warming on the European sea bass *D. labrax*, while ocean warming alone would negatively influence the larval growth, development, and swimming performance (Cominassi et al., 2019). A short-term (7 days) exposure to experimental warming and acidification can negatively and additively impact antipredator defense strategies in two species of China mussels, namely *M. coruscus* and *M. edulis* (Kong, Clements, et al., 2019). No significant interactive effect of treatment pH and temperature was detected on the heart rate, calcium, and L-lactate in the subadult American lobsters *Homarus americanus* (Harrington & Hamlin, 2019). The green sea urchin *Lytechinus variegatus* exposed to both OA and warming stressors showed decreased fertilization rates and accelerated larval development (due to increased respiration), in which the more asymmetric larvae with smaller size indicated additive interactions (Lenz et al., 2019). Studies by Lu et al. (2018) found that the ongoing OA and warming may additively interfere with the calcification physiology of the blue mussel *M. edulis*. Elevated pCO_2 did not compensate for the negative effects of thermal stress on the mitochondrial function of the Atlantic herring *C. harengus* (Leo et al., 2018). OA and ocean warming were observed to have no interactive effects on the antioxidant enzyme activity (catalase and glutathione S-transferase), lipid peroxidation, and heat shock response of the octocoral *Veretillum cynomorium* (Lopes et al., 2018). No interactive effects of elevated pCO_2 and higher seawater temperature were revealed on the survival and settlement of planula larvae in the moon jellyfish *Aurelia coerulea* (Dong & Sun, 2018). Increased temperature and reduced pH had significant positive impacts on the fitness of the zooplanktonic appendicularian *Oikopleura dioica*, with no interactions, due to increased fecundity and shortened generation time (Bouquet et al., 2018). In the polar pteropod *Limacina helicina antarctica*, it was shown that the survivorship was mainly influenced by acidification, while warming was more likely sublethal and did not reinforce the mortality when combined with acidification (Gardner et al., 2018). Multigenerational exposure to OA and warming caused a counteractive, but not interactive, effect on the juvenile developmental rate, reactive oxygen species production and mitochondrial density, average reproductive body size, fecundity, and fluctuations in mitochondrial capacity in the emerging model marine polychaete *Ophryotrocha labronica* (Gibbin et al., 2017). Reduced growth rates and increased oxygen consumption under high pCO_2 may have been additionally exacerbated by elevated temperature in the sea urchin *H. erythrogramma* (Carey et al., 2016).

Ultimately, it is important to estimate the energetic balance of complex climate conditions in the future, especially to predict the acclimation mechanisms and population dynamics of marine organisms upon the coexposure to OA and warming. Thus more attention must be paid toward the development of bioenergetic frameworks to understand the effects of single and combined exposure to OA and ocean warming on specific species during different life stages and to predict future ecosystem-level changes.

Interaction of ocean acidification with hypoxia

Oxygen is fundamental for biological and biogeochemical processes in the ocean. A decline in oxygen concentration will cause major changes in ocean productivity, biodiversity, and biogeochemical cycles. Generally, hypoxia is defined as the condition in which the concentration of dissolved oxygen (DO) in water is less than 2.0 mg/L (Diaz & Rosenberg, 1995; Gray et al., 2002). As a worldwide environmental phenomenon, hypoxia has become common in various water ecosystems, such as lakes, estuaries, near-shore waters, and deep seas (Gray et al., 2002; Kidwell et al., 2009). The number of hypoxic water bodies in the world has increased to more than 400, with a total area of 245,000 km^2, since the year 1960 (Diaz & Rosenberg, 2008). Hypoxia is driven by variable inducers in different water environments, including human activities and physical or biological factors. Global warming is expected to contribute to hypoxia, both directly, as the solubility of oxygen in warmer waters decreases, and indirectly, by inducing changes in ocean dynamics that reduce ocean ventilation. Low water flow rate or the gas exchange occurring at the limited water–air interface is another vital reason. The degree of hypoxia can be exacerbated when the water body has vertical stratification, such as salt or thermocline (Diaz & Rosenberg, 1995; Rabouille et al., 2008). In addition, outbreaks of eutrophication in water bodies also aggravate hypoxia. Degradation of large amounts of organic matter input brought about by human activities and the excessive organic matter produced by primary productivity consume a large amount of oxygen in the water body, which has led to more frequent and severe subsea hypoxia globally (Diaz & Rosenberg, 1995). As recorded, oxygen content in global seawater is in overall decline, with a dramatic increase in extreme (hypoxic) events (Stramma et al., 2010). Ocean hypoxia has caused a range of far-reaching effects on marine ecosystems, from the molecular level to the whole ecosystem (Diaz & Rosenberg, 1995; Wu, 2002). Molecular sensors for oxygen have been widely recognized in the cells of marine organisms (Wenger, 2000), including NAD(P)H oxidase, which generates peroxide, reactive oxygen species, mitochondrial cytochrome a3, cytochrome c oxidase, and so on. It has been found that hypoxia would downregulate energy demand and energy supply pathways (e.g., protein synthesis, protein degradation, glucose synthesis, urea synthesis, and maintenance of electrochemical gradients) by affecting these molecular sensors directly in marine organisms to maintain energy balance in the cells under extremely low levels of ATP turnover (Wu, 2002). Effects induced by hypoxia, including the downregulation of protein synthesis and the downregulation and/or modification of certain regulatory enzymes in the anaerobic and aerobic pathways, lead to metabolic depression in marine organisms and eventually manifest as growth reduction (Storey, 1988). For example, growth reductions have been observed in the ophiuroid echinoderm *Amphiura filiformis* (between 2.7 and 1.8 mg O_2/L) and the bivalve mollusks *Crassostrea virginica* and *M. edulis* (between 1.5 and 0.6 mg O_2/L) (Diaz & Rosenberg, 1995). In addition, the species composition of marine ecosystems was also significantly altered by hypoxia. Dauer (1993) found alterations in biomass distribution among species groups in hypoxic areas, with the presence of less biomass of deep-dwelling species and equilibrium species but more biomass consisting of opportunistic species.

OA and hypoxia are very likely occurring together in coastal areas, considering the frequent human activities and the 400 plus hypoxic zones identified in coastal zones across the planet (Diaz & Rosenberg, 2008). The two environmental factors commonly display high dynamics similarity in coastal ecosystems because of the respiration processes of marine organisms (Cai et al., 2011; Feely et al., 2011; Melzner et al., 2013).

In temperate coastal zones, the two conditions occur simultaneously in warmer months, perhaps because respiration rates are maximal and thermal stratification is most likely occurring for this reason. In addition, they both occur at regions near large coastal cities that receive excessive nutrient loads (Melzner et al., 2013; Wallace et al., 2014) or within eutrophic river plumes (Cai et al., 2011). Estuaries have long been known as net heterotrophic ecosystems owing to both allochthonous (imported) and autochthonous (self-produced) sources of organic carbon; hence, on a net annual basis, they produce CO_2 and consume O_2. Besides, specific estuarine habitats such as salt marshes and mangroves are naturally enriched in organic carbon due to respiration, which creates hypoxic and acidified conditions, particularly within warmer waters (Baumann et al., 2014). Considering the coexistence of OA and hypoxia in many coastal regions, the two processes will lead to the induction of interactive effects on marine ecosystems in the foreseeable future. Understanding the interactive effects between the two is therefore a key priority for ecotoxicology and environmental risk assessment under environmentally realistic scenarios. During the past decade, the interactive effects of OA and hypoxia have garnered significant attention among scientists. To summarize the peer-reviewed papers that explicitly investigated interactive effects between OA and hypoxia, we searched on the ISI Web of Science using a combination of the search terms "marine," "ocean," "acidification," "elevated pCO_2," "high pCO_2," "deoxygenation," "hypoxia," "O_2 reduction," and "O_2 depletion." As a result, a total of 36 published articles in the last two decades (2000−20) were accessed, and a dataset comprising 35 species was acquired, most of which are key species or model organisms.

Most of the studied organisms with coexposure to OA and hypoxia belong to invertebrates (Fig. 5.3A), including corals, bivalves, gastropods, and crustaceans. Due to the consistency of the driven factors, OA and hypoxia often occur simultaneously in coastal areas. Invertebrates may be the most sensitive organisms in this area, attributing to their life habit, especially during the early life stages (Franklin, 2008). Thus most recent studies have considered the survival, metabolism, and growth in the early stages (embryonic, larval, and juvenile) of marine organisms (Fig. 5.3B and C). A strong response in early-stage development may be attributed to the higher energy demand during this stage. In addition, the effects of the two stressors on other physiological processes, such as photosynthesis, reproduction, oxidative stress, calcification, immune, acid−base balance, and behaviors (Fig. 5.3C), were also found in the database.

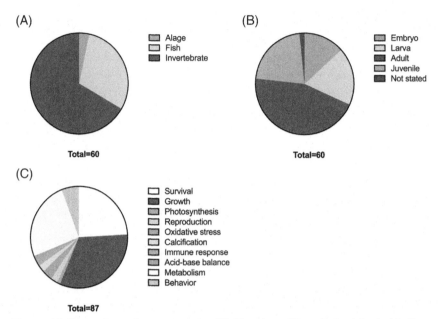

Figure 5.3 Proportions of group species (A), life stages (B), and physiological indices (C) in the studies related to interactions between ocean acidification and hypoxia.

As a result of the differences in the physiological and ecological characteristics, individual responses of marine organisms to OA and hypoxia vary widely. A number of studies indicated that OA and hypoxia have additive negative effects on the growth, survival, and metamorphosis of marine organisms (Fig. 5.4). For instance, concurrent and independent OA and hypoxia were found to significantly depress the growth and survival rates of the hard clams *Mercenaria mercenaria* (Gobler et al., 2014). The growth rates of the juvenile red abalones *Haliotis rufescens* reduced significantly upon exposure to low oxygen and low pH treatments independently, while additive growth rates were found in the combined exposure treatment (Kim et al., 2013). Studies by Tomasetti et al. (2018) found that low pH can act as a secondary stressor, producing additively negative effects on the survival of the blue crabs *Callinectes sapidus*. Although the combination of OA and hypoxia delayed the hatching in embryos of the sheepshead minnows *Cyprinodon variegatus*, no interactions between the two stressors were detected (Enzor et al., 2020). Additive effects of OA and hypoxia were evidently found for the clearance rate, absorption efficiency, respiration rate, and excretion rate of the blue mussel *M. edulis*, and the low pH (pH 7.3) appeared to elicit a stronger effect than the

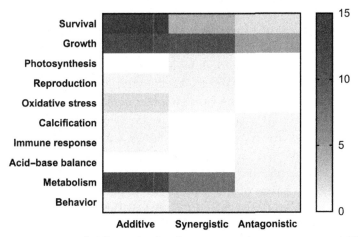

Figure 5.4 Heatmap of different kinds of interactions between ocean acidification and hypoxia across different physiological responses.

hypoxia (2.0 mg/L) (Gu et al., 2019). Physiologically, the rockfish genus *Sebastes* showed acute alterations in the cellular metabolic enzyme activity and exploration behavior after 1 week of acclimation to reduced pH and hypoxia, but no interactions were observed (Davis et al., 2018). Hypoxia was not found to interact with OA in affecting the variation in the fertilization success of the red abalone *H. rufescens* according to the result of the GLM model (Boch et al., 2017). No clear interactive effects of pH and hypoxia on the growth or survival of the juvenile weakfish *Cynoscion regalis* were observed despite the pH being as low as 6.86, indicating that juvenile weakfish have substantial tolerance to dial cycles of oxygenation and acidification encountered in shallow estuarine nursery habitats (Lifavi et al., 2017). With hypoxia, cycling treatments with low mean daily pH ($\sim$ 6.87) and high pCO_2 ($\sim$ 10,000 µatm) had neither an independent nor an interactive effect on the growth of the juvenile summer flounder *Paralichthys dentatus* (Davidson et al., 2016). OA and hypoxia in tandem might result in enhanced larval mortality of marine organisms due to a metabolism-related mechanism (e.g., reducing the aerobic scope to negative values and capacity to provide oxygen to the tissues). Anyway, such additive effects of OA and hypoxia on marine organisms would be expected to influence the distribution and abundance of marine organism populations in the future.

It is common to find the synergistic effect of OA and hypoxia on the physiological responses of marine organisms in recent studies (Fig. 5.4).

In a study addressing the effect of seawater acidification and hypoxia on the early life stages of the bivalves *A. irradians* and the scallop clam *M. mercenaria*, it was found that acidification exposure reduced the viability of scallop larvae, hypoxia exposure inhibited the growth and metamorphosis, and the alternating exposure to acidification and hypoxia caused a more negative response in the larvae (Gobler et al., 2014). Similarly, individuals in the early life stages of calcifying bivalves such as the hard clams *M. mercenaria* and the bay scallops *A. irradians* exhibited synergistic negative responses to OA and hypoxia, including slower growth rates, impaired survival, and inability to metamorphose (Gobler et al., 2014). Additionally, the larvae of the ecologically important forage fish *Menidia* spp. displayed dramatic declines in posthatch survival when exposed to both low pH and low DO (Baumann et al., 2014). Young & Gobler (2020) revealed that OA and hypoxia can synergistically disrupt herbivory of the gastropod *Lacuna vincta* on macroalgae, thus perhaps negatively affecting its survival. The combination of the two stressors affects the digestive enzyme performance more drastically compared to the effects of a single stressor alone, indicating a case of synergistic interaction (Khan et al., 2020). Significant synergistic interaction between OA and hypoxia was recorded in studies on the embryos of the coastal forage fish *Menidia menidia*, in which the oxygen consumption rate was unaffected by the pO_2 level at ambient pCO_2 but decreased with declining pO_2 under elevated pCO_2 (Schwemmer et al., 2020). Reduced pH and DO concentration synergistically reduced the growth rates of the abalone *Haliotis* spp. (Aalto et al., 2020). Synergistic effects of OA and hypoxia on the cleavage rate of fertilized eggs and the deformation rate of mussel larvae were observed in the thick shell mussel *M. coruscus* (Wang et al., 2020). Exposure to OA and hypoxia appears to synergistically induce transcriptomic response related to the ionoregulatory and hypoxia-responsive genes in the blue rockfish *Sebastes mystinus*, which was not observed upon independent exposure to the stressors alone (Cline et al., 2020). Cross et al. (2019) reported that static elevated pCO_2 did not affect the size-at-hatch and postlarval growth rates of the coastal forage fish *M. menidia*. Nevertheless, a synergistic negative effect was observed on embryonic survival under hypoxic conditions. Synergistically negative effects were observed on the growth, carbon fixation, nitrate uptake, and photosynthetic efficiency (Fv/Fm) of the harmful dinoflagellate *Amphidinium carterae* under conditions of OA combined with hypoxia (Bausch et al., 2019). Exposure to OA and hypoxia had a synergistically negative effect on the developmental traits (embryo morphology) in the embryos of the blue mussel *M. edulis* (Kong, Jiang, et al., 2019). In a study by Fontanini et al. (2018) a total of 11 species (crustacean, chordate, echinoderm, and bivalve) of marine ecosystems in Skagerrak were exposed to four different treatments (varying pCO_2 of 450−1300 μatm and dissolved O_2 concentrations of 2−3.5 and 9−10 mg/L) for 6 days, and highly species-specific respiratory responses to OA and hypoxia were found. Synergistic effects of OA and hypoxia on respiration were found in *Tarebia granifera*, *Ophiocomina nigra*, *Ophiothrix fragilis*, and *Asterias rubens* (Fontanini et al., 2018). Significant synergistic interactions between DO and pH were observed to impact the activities of superoxide dismutase and catalase in the gills and hemolymph of the hard-shelled mussel *M. coruscus* (Sui et al., 2017). There was a synergistically significant negative effect of pH and DO on survival of the hard clam *M. mercenaria* and larvae of the bay scallops *A. irradians* (Clark & Gobler, 2016). The significant negative effect of elevated pCO_2 on the fertilization success of the sea urchin *P. lividus* was aggravated when combined with a simulated hypoxic event (Graham et al., 2016). Concurrent stressors such as hypoxia and acidification can synergistically decrease an organism's aerobic scope (or fraction of energy available beyond the basal metabolic requirements) (Sokolova, 2013). Combined exposure to OA and hypoxia might reduce the aerobic scope of marine organisms to negative values and enhance larval mortality. It is widely acknowledged that low DO concentrations limit oxygen supply to tissues. The aerobic scope could be intensified by low pH exposures of long duration, due to the more energy demand induced by the activated physiological process of marine organisms, like acid−base regulatory pathways. When marine organisms are exposed to extreme stress like coexposure to OA and hypoxia, coping strategies such as metabolic depression and shifts to anaerobic pathways are probably initiated. However, these defenses have limited capacity under intensifying physiological stress and the inability to meet the energy demand for basal metabolic needs can eventually lead to a negative aerobic scope and death (Pörtner et al., 2005).

Compared to additive and synergistic effects, antagonistic effects of OA and hypoxia were not observed as often. Nevertheless, they did exist in several previous studies on certain physiological processes of marine organisms. In a study on the widespread fouling barnacle, *Balanus amphitrite*, the juvenile growth of the barnacle was not affected by single stressor treatments, except for the fact that the juvenile CI represented by the proportion of organic weight to inorganic weight was influenced by coexposure to OA and hypoxia (Campanati et al., 2016). This could be attributed to the less or weaker calcified structures developed under the interaction of the two stressors. In the larvae of *Hydroides elegans*, it has been reported that hypoxia can alleviate the reduced calcification rates induced by OA, represented by the upregulated expression of calreticulin (Mukherjee et al., 2013). Calreticulin is a conserved essential endoplasmic reticulum protein expressed in all eukaryotic organisms (Jia et al., 2009). This protein plays key roles in diverse cellular processes such as molecular chaperone/protein folding and calcium-related metabolism (Michalak et al., 2002). Regulation in calreticulin isoforms is speculated to keep balance in the calcium homeostasis of marine organisms under combined exposure to OA and hypoxia. Antagonistic interactions between OA and hypoxia were found to have altered the hatch times of three species of forage fish endemic to the United States East Coast: *M. menidia*, *M. beryllina*, and *C. variegatus* (Morrell & Gobler, 2020). In a realistic environment, OA during a hypoxic event increased the hypoxic tolerance of the European sea bass *D. labrax*, which is partly attributed to the enhanced O_2 uptake ability of fish due to the increased hemoglobin$-O_2$ affinity (Montgomery et al., 2019). After 4 weeks of exposure to a combination of reduced pH and DO concentration, the growth rates of three North Atlantic bivalves, namely *A. irradians*, *C. virginica*, and *M. edulis*, were found to be higher than those upon exposure to either single stressor independently, as predicted, indicating that some anaerobic metabolic pathways may function optimally under hypercapnia (Stevens & Gobler, 2018). OA was found to generally ameliorate the effects on immune traits (phagocytosis and reactive oxygen species production) and metabolic traits (respiration rate, ammonium excretion rate) of the blue mussel *M. coruscus* induced by hypoxia (Sui, Kong, Huang, et al., 2016; Sui, Kong, Shang, et al., 2016). Significant interactive effects of OA and hypoxia were found on the percentage of dead hemocytes, phagocytosis, and apoptosis of the eastern oyster *C. virginica* (Keppel et al., 2015). Although positive growth was observed in all treatments, exposure to the combination of OA and hypoxia was found to have a significant antagonistic effect on the relative growth of the juvenile bivalve *Macoma balthica* (Jansson et al., 2015). Sui et al. (2015) highlight the significant antagonistic effects of OA and hypoxia on the antipredatory responses of the adult mussel *M. coruscus* under the coexposure condition.

Collectively, the effects of low oxygen and acidification on the bivalves during the early life stage, as well as on other marine organisms, are much severe than those induced by a single stressor alone, as predicted. Thus these interactions must be considered when assessing how animals in the ocean respond to these conditions nowadays and the climate change scenarios in the future.

Interaction of ocean acidification with salinity fluctuation

Salinity is the measure of the grams of solute (mainly including Na^+, Cl^-, SO_4^{2-}, Mg^{2+}, Ca^+, K^+, and Br^-) in a kilogram of water. Benefitting from the development of techniques, metrics to quantify salinity have gone through a series of changes in the last 50 years (Russell, 2013). Recently, "absolute salinity," which use ppt as the unit, was advocated by the Scientific Committee on Oceanic Research (SCOR), the International Association for the Physical Sciences of the Oceans (IAPSO), and the Intergovernmental Oceanographic Commission (IOC) for estimating salinity instead of the practical salinity scale. Salinity fluctuations are common in estuaries but might also occur in marine ecosystems as a result of climate change. With warmer air and increasing seawater temperature, sea ice coverage has been declining, leading to a decrease in salinity, especially in the coastal areas at high latitudes (Dickson et al., 2002; Massom & Stammerjohn, 2010). For instance, a sharp decline in salinity has been observed in the coastal areas of the Norwegian Sea during the last few decades (Blindheim et al., 2000). On the other hand, rainfall or hydrological factors associated with climate change can also alter the salinity of surface seawater. Due to the various drivers like the ones mentioned above, a body of studies has detected seawater salinity fluctuation around the world. For instance, both the complete set from 2005 to 2015 (Roemmich—GilsonScripps Institution of Oceanography, https://sio-argo.ucsd.edu/RG/_Climatology.html) and a subset of objective analysis products from 2004 to 2015 (International Pacific Research Center, https://apdrc.soest.hawaii.edu/projects/Argo) have shown a general decrease in salinity in the northern North Atlantic, especially in the central North Atlantic and western subpolar gyre including the Labrador Sea (Tesdal et al., 2018). In addition, surface salinity in the Northwest Pacific Subtropical Gyre decreased about 0.10 psu from 1987 to 2012 with a freshening trend of −0.0042 psu per year (Nan et al., 2015).

Exposure to reduced salinity increases osmotic gradients across the body surface, which results in the passive influx of water and the loss of ions; thus salinity change is a considerable threat to marine organisms from the perspective of cellular responses (homeostatic responses) to alternations in the community and the ecosystem (distribution and dispersion patterns) (Henry et al., 2012). Species across classes differ in their abilities to regulate ions when they are faced with osmotic challenges. Marine organisms are classified into osmoconformers and osmoregulators depending on their osmoregulation ability. Osmoconformers mainly include many bivalves (Carregosa et al., 2014; Shumway, 1977), polychaetes (Freitas et al., 2015; Shumway, 1977), crustaceans (McAllen et al., 1998; Svetlichny et al., 2012), and echinoderms (Castellano et al., 2016). These species maintain internal medium isosmotic to the environment and minimize water fluxes across membranes. When exposed to salinity changes, osmoconformers do not invest energy into transport mechanisms, therefore the osmolality of their internal medium fluctuates according to the osmolality of the environment. They are not able to perform osmotic regulation of their extracellular fluid and rely solely on isosmotic intracellular regulation. This involves (1) increasing or decreasing the concentrations of osmotically active solutes [e.g., ninhydrin-positive substances, K^+ and free amino acids (FAA)] to achieve cell volume regulation and (2) modifying membrane-bound transporters (Gilles, 1987; Pequeux, 1995). Osmosensing is achieved through a wide variety of internal mechanisms (e.g., Ca^{2+} gradients, transient receptor potential ion channels, cell volume sensors) and is often controlled by specific hormones (Dietmar, 2007). Opposite to osmoconformers are osmoregulators, which are species that carry out anisosmotic extracellular regulation when exposed to extracellular osmolality changes. When exposed to dilute seawater or freshwater, these organisms initiate a series of mechanisms (energetically costly) that allow them to maintain the extracellular fluids at a higher osmolality than the surrounding medium. This is thought to represent a selective advantage when dealing with fluctuating salinities (e.g., estuaries) (Barnes, 1967). Normally, the osmoregulation mechanisms mainly involve the controlling of ionic fluxes (mostly those of Na^+ and Cl^- ions), the mobilization of organic osmolytes, and a decrease in body surface permeability (Henry et al., 2012; Rivera-Ingraham & Lignot, 2017). This control involves both limiting and compensatory processes (e.g., control of membrane permeability or epithelial leaks and active pumping, respectively).

The concentrations of many ions in seawater were also affected by OA. OA causes increases in hydrogen ion concentration (H^+) and bicarbonate ions (HCO_3^-) and reductions in carbonate ions (CO_3^{2-}) (Hopkins et al., 2020). OA acts in concert with salinity and modifies the physical–chemical environment of pelagic and benthic communities, leading to severe stress on the osmoregulation of marine organisms. We searched on the ISI Web of Science for peer-reviewed articles for studies that explicitly investigated the interactive effects between OA and salinity fluctuation using a combination of the key words "marine," "ocean," "acidification," "elevated pCO_2," "high pCO_2," "salinity fluctuation," "salinity variation," and "decreased salinity." A total of 17 scientific research articles were published in the last two decades (2000–2020), which allow us to produce a dataset comprising 18 species, most of which are key species or model organisms in their respective ecosystems.

According to this dataset, most subjects of studies related to coexposure to OA and salinity belong to invertebrates (Fig. 5.5A), including bivalves and gastropods. Half of these studies focused on the physiological process of survival and growth in the early stages (embryonic, larval, and

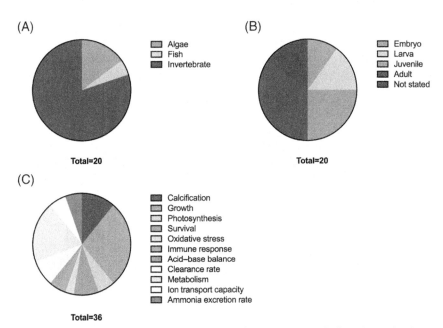

Figure 5.5 Proportions of group species (A), life stages (B), and physiological indices (C), in the studies related to interactions between ocean acidification and heavy salinity fluctuation.

juvenile) of marine organisms (Fig. 5.5B), and studies in recent years have mainly shown that the combined exposure to OA and salinity fluctuation was used to evaluate the physiological responses related to growth, metabolism, and survival of marine organisms (Fig. 5.5C). This accordance in the results is perhaps due to the higher energy demand during early-stage development. In addition, some effects of the two stressors on calcification, photosynthesis, oxidative stress, immune response, acid−base balance, clearance rate, ion transport capacity, and ammonia excretion rate were also revealed in the database (Fig. 5.5C).

Although the total number is small, all of the three types of interactive effects (additive, synergistic, and antagonistic) were observed in the previous studies that focus on the interactive effects between OA and salinity fluctuation on marine organisms (Fig. 5.6).

According to the summary of results of the recent studies, synergistic effects of OA and salinity fluctuation on marine organisms were commonly found. For instance, a significant synergistic effect was observed on the ammonia excretion rate of the European sea bass *D. labrax* (Shrivastava et al., 2019). In addition, under the condition of the combination of OA and lower salinity, the levels of pCO_2 had a synergistic effect, with the lowest clearance rate recorded in the economically important filter feeders *M. edulis* and *Ciona intestinalis* (Rastrick et al., 2018). Moreover, low pH and salinity were observed to lead to a stronger negative effect on hemocyte mortality, esterase activity, reactive oxygen species production, and total hemocyte count in vivo in the blue mussel

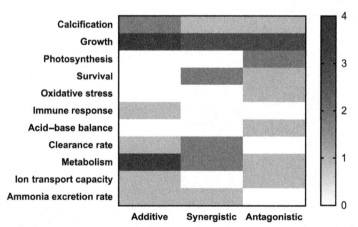

Figure 5.6 Heatmap of different kinds of interactions between ocean acidification and salinity across different physiological responses.

M. coruscus (Wu, Xie, et al., 2018). Additionally, the Sydney rock oyster *S. glomerata* acclimated to elevated pCO_2 showed significant metabolic depression and extracellular acidosis with acute exposure to reduced salinity, especially in the treatment with highest pCO_2 of 1500 µatm (Parker et al., 2017). There was some evidence of a synergistic effect between elevated pCO_2 and reduced salinity on the embryonic development of the amphipod *Echinogammarus marinus*, indicating that the life stage plays a pivotal role in the future survival of this species (Egilsdottir et al., 2009). Reduced salinity and decreased pH synergistically delayed the pre- and postsettlement growth of the Pacific Oyster *C. gigas* (Ko et al., 2014). The larval mortality of two congeneric gastropods, namely the intertidal *Nassarius festivus* and the subtidal *Nassarius conoidalis*, was affected by both exposure to OA (~ 1250 ppm) and reduced salinity (~ 10 psu) independently and the synergistic effects between them (Zhang et al., 2014). Low salinity led to stronger changes in the juvenile eastern oyster *C. virginica* than a high level of pCO_2, whereas the combination of the two induced more severe effects on the shell properties (Vickers microhardness and crack radius) of these mollusks than each of the factors alone (Dickinson et al., 2013).

OA and salinity perhaps have opposite effects on acid–base balance and ion transport capacity in some marine invertebrates; therefore some antagonistic effects between the two stressors were detected. Studies by Xu et al. (2020) found that OA and salinity fluctuation, both separately and antagonistically, affected the pigment content, the photosynthesis, and the ratio of photosynthesis to calcification in the coccolithophorid *Emiliania huxleyi*. In response to elevated pCO_2, the degree of compensation for blood pH in the European sea bass *D. labrax* was much more efficient in normal seawater than in seawater with lower salinity (Shrivastava et al., 2019).

Gao et al. (2019) demonstrated that low salinity can reduce growth and carbon and nitrogen assimilation in the green tide alga *Ulva linza*, but reduced pH can offset or alleviate the negative effect of low salinity by enhancing carbon and nitrogen assimilation. OA was found to significantly and antagonistically interact with reduced salinity and impact the measurements of oxygen uptake in the juvenile mussel *Mytilus chilensis* (Duarte et al., 2018). Effects of OA on autofluorescence and cell size of the marine microalga *Phaeodactylum tricornutum* seem to be mitigated under high salinity conditions (Bautista-Chamizo et al., 2018). The clams *Ruditapes philippinarum* were able to maintain their physiological status

(CI, Na$^+$ and K$^+$ concentrations) and biochemical performance (oxidative stress—related biomarkers) under low pH (7.3) and alternated salinities, suggesting that the low pH tested may mask the negative effects of salinity (Velez et al., 2016). Moderate hypercapnia ($\sim$800 μatm pCO$_2$) appeared to stimulate shell and tissue growth and reduce the mortality of the juvenile clams *M. mercenaria*; however, exposure to low salinity abates these effects (Dickinson et al., 2013).

A few studies reported the additive effect of OA and salinity fluctuation on the physiological process of marine organisms. Laboratory exposure to elevated pCO$_2$ ($\sim$1000 μatm) and reduced salinity ($\sim$23) for 1 month increased gill Na$^+$/K$^+$-ATPase activities and reduced cellular energy allocation in the circumpolar arctic/subarctic amphipods *Gammarus setosus*, but the stressor treatments increased metabolic rates in the higher salinity population, while no interactions between the two environmental factors were detected (Brown et al., 2020). Growth rate and calcification rate in the coccolithophorids *E. huxleyi* significantly responded to only salinity or OA treatment alone, with no interactions between the two detected (Xu et al., 2020). Salinity fluctuation was closely linked to the larval growth of the Olympia oyster *Ostrea lurida*; however, the larvae were tolerant to OA at the same scale (Lawlor & Arellano, 2020). One study on the combined effects of pH and salinity on the properties related to shell formation processes (total shell organic matter, growth rate, and shell microstructure) of the edible mussel *M. chilensis* showed that the shells were more affected by lower pH conditions, while there were no interactive effects observed between the two stressors (Grenier et al., 2020).

Interaction of ocean acidification with heavy metal pollution

Thousands of substances or materials have been introduced to the marine environment by human activities (agriculture, coastal tourism and recreation, port and harbor activities, urban and industrial development, mining, fisheries, and aquaculture). Once the concentrations are higher than certain threshold values, these elements might present negative effects on the biological components of ecosystems and therefore become pollutants (Ricardo, 2018). Heavy metals are typical hazardous substances in coastal ecosystems, induced by discharges or solid wastes of domestic sewage, industrial effluents through the runoff of rivers (point-sources), or exchanges between the ocean and the atmosphere (Turner, 2010; Wilhelmsson et al., 2013). Normally, zinc (Zn), copper (Cu), chromium (Cr), lead (Pb), arsenic (As), mercury (Hg), and cadmium (Cd) are considered major metal pollutants according to most of the available literature. Concentrations of Cu in Chinese offshore marine waters were found to range from 0.1 to 43.2 µg/L (Jin et al., 2015). It has been reported that Cd is one of the dominant metal pollutants in the Bohai Sea, and the concentration of Cd could reach as high as 5 µg/L in the coastal waters of the Bohai Sea (Gao et al., 2014). Heavy metal pollutants are characterized as nondegradable, meaning that they remain in the nearshore water for relatively long periods, leading to severe toxic effects on marine organisms. The toxic effects encompass an increase in the energy demand, oxidative stress, immunosuppressed reactions, and impaired development and reproduction (Adiele et al., 2011; Coteur et al., 2005; Fonseca et al., 2009; Jezierska et al., 2009).

OA can significantly modulate the impacts of pollution on estuarine and coastal ecosystems (Millero et al., 2009). Metals that form strong complexes with hydroxide and carbonate will undergo significant changes in speciation as the pH of seawater decreases (Millero et al., 2009). Moreover, a decrease in the concentration of OH^- and CO_3^{2-} ions can also affect the solubility, adsorption, toxicity, and rates of redox processes of metals in seawater (Millero et al., 2009; Tatara et al., 1997). The changes in metal characteristics in seawater driven by OA may perhaps alter the impacts of metals on marine organisms and ecosystems. For summarization, we searched on the ISI Web of Science for peer-reviewed papers that explicitly investigated the interactive effects between OA and heavy metal pollutions, using keywords like "ocean acidification copper" and another six kinds of metals combined with "ocean acidification". It was found that a total of 43 peer-reviewed studies were carried out in the last two decades (2000−20), involving 104 various biological and physiological responses of organisms from 32 species.

Nearly 90% of the studies have investigated the toxicity of Cu, Zn, and Cd under OA conditions (Fig. 5.7A). Cu was the most commonly considered heavy metal according to our summary, probably because of changing character of Cu in seawater driven by OA. The decrease of CO_3^{2-} levels in seawater will result in Cu forming more strong complex with hydroxide and carbonate, and an increase in free and Cu^{2+}

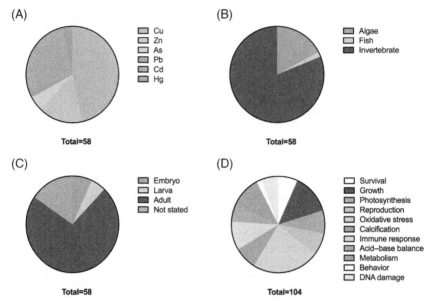

Figure 5.7 Proportions of metal elements (A), group species (B), life stages (C), and physiological indices (D) in the studies related to interactions between ocean acidification and heavy metal pollution.

concentration (from the present-day ∼8% to ∼32% by the year 2250), which perhaps increase its toxicity to marine organisms (Millero et al. 2009). In addition, Zn is an essential cofactor in a number of biochemical processes, such as apoptosis. The concentration of Cd, which is highly toxic to living systems, has maintained at a high level in the marine environment since the 1980s. Nevertheless, Cd speciation is less related to seawater pH and the biological functions of Cd in animals are yet to be known (Bertin & Averbeck, 2006; Das et al., 2014). On the other hand, more than half of the studied subjects were adult invertebrates (Fig. 5.7B) and adult individuals (Fig. 5.7C). Invertebrates, which live at the bottom of coastal areas, usually have weaker motility and are more susceptible to heavy metal pollution. In addition, filter-feeding of some adult bivalve mollusks accelerates metal accumulation in tissues, leading to severe physiological responses. Impacts on the suborganismal and organismal levels were most commonly studied, perhaps due to the possibility of testing the interactive effects of OA and heavy metal pollution under experimental settings. The key physiological functions of concern include different indicators related to survival, growth, photosynthesis, reproduction, oxidative stress, calcification, immune response, acid—base balance, metabolism, behaviors, and DNA damage (Fig. 5.7D).

We summarized the current knowledge concerning the interactive effects of OA and some typical heavy metal pollutants and tried to discuss the potential mechanisms involved in response to these stressor interactions. According to our results from summarization, interactive synergisms and antagonisms were more common than response-additive toxicity (Fig. 5.8).

The synergistic effects of OA and heavy metal pollution have been found on a range of physiological responses in marine organisms. For example, OA and Cu have synergistic effects on reducing fertilization success and inhibiting larval calcification in the abalone *Haliotis discus hannai* (Guo et al., 2020). Similarly, low pH alone negatively impacted the development of embryos and trochophore larvae of *Pomatoceros lamarckii*, causing high mortality and body asymmetry, while coexposure to low concentrations of Cu and low pH further reduced the larval survival (Lewis et al., 2013). Exposure to OA combined with Cd negatively affected the antioxidative enzyme activity and apoptosis and increased DNA damage in the hemocytes of the oyster *C. gigas* (Cao et al., 2018). In the clam *M. mercenaria* and the oyster *C. virginica*, it was shown that inhibition of phagocytic activity, adhesion capacity, and lysozyme activity due to Cd exposure were enhanced by OA conditions (Ivanina et al., 2014). The Atlantic cod achieved the ability to fully compensate for the

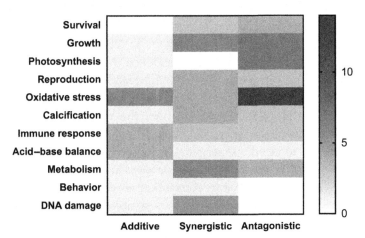

Figure 5.8 Heatmap of different kinds of interactions between ocean acidification and heavy metal pollution across different physiological responses.

extracellular acidosis to resist OA. But this compensatory process was inhibited by Cu exposure perhaps due to the inhibitory effects of this metal on Na^+/K^+-ATPase, HCO_3^-/Cl^- exchanger, and carbonic anhydrase (Larsen et al., 1997). The combined exposure to OA and metal stress is likely to increase the energy costs of the protein synthesis and turnover, as well as the ATP requirements for acid–base homeostasis. These will result in greater energy demand for basal maintenance and may reduce the aerobic scope for fitness-related functions, such as activity, reproduction, and growth (Sokolova, 2013). Synergistic inhibitory effects of water acidity and metallic divalent cations (Zn^{2+} and Cu^{2+}) on ^{45}Ca uptake were revealed in the apical plasma membrane vesicles in the gills of the lobster branchiostegite *H. americanus* (An et al., 2019). Significant synergistic interactions between pH and nano-ZnO were observed to affect the calcification (cumulative length of byssus thread and cumulative volume of byssus thread) of the mussels *M. coruscus*, in which exposure to the two stressors reduces the adhesion strength of the byssus thread and may increase the probability of the mussels being preyed on (Shang et al., 2019). Similarly, metabolism (respiration rate and clearance rate) and immunity (total hemocyte count and reactive oxygen species production) of the mussels *M. coruscus* were also synergistically affected under the combined exposure to ZnO nanoparticles and OA (Shang et al., 2018; Wu, Cui, et al., 2018). OA and copper are likely to exert synergistic effects on the mortality of the embryos of the sea urchin *Heliocidaris crassispina* (Dorey et al., 2018). OA and copper synergistically interact to reduce the size and lipid content of the eggs of the Sydney rock oyster *S. glomerata*, indicating greater energetic demands in the oysters under stressful conditions (Scanes et al., 2018). Huang, Jiang, et al. (2018) showed that Cu in high concentrations (50 µg/L) could influence the hemocyte parameters (lysosomal content, phagocytic activity, and reactive oxygen species production) of the estuarine oyster *Crassostrea rivularis* stronger than the stressor of low pH (pH 7.7); meanwhile, the two stressors showed a concentration-dependent synergistic negative effect. Coexposure to copper and low pH induced a synergistic response of the glutathione reductase activity in the Staghorn coral *Acropora cervicornis* coral and the cauliflower coral *P. damicornis* zooxanthellae (Bielmyer-Fraser et al., 2018). Ca^{2+}-ATPase activity in the symbiont-bearing foraminifer *Amphistegina gibbosa* was synergistically inhibited upon exposure to OA and copper together (Marques et al., 2017). In the mussel *M. edulis*, DNA damage was only observed upon exposure to OA and copper combined, resulting in a 1.9-fold increase in DNA damage, from 14% DNA damage in the ambient treatment to 27% upon combined exposure (Lewis et al., 2016). The toxicity of copper on the polychaete *Arenicola marina* increased the DNA damage of its sperms and inhibited its early larval survivorship, and OA aggravated these effects synergistically (Campbell et al., 2014). The decline in naupliar production of the benthic copepod *Tisbe battagliai* under low pH conditions was greater with the addition of copper when compared to the low pH alone, revealing a negative synergistic impact between OA (pH 7.67) and environmentally relevant copper concentrations (20 g/μL) (Fitzer et al., 2013). The mortality rates of the bivalve *M. edulis L.* were found to increase with reduced pH in all heavy metal (Cd, Pb, Cu) treatments, which demonstrates OA can aggravate the toxicity of heavy metals pollution on marine organisms (Han et al., 2014). Lysosomal membrane stability and the onset of genotoxic damages in the smooth scallop *Flexopecten glaber* showed high sensitivity with possible synergistic effects between OA and Cd (Nardi, Benedetti, Fattorini, et al., 2018). In summary, oxidative stress, impaired immunity, disturbances in acid−base balance, more energy demand, and even mortality could occur in marine organisms upon coexposure to OA and heavy metals, which might be related to the increased bioavailability and accumulation of trace metals in marine organisms. Increased Cd accumulation was found in both the gills and digestive glands of *C. gigas* upon exposure to elevated pCO_2 (Cao et al., 2018). Similarly, the accumulation of Mn and Fe in the polychaete *Nereis diversicolor* (Rodríguez-Romero et al., 2014) and the accumulation of Zn, Pb, Cu, Ni, Cr, Hg in the carpet clams *R. philippinarum* (López et al. 2010) were all increased under ocean acidification conditions. Similarly, OA enhanced Cd and Cu accumulation in mantle tissues in the marine bivalves *C. virginica* and *M. mercenaria* (Götze et al., 2014). Sea anemones *Exaiptasia pallida* exposed to 1000 ppm CO_2 had higher tissue Cu concentrations than those exposed to 400 ppm CO_2 (Siddiqui & Bielmyer-Fraser, 2015). Cd uptake was significantly increased with decreasing pH in the juvenile Mediterranean mussel *M. galloprovincialis* (Sezer et al., 2020), which could be attributed to the fact that reduced pH can increase the solubility of metals and may cause desorption of the metals from the sediments and organic ligands, leading to an elevated flux of the dissolved metals into the water column (De Orte, Lombardi, et al., 2014; de Orte, Sarmiento, et al., 2014). It has also been reported that the Cd influx through the Ca channel in marine organisms could increase along the elevated Cd^{2+}/Ca^{2+} concentrations in low-pH seawater

(Shi et al., 2016). Apart from the changes in the metal characteristics in low-pH seawater, the uptake of metals in aquatic organisms is also related to the metabolic rates, due to the presence of a positive correlation between the uptake of waterborne metals in the gills and the ventilation rates (driven by the oxygen demand of an organism) and that between the uptake of dietary metals and uptake and assimilation of food rates (driven by the organism's nutrient demand).

Antagonistic effects of OA and heavy metal pollution were often found in the studies on microalgae. For instance, OA mitigated the Cd toxicity on the antioxidative systems in the marine diatom *Skeletonema costatum* (Dong, Wang, et al., 2020). Similarly, OA alleviated the inhibition of Cd to the relative growth rate and superoxide dismutase activity in *Pyropia haitanensis* (Ma et al., 2018). The acceleration of the growth rate is perhaps due to the enhanced photosynthesis, resulting from the exceeded contents of *chlorophyll a* due to the increased synthesis of chlorophyll-Cd^{2+} (Küpper et al., 2002). Ma et al. (2020) found that pCO_2 and Zn had a significant antagonistic effect on the growth rate and net photosynthetic rate of the red alga *P. yezoensis*. The outcomes of chlorophyll α content and Ca^{2+}-ATPase activity in the reef calcifier *A. gibbosa* decreased upon coexposure to OA and copper, indicating an overall synergism between the two stressors (Marques et al., 2020). Antagonistic interactions between copper and reduced pH were observed on the photosynthesis of the seagrass *Zostera noltei* (de los Santos et al., 2019). After chronic exposure (35 days), exposure to the combination of OA and copper has shown antagonistic effects on the carbonic anhydrase activity in the zooxanthellate scleractinian coral *Mussismilia harttii* (Marangoni et al., 2019). Gao et al. (2017) demonstrated that a modest increase in pCO_2 (pH 7.80, corresponding to 1000 µatm pCO_2) can alleviate the toxicity of Cu on the respiratory rate and the relative electron transport rate of the green tide alga *U. prolifera*, whereas a further increase in pCO_2 (pH 7.69, corresponding to 1400 µatm pCO_2) exacerbates the toxicity when compared to the control (pH 8.16, corresponding to 390 µatm). The alleviation of the impacts of additive Cd toxicity on *P. tricornutum* by OA has not only been revealed in well-controlled indoor experiments but also been observed in the outdoor mesocosm experiments that reflect more natural growth conditions (Zhang et al., 2020). The combination of OA and Cd groups inhibited the superoxide dismutase activities in the marine diatom *P. tricornutum* compared to their counterparts with exposure to the single Cd stressor, indicating the antagonistic effects of the two factors (Dong, Zhu, et al., 2020). Cd^{2+} dramatically suppressed the maximum net photosynthesis oxygen evolution rate and the minimum saturation irradiance in the maricultivated macroalga *P. haitanensis* that was grown under ambient conditions. However, such suppression by Cd^{2+} was much decreased when the macroalga was cultured under elevated $p\text{CO}_2$ conditions (Ma et al., 2018). The enhanced antioxidant ability in the microalgae might be attributed to the increased carotenoid contents as a stress response to Cd^{2+} (Strzałka et al., 2003). In addition, antagonistic effects of OA and heavy metal pollution were also found in some marine animals. Notably, negative effects of metals (Cu and Cd) on reactive oxygen species production were mitigated by low pH in the marine bivalves *C. virginica* and *M. mercenaria*, which has potentially affected the mitochondrial function (including mitochondrial coupling and ATP synthesis capacity) by influencing the metal transport and binding (Ivanina et al., 2015; Ivanina & Sokolova, 2013). OA can alleviate the toxicity of Cu and Zn to inhibit the fertilization of the oyster *Crassostrea angulate* (Guo et al., 2020). Antagonistic toxicities were observed upon exposure to the combination of OA and the heavy metal Cu or Cd toward the mortality rates of the coastal harpacticoid copepods *Amphiascoides atopus* (Pascal et al., 2010). In the two closely related oyster species *C. angulata* and *C. gigas*, the increased glutathione-S-transferase activity induced by arsenic exposure was suppressed under the condition of OA and arsenic combined, indicating an antagonistic effect between OA and warming (Moreira et al., 2016). *Hediste diversicolor* was negatively and antagonistically affected by exposure to conditions of Hg and low pH, which caused oxidative stress and metabolic changes in the polychaetes of the organism (Freitas et al., 2017). Similarly, OA also alleviated the Hg toxicity toward reproductive performance (inhibitory effects on a number of nauplii/clutch and fecundity) of the marine copepod *Tigriopus japonicus* (Li et al., 2017). In the larvae of the flounder *P. olivaceus*, OA not only affected directly the antioxidant defenses (catalase activity and glutathione peroxidase activity) but also antagonistically interacted with Cd to regulate these defense responses (Cui et al., 2020). Under CO_2-driven OA conditions, the mitigated toxicity of Cd induced oxidative stress and inhibited growth in the marine diatom *S. costatum* (Dong, Wang, et al., 2020). Long-term exposure to CO_2-driven acidification mitigated the as toxicity on reproduction of the harpacticoid copepod *T. japonicus*, which could be explained by the enhanced lysosome—autophagy pathway proteomes that are responsible for repairing and removing damaged proteins and enzymes under stress

(Wang et al., 2017). As the most important intracellular process used by eukaryotic cells to sequester and degrade cytoplasm portions and organelles via the lysosomal pathway, autophagy is one of the most useful pathways in the defense against cadmium toxicity (Bargagli, 2000; Chiarelli et al., 2016; Klionsky & Emr, 2000) and oxidative stress (Moore et al., 2006) (Moore, Allen). Although the lysosome—autophagy pathway was not observed in the other studies mentioned before, it could be a potential mechanism leading to the alleviation of the oxidative stress of OA on heavy metal toxicity.

There are also a number of studies showing no interactions between the two stressors, namely OA and heavy metal pollution. For example, independent exposure to high pCO_2 and high Zn concentration altered malondialdehyde (MDA) concentrations in the red alga *P. yezoensis*, but no interactions between the two stressors were detected (Ma et al., 2020). OA and nano-ZnO in tandem might result in enhanced antioxidant enzyme activities in the mussels *M. coruscus*, but no interactions between the stressors were detected (Huang, Liu, et al., 2018). No significant interactive effect between pH and copper was found on superoxide dismutase activity, lipid peroxidation, and DNA damage in the king ragworm *Alitta virens* (Nielson et al., 2019). OA additively increases copper toxicity toward the respiration rate, activities of antioxidant enzymes (superoxide dismutase and glutathione transferase), lipid peroxidation, and hexokinase activities in the Pacific Oyster *C. gigas* (Cao et al., 2019). Interactions between OA and Cu were not detected in clams *M. mercenaria* and oysters *C. virginica*, although exposure to Cu stimulated the key immune parameters of the two species, including an increased number of circulating hemocytes, higher phagocytosis and adhesion ability of hemocytes, and enhanced antiparasitic and antibacterial properties of the hemolymph (Ivanina et al., 2016).

In the future, studies about interactions of OA and heavy metal pollution should focus on three critical physiological functions, namely acid—base regulation, oxidative stress, and energy homeostasis, in ecologically and economically important marine species.

References

Aalto, E. A., Barry, J. P., Boch, C. A., Litvin, S. Y., Micheli, F., Woodson, C. B., & De Leo, G. A. D. (2020). Abalone populations are most sensitive to environmental stress effects on adult individuals. *Marine Ecology Progress Series, 643*, 75—85. Available from https://doi.org/10.3354/meps13320.

Adiele, R. C., Stevens, D., & Kamunde, C. (2011). Cadmium- and calcium-mediated toxicity in rainbow trout (*Oncorhynchus mykiss*) in vivo: Interactions on fitness and mitochondrial endpoints. *Chemosphere, 85*(10), 1604–1613. Available from https://doi.org/10.1016/j.chemosphere.2011.08.007.

Allen, R., Hoffmann, L. J., Law, C. S., & Summerfield, T. C. (2020). Subtle bacterioplankton community responses to elevated CO_2 and warming in the oligotrophic South Pacific gyre. *Environmental Microbiology Reports, 12*(4), 377–386. Available from https://doi.org/10.1111/1758-2229.12844.

Alma, L., Kram, K. E., Holtgrieve, G. W., Barbarino, A., Fiamengo, C. J., & Padilla-Gamiño, J. L. (2020). Ocean acidification and warming effects on the physiology, skeletal properties, and microbiome of the purple-hinge rock scallop. *Comparative Biochemistry and Physiology Part A: Molecular & Integrative Physiology, 240*, 110579. Available from https://doi.org/10.1016/j.cbpa.2019.110579.

An, D., Husovic, A., Ali, L., Weddle, E., Nagle, L., & Ahearn, G. A. (2019). Ocean acidification: Synergistic inhibitory effects of protons and heavy metals on ^{45}Ca uptake by lobster branchiostegite membrane vesicles. *Journal of Comparative Physiology B: Biochemical, Systemic, and Environmental Physiology, 189*(5), 513–521. Available from https://doi.org/10.1007/s00360-019-01227-7.

Andersson, A. J., Bates, N. R., & Mackenzie, F. T. (2007). Dissolution of carbonate sediments under rising pCO_2 and ocean acidification: Observations from Devil's Hole, Bermuda. *Aquatic Geochemistry, 13*(3), 237–264. Available from https://doi.org/10.1007/s10498-007-9018-8.

Araújoa, J. E., Madeira, D., Vitorino, R., Repolho, T., Rosa, R., & Diniz, M. (2018). Negative synergistic impacts of ocean warming and acidification on the survival and proteome of the commercial sea bream, *Sparus aurata. Journal of Sea Research, 139*, 50–61. Available from https://doi.org/10.1016/j.seares.2018.06.011.

Bahr, K. D., Tran, T., Jury, C. P., & Toonen, R. J. (2020). Abundance, size, and survival of recruits of the reef coral *Pocillopora acuta* under ocean warming and acidification. *PLoS One, 15*(2), e0228168. Available from https://doi.org/10.1371/journal.pone.0228168.

Bailey, R. M., & van der Grient, J. M. A. (2020). OSIRIS: A model for integrating the effects of multiple stressors on marine ecosystems. *Journal of Theoretical Biology, 493*, 110211. Available from https://doi.org/10.1016/j.jtbi.2020.110211.

Bargagli, R. (2000). Trace metals in Antarctica related to climate change and increasing human impact. *Reviews of Environmental Contamination and Toxicology, 166*, 129–173.

Barnes, R. S. K. (1967). Osmotic behaviour of a number of grapsoid crabs with respect to their differential penetration of an estuarine system. *Journal of Experimental Biology, 47* (3), 535–551.

Baumann, H., Wallace, R. B., Tagliaferri, T., & Gobler, C. J. (2014). Large natural pH, CO_2 and O_2 fluctuations in a temperate tidal salt marsh on diel, seasonal, and interannual time scales. *Estuaries and Coasts, 38*(1), 220–231. Available from https://doi.org/10.1007/s12237-014-9800-y.

Bausch, A. R., Juhl, A. R., Donaher, N. A., & Cockshutt, A. M. (2019). Combined effects of simulated acidification and hypoxia on the harmful dinoflagellate *Amphidinium carterae. Marine Biology, 166*(6), 80. Available from https://doi.org/10.1007/s00227-019-3528-y.

Bautista-Chamizo, E., Sendra, M., Cid, Á., Seoane, M., Romano de Orte, M., & Riba, I. (2018). Will temperature and salinity changes exacerbate the effects of seawater acidification on the marine microalga *Phaeodactylum tricornutum? Science of the Total Environment, 634*, 87–94. Available from https://doi.org/10.1016/j.scitotenv.2018.03.314.

Behbehani, M., Uddin, S., Dupont, S., Sajid, S., Al-Musalam, L., & Al-Ghadban, A. (2019). Response of corals acropora pharaonis and porites lutea to changes in pH and temperature in the gulf. *Sustainability (Switzerland), 11*(3156), 1–7. Available from https://doi.org/10.3390/su11113156.

Bennett, H. M., Altenrath, C., Woods, L., Davy, S. K., Webster, N. S., & Bell, J. J. (2017). Interactive effects of temperature and pCO_2 on sponges: From the cradle to the grave. *Global Change Biology, 23*(5), 2031–2046. Available from https://doi.org/10.1111/gcb.13474.

Bertin, G., & Averbeck, D. (2006). Cadmium: Cellular effects, modifications of biomolecules, modulation of DNA repair and genotoxic consequences (a review). *Biochimie, 88*(11), 1549–1559. Available from https://doi.org/10.1016/j.biochi.2006.10.001.

Bielmyer-Fraser, G. K., Patel, P., Capo, T., & Grosell, M. (2018). Physiological responses of corals to ocean acidification and copper exposure. *Marine Pollution Bulletin, 133*, 781–790. Available from https://doi.org/10.1016/j.marpolbul.2018.06.048.

Blindheim, J., Borovkov, V., Hansen, B., Malmberg, S. A., Turrell, W. R., & Østerhus, S. (2000). Upper layer cooling and freshening in the Norwegian Sea in relation to atmospheric forcing. *Deep-Sea Research Part I: Oceanographic Research Papers, 47*(4), 655–680. Available from https://doi.org/10.1016/S0967-0637(99)00070-9.

Boch, C. A., Litvin, S. Y., Micheli, F., De Leo, G., Aalto, E. A., Lovera, C., Woodson, C. B., Monismith, S., & Barry, J. P. (2017). Effects of current and future coastal upwelling conditions on the fertilization success of the red abalone (*Haliotis rufescens*). *ICES Journal of Marine Science, 74*(4), 1125–1134. Available from https://doi.org/10.1093/icesjms/fsx017.

Bopp, L., Resplandy, L., Orr, J. C., Doney, S. C., Dunne, J. P., Gehlen, M., Halloran, P., Heinze, C., Ilyina, T., Séférian, R., Tjiputra, J., & Vichi, M. (2013). Multiple stressors of ocean ecosystems in the 21st century: Projections with CMIP5 models. *Biogeosciences, 10*, 6225–6245. Available from https://doi.org/10.5194/bg-10-6225-2013.

Bouquet, J. M., Troedsson, C., Novac, A., Reeve, M., Lechtenbörger, A. K., Massart, W., Skaar, K. S., Aasjord, A., Dupont, S., & Thompson, E. M. (2018). Increased fitness of a key appendicularian zooplankton species under warmer, acidified seawater conditions. *PLoS One, 13*(1), e0190625. Available from https://doi.org/10.1371/journal.pone.0190625.

Boyd, P. W., & Hutchins, D. A. (2012). Understanding the responses of ocean biota to a complex matrix of cumulative anthropogenic change. *Marine Ecology Progress Series, 470*, 125–135. Available from https://doi.org/10.3354/meps10121.

Braby, C. E., & Somero, G. N. (2006). Following the heart: Temperature and salinity effects on heart rate in native and invasive species of blue mussels (genus *Mytilus*). *Journal of Experimental Biology, 209*(13), 2554–2566. Available from https://doi.org/10.1242/jeb.02259.

Brothers, C. J., Harianto, J., McClintock, J. B., & Byrne, M. (2016). Sea urchins in a high-CO_2 world: The influence of acclimation on the immune response to ocean warming and acidification. *Proceedings of the Royal Society B: Biological Sciences, 283* (1837). Available from https://doi.org/10.1098/rspb.2016.1501.

Brown, J., Whiteley, N. M., Bailey, A. M., Graham, H., Hop, H., & Rastrick, S. P. S. (2020). Contrasting responses to salinity and future ocean acidification in arctic populations of the amphipod *Gammarus setosus. Marine Environmental Research, 162*, 105176. Available from https://doi.org/10.1016/j.marenvres.2020.105176.

Byrne, M., Ho, M., Selvakumaraswamy, P., Nguyen, H. D., Dworjanyn, S. A., & Davis, A. R. (2009). Temperature, but not pH, compromises sea urchin fertilization and early development under near-future climate change scenarios. *Proceedings of the Royal Society B: Biological Sciences, 276*(1663), 1883–1888. Available from https://doi.org/10.1098/rspb.2008.1935.

Cabrerizo, M. J., Álvarez-Manzaneda, M. I., León-Palmero, E., Guerrero-Jiménez, G., de Senerpont Domis, L. N., Teurlincx, S., & González-Olalla, J. M. (2020). Warming and CO_2 effects under oligotrophication on temperate phytoplankton communities. *Water Research*, *173*, 115579. Available from https://doi.org/10.1016/j.watres.2020.115579.

Cai, W. J., Hu, X., Huang, W. J., Murrell, M. C., Lehrter, J. C., Lohrenz, S. E., Chou, W. C., Zhai, W., Hollibaugh, J. T., Wang, Y., Zhao, P., Guo, X., Gundersen, K., Dai, M., & Gong, G. C. (2011). Acidification of subsurface coastal waters enhanced by eutrophication. *Nature Geoscience*, *4*(11), 766−770. Available from https://doi.org/10.1038/ngeo1297.

Caldeira, K., & Wickett, M. (2005). Ocean model predictions of chemistry changes from carbon dioxide emissions to the atmosphere and ocean. *Journal of Geophysical Research C: Oceans*, *110*(9), 1−12. Available from https://doi.org/10.1029/2004JC002671.

Campanati, C., Yip, S., Lane, A., & Thiyagarajan, V. (2016). Combined effects of low pH and low oxygen on the early-life stages of the barnacle *Balanus amphitrite*. *ICES Journal of Marine Science*, *73*(3), 791−802. Available from https://doi.org/10.1093/icesjms/fsv221.

Campbell, A. L., Mangan, S., Ellis, R. P., & Lewis, C. (2014). Ocean acidification increases copper toxicity to the early life history stages of the polychaete arenicola marina in artificial seawater. *Environmental Science & Technology*, *48*(16), 9745−9753. Available from https://doi.org/10.1021/es502739m.

Cao, R., Liu, Y., Wang, Q., Dong, Z., Yang, D., Liu, H., Ran, W., Qu, Y., & Zhao, J. (2018). Seawater acidification aggravated cadmium toxicity in the oyster *Crassostrea gigas*: Metal bioaccumulation, subcellular distribution and multiple physiological responses. *Science of the Total Environment*, *642*, 809−823. Available from https://doi.org/10.1016/j.scitotenv.2018.06.126.

Cao, R., Zhang, T., Li, X., Zhao, Y., Wang, Q., Yang, D., Qu, Y., Liu, H., Dong, Z., & Zhao, J. (2019). Seawater acidification increases copper toxicity: A multi-biomarker approach with a key marine invertebrate, the Pacific Oyster *Crassostrea gigas*. *Aquatic Toxicology*, *210*, 167−178. Available from https://doi.org/10.1016/j.aquatox.2019.03.002.

Carey, N., Harianto, J., & Byrne, M. (2016). Sea urchins in a high-CO_2 world: Partitioned effects of body size, ocean warming and acidification on metabolic rate. *Journal of Experimental Biology*, *219*(8), 1178−1186. Available from https://doi.org/10.1242/jeb.136101.

Carregosa, V., Figueira, E., Gil, A. M., Pereira, S., Pinto, J., Soares, A. M. V. M., & Freitas, R. (2014). Tolerance of *Venerupis philippinarum* to salinity: Osmotic and metabolic aspects. *Comparative Biochemistry and Physiology Part A Molecular & Integrative Physiology*, *171*, 36−43. Available from https://doi.org/10.1016/j.cbpa.2014.02.009.

Castellano, G. C., Souza, M. M., & Freire, C. A. (2016). Volume regulation of intestinal cells of echinoderms: Putative role of ion transporters (Na^+/K^+-ATPase and NKCC). *Comparative Biochemistry and Physiology Part A: Molecular & Integrative Physiology*, *201*, 124−131. Available from https://doi.org/10.1016/j.cbpa.2016.07.006.

Celis-Plá, P. S. M., Martínez, B., Korbee, N., Hall-Spencer, J. M., & Figueroa, F. L. (2017). Ecophysiological responses to elevated CO_2 and temperature in *Cystoseira tamariscifolia* (Phaeophyceae). *Climatic Change*, *142*(1−2), 67−81. Available from https://doi.org/10.1007/s10584-017-1943-y.

Chan, W. Y., & Eggins, S. M. (2017). Calcification responses to diurnal variation in seawater carbonate chemistry by the coral *Acropora formosa*. *Coral Reefs*, *36*(3), 763−772. Available from https://doi.org/10.1007/s00338-017-1567-8.

Chiarelli, R., Martino, C., Agnello, M., Bosco, L., & Roccheri, M. C. (2016). Autophagy as a defense strategy against stress: Focus on *Paracentrotus lividus* sea urchin embryos exposed to cadmium. *Cell Stress and Chaperones, 21*(1), 19−27. Available from https://doi.org/10.1007/s12192-015-0639-3.

Clark, H. R., & Gobler, C. J. (2016). Diurnal fluctuations in CO_2 and dissolved oxygen concentrations do not provide a refuge from hypoxia and acidification for early-life-stage bivalves. *Marine Ecology Progress Series, 558*, 1−14. Available from https://doi.org/10.3354/meps11852.

Cline, A. J., Hamilton, S. L., & Logan, C. A. (2020). Effects of multiple climate change stressors on gene expression in blue rockfish (*Sebastes mystinus*). *Comparative Biochemistry and Physiology Part A: Molecular & Integrative Physiology, 239*, 110580. Available from https://doi.org/10.1016/j.cbpa.2019.110580.

Coello-Camba, A., Agustí, S., Holding, J., Arrieta, J. M., & Duarte, C. M. (2014). Interactive effect of temperature and CO_2 increase in Arctic phytoplankton. *Frontiers in Marine Science, 1*, 49. Available from https://doi.org/10.3389/fmars.2014.00049.

Cominassi, L., Moyano, M., Claireaux, G., Howald, S., Mark, F. C., Zambonino-Infante, J. L., Le Bayon, N., & Peck, M. A. (2019). Combined effects of ocean acidification and temperature on larval and juvenile growth, development and swimming performance of European sea bass (*Dicentrarchus labrax*). *PLoS One, 14*(9). Available from https://doi.org/10.1371/journal.pone.0221283.

Cominassi, L., Moyano, M., Claireaux, G., Howald, S., Mark, F. C., Zambonino-Infante, J. L., & Peck, M. A. (2020). Food availability modulates the combined effects of ocean acidification and warming on fish growth. *Scientific Reports, 10*(1), 2338. Available from https://doi.org/10.1038/s41598-020-58846-2.

Connell, S. D., & Russell, B. D. (2010). The direct effects of increasing CO_2 and temperature on non-calcifying organisms: Increasing the potential for phase shifts in kelp forests. *Proceedings of the Royal Society B: Biological Sciences, 277*(1686), 1409−1415. Available from https://doi.org/10.1098/rspb.2009.2069.

Coteur, G., Gillan, D., Pernet, P., & Dubois, P. (2005). Alteration of cellular immune responses in the seastar *Asterias rubens* following dietary exposure to cadmium. *Aquatic Toxicology, 73*(4), 418−421. Available from https://doi.org/10.1016/j.aquatox.2005.04.003.

Crain, C. M., Kroeker, K., & Halpern, B. S. (2008). Interactive and cumulative effects of multiple human stressors in marine systems. *Ecology Letters, 11*(12), 1304−1315. Available from https://doi.org/10.1111/j.1461-0248.2008.01253.x.

Cross, E. L., Murray, C. S., & Baumann, H. (2019). Diel and tidal pCO_2 × O_2 fluctuations provide physiological refuge to early life stages of a coastal forage fish. *Scientific Reports, 9*(1), 18146. Available from https://doi.org/10.1038/s41598-019-53930-8.

Cui, W., Cao, L., Liu, J., Ren, Z., Zhao, B., & Dou, S. (2020). Effects of seawater acidification and cadmium on the antioxidant defense of flounder *Paralichthys olivaceus* larvae. *Science of the Total Environment, 718*, 137234. Available from https://doi.org/10.1016/j.scitotenv.2020.137234.

Darling, E. S., & Côté, I. M. (2008). Quantifying the evidence for ecological synergies. *Ecology Letters, 11*(12), 1278−1286. Available from https://doi.org/10.1111/j.1461-0248.2008.01243.x.

Das, S., Raj, R., Mangwani, N., Dash, H. R., & Chakraborty, J. (2014). *Heavy metals and hydrocarbons: Adverse effects and mechanism of toxicity. Microbial biodegradation and bioremediation* (pp. 24−54). Elsevier Inc. Available from https://doi.org/10.1016/B978-0-12-800021-2.00002-9.

Dauer, D. M. (1993). Biological criteria, environmental health and estuarine macrobenthic community structure. *Marine Pollution Bulletin, 26*(5), 249−257. Available from https://doi.org/10.1016/0025-326X(93)90063-P.

Davidson, M. I., Targett, T. E., & Grecay, P. A. (2016). Evaluating the effects of diel-cycling hypoxia and pH on growth and survival of juvenile summer flounder

Paralichthys dentatus. *Marine Ecology Progress Series*, *556*, 223−235. Available from https://doi.org/10.3354/meps11817.

Davis, B. E., Flynn, E. E., Miller, N. A., Nelson, F. A., Fangue, N. A., & Todgham, A. E. (2018). Antarctic emerald rockcod have the capacity to compensate for warming when uncoupled from CO_2-acidification. *Global Change Biology*, *24*(2), e655−e670. Available from https://doi.org/10.1111/gcb.13987.

de los Santos, C. B., Arenas, F., Neuparth, T., & Santos, M. M. (2019). Interaction of short-term copper pollution and ocean acidification in seagrass ecosystems: Toxicity, bioconcentration and dietary transfer. *Marine Pollution Bulletin*, *142*, 155−163. Available from https://doi.org/10.1016/j.marpolbul.2019.03.034.

de Orte, M. R., Lombardi, A. T., Sarmiento, A. M., Basallote, M. D., Rodriguez-Romero, A., Riba, I., & Del Valls, A. (2014). Metal mobility and toxicity to microalgae associated with acidification of sediments: CO_2 and acid comparison. *Marine Environmental Research*, *96*, 136−144. Available from https://doi.org/10.1016/j.marenvres.2013.10.003.

de Orte, M. R., Sarmiento, A. M., Basallote, M. D., Rodríguez-Romero, A., Riba, I., & delValls, A. (2014). Effects on the mobility of metals from acidification caused by possible CO_2 leakage from sub-seabed geological formations. *Science of the Total Environment*, *470−471*, 356−363. Available from https://doi.org/10.1016/j.scitotenv.2013.09.095.

Deutsch, C., Ferrel, A., Seibel, B., Portner, H.-O., & Huey, R. B. (2015). Climate change tightens a metabolic constraint on marine habitats. *Science (New York, N.Y.)*, *348* (6239), 1132−1135. Available from https://doi.org/10.1126/science.aaa1605.

Di Santo, V. (2019). Ocean acidification and warming affect skeletal mineralization in a marine fish. *Proceedings of the Royal Society B: Biological Sciences*, *286*(1894). Available from https://doi.org/10.1098/rspb.2018.2187.

Diaz, R. J., & Rosenberg, R. (1995). Marine benthic hypoxia: A review of its ecological effects and the behavioural responses of benthic macrofauna. *Oceanography and Marine Biology: An Annual Review*, *33*, 245−303.

Diaz, R. J., & Rosenberg, R. (2008). Spreading dead zones and consequences for marine ecosystems. *Science (New York, N.Y.)*, *321*(5891), 926−929. Available from https://doi.org/10.1126/science.1156401.

Dickinson, G. H., Matoo, O. B., Tourek, R. T., Sokolova, I. M., & Beniash, E. (2013). Environmental salinity modulates the effects of elevated CO_2 levels on juvenile hard-shell clams, *Mercenaria mercenaria*. *Journal of Experimental Biology*, *216*(14), 2607−2618. Available from https://doi.org/10.1242/jeb.082909.

Dickson, B., Yashayaev, I., Meincke, J., Turrell, B., Dye, S., & Holfort, J. (2002). Rapid freshening of the deep North Atlantic Ocean over the past four decades. *Nature*, *416* (6883), 832−837. Available from https://doi.org/10.1038/416832a.

Dietmar, K. (2007). Osmotic stress sensing and signaling in animals. *FEBS Journal*, *274* (22), 5781. Available from https://doi.org/10.1111/j.1742-4658.2007.06097.x.

Dong, F., Wang, P., Qian, W., Tang, X., Zhu, X., Wang, Z., Cai, Z., & Wang, J. (2020). Mitigation effects of CO_2-driven ocean acidification on Cd toxicity to the marine diatom *Skeletonema costatum*. *Environmental Pollution*, *259*, 113850. Available from https://doi.org/10.1016/j.envpol.2019.113850.

Dong, F., Zhu, X., Qian, W., Wang, P., & Wang, J. (2020). Combined effects of CO_2-driven ocean acidification and Cd stress in the marine environment: Enhanced tolerance of *Phaeodactylum tricornutum* to Cd exposure. *Marine Pollution Bulletin*, *150*, 110594. Available from https://doi.org/10.1016/j.marpolbul.2019.110594.

Dong, Z., & Sun, T. (2018). Combined effects of ocean acidification and temperature on planula larvae of the moon jellyfish *Aurelia coerulea*. *Marine Environmental Research*, *139*, 144−150. Available from https://doi.org/10.1016/j.marenvres.2018.05.015.

Dorey, N., Maboloc, E., & Chan, K. Y. K. (2018). Development of the sea urchin *Heliocidaris crassispina* from Hong Kong is robust to ocean acidification and copper contamination. *Aquatic Toxicology*, *205*, 1−10. Available from https://doi.org/10.1016/j.aquatox.2018.09.006.

Duarte, C., Navarro, J. M., Quijón, P. A., Loncon, D., Torres, R., Manríquez, P. H., Lardies, M. A., Vargas, C. A., & Lagos, N. A. (2018). The energetic physiology of juvenile mussels, *Mytilus chilensis* (Hupe): The prevalent role of salinity under current and predicted pCO_2 scenarios. *Environmental Pollution*, *242*(Pt A), 156−163. Available from https://doi.org/10.1016/j.envpol.2018.06.053.

Egilsdottir, H., Spicer, J. I., & Rundle, S. D. (2009). The effect of CO_2 acidified sea water and reduced salinity on aspects of the embryonic development of the amphipod *Echinogammarus marinus* (Leach). *Marine Pollution Bulletin*, *58*(8), 1187−1191. Available from https://doi.org/10.1016/j.marpolbul.2009.03.017.

Enzor, L. A., Hankins, C., Hamilton-Frazier, M., Moso, E., Raimondo, S., & Barron, M. G. (2020). Elevated pCO_2 and hypoxia alter the acid − base regulation of developing sheepshead minnows *Cyprinodon variegatus*. *Marine Ecology Progress Series*, *636*, 157−168. Available from https://doi.org/10.3354/meps13220.

Feely, R. A., Alin, S., Sabine, C. L., & Newton, J. (2011). The combined effects of ocean acidification, mixing, and respiration on pH and carbonate saturation in an urbanized estuary. *Journal of Phycology*, *47*, S2.

Fitzer, S. C., Caldwell, G. S., Clare, A. S., Upstill-Goddard, R. C., & Bentley, M. G. (2013). Response of copepods to elevated pCO_2 and environmental copper as co-stressors - A multigenerational study. *PLoS One*, *8*(8), e71257. Available from https://doi.org/10.1371/journal.pone.0071257.

Flynn, E. E., Bjelde, B. E., Miller, N. A., & Todgham, A. E. (2015). Ocean acidification exerts negative effects during warming conditions in a developing Antarctic fish. *Conservation Physiology*, *3*(1), cov033. Available from https://doi.org/10.1093/conphys/cov033.

Fonseca, V., Serafim, A., Company, R., Bebianno, M. J., & Cabral, H. (2009). Effect of copper exposure on growth, condition indices and biomarker response in juvenile sole *Solea senegalensis*. *Scientia Marina*, *73*(1), 51−58. Available from https://doi.org/10.3989/scimar.2009.73n1051.

Fontanini, A., Steckbauer, A., Dupont, S., & Duarte, C. M. (2018). Variable metabolic responses of Skagerrak invertebrates to low O_2 and high CO_2 scenarios. *Biogeosciences*, *15*(12), 3717−3729. Available from https://doi.org/10.5194/bg-15-3717-2018.

Franklin, C. (2008). Climate change and conservation physiology. *Comparative Biochemistry and Physiology Part A: Molecular & Integrative Physiology*, *150*(3 suppl), S167. Available from https://doi.org/10.1016/j.cbpa.2008.04.438.

Franzova, V. A., MacLeod, C. D., Wang, T., & Harley, C. D. G. (2019). Complex and interactive effects of ocean acidification and warming on the life span of a marine trematode parasite. *International Journal for Parasitology*, *49*(13−14), 1015−1021. Available from https://doi.org/10.1016/j.ijpara.2019.07.005.

Freitas, R., de Marchi, L., Moreira, A., Pestana, J. L. T., Wrona, F. J., Figueira, E., & Soares, A. M. V. M. (2017). Physiological and biochemical impacts induced by mercury pollution and seawater acidification in *Hediste diversicolor*. *Science of the Total Environment*, *595*, 691−701. Available from https://doi.org/10.1016/j.scitotenv.2017.04.005.

Freitas, R., Pires, A., Velez, C., Almeida, Â., Wrona, F. J., Soares, A. M. V. M., & Figueira, E. (2015). The effects of salinity changes on the Polychaete *Diopatra neapolitana*: Impacts on regenerative capacity and biochemical markers. *Aquatic Toxicology*, *163*, 167−176. Available from https://doi.org/10.1016/j.aquatox.2015.04.006.

Friedlingstein, P., Jones, M. W., O'Sullivan, M., Andrew, R. M., Hauck, J., Peters, G. P., Peters, W., Pongratz, J., Sitch, S., Le Quéré, C., DBakker, O. C. E., Canadell1, J. G., Ciais1, P., Jackson, R. B., Anthoni1, P., Barbero, L., Bastos, A., Bastrikov, V., Becker, M., & Zaehle, S. (2019). Global carbon budget 2019. *Earth System Science Data*, *11*(4), 1783−1838. Available from https://doi.org/10.5194/essd-11-1783-2019.

Gao, G., Liu, Y., Li, X., Feng, Z., Xu, Z., Wu, H., & Xu, J. (2017). Expected CO_2-induced ocean acidification modulates copper toxicity in the green tide alga *Ulva prolifera*. *Environmental and Experimental Botany*, *135*, 63−72. Available from https://doi.org/10.1016/j.envexpbot.2016.12.007.

Gao, G., Qu, L., Xu, T., Burgess, J. G., Li, X., Xu, J., & Norkko, J. (2019). Future CO_2-induced ocean acidification enhances resilience of a green tide alga to low-salinity stress. *ICES Journal of Marine Science*, *76*(7), 2437−2445. Available from https://doi.org/10.1093/icesjms/fsz135.

Gao, G., Shi, Q., Xu, Z., Xu, J., Campbell, D. A., & Wu, H. (2018). Global warming interacts with ocean acidification to alter PSII function and protection in the diatom *Thalassiosira weissflogii*. *Environmental and Experimental Botany*, *147*, 95−103. Available from https://doi.org/10.1016/j.envexpbot.2017.11.014.

Gao, X., Zhou, F., & Chen, C. T. A. (2014). Pollution status of the Bohai Sea: An overview of the environmental quality assessment related trace metals. *Environment International*, *62*, 12−30. Available from https://doi.org/10.1016/j.envint.2013.09.019.

García, E., Clemente, S., & Hernández, J. C. (2015). Ocean warming ameliorates the negative effects of ocean acidification on *Paracentrotus lividus* larval development and settlement. *Marine Environmental Research*, *110*, 61−68. Available from https://doi.org/10.1016/j.marenvres.2015.07.010.

Gardner, J., Manno, C., Bakker, D. C. E., Peck, V. L., & Tarling, G. A. (2018). Southern Ocean pteropods at risk from ocean warming and acidification. *Marine Biology*, *165*(1). Available from https://doi.org/10.1007/s00227-017-3261-3.

Garzke, J., Hansen, T., Ismar, S. M. H., & Sommer, U. (2016). Combined effects of ocean warming and acidification on copepod abundance, body size and fatty acid content. *PLoS One*, *11*(5). Available from https://doi.org/10.1371/journal.pone.0155952.

Gattuso, J.-P., Magnan, A., Billé, R., Cheung, W. W. L., Howes, E. L., Joos, F., Allemand, D., Bopp, L., Cooley, S. R., Eakin, C. M., Hoegh-Guldberg, O., Kelly, R. P., Pörtner, H.-O., Rogers, A. D., Baxter, J. M., Laffoley, D., Osborn, D., Rankovic, A., Rochette, J., Sumaila, U. R., Treyer, S., & Turley, C. (2015). Contrasting futures for ocean and society from different anthropogenic CO_2 emissions scenarios. *Science (New York, N.Y.)*, *349*(6243), aac4722. Available from https://doi.org/10.1126/science.aac4722.

Gibbin, E. M., Chakravarti, L. J., Jarrold, M. D., Christen, F., Turpin, V., N'Siala, G. M., Blier, P. U., & Calosi, P. (2017). Can multi-generational exposure to ocean warming and acidification lead to the adaptation of life history and physiology in a marine metazoan? *Journal of Experimental Biology*, *220*(4), 551−563. Available from https://doi.org/10.1242/jeb.149989.

Gilles, R. (1987). Volume regulation in cells of euryhaline invertebrates. *Current Topics in Membranes and Transport*, *30*, 205−247. Available from https://doi.org/10.1016/S0070-2161(08)60372-X.

Gobler, C. J., DePasquale, E. L., Griffith, A. W., & Baumann, H. (2014). Hypoxia and acidification have additive and synergistic negative effects on the growth, survival, and metamorphosis of early life stage bivalves. *PLoS One*, *9*(1). Available from https://doi.org/10.1371/journal.pone.0083648.

Goncalves, P., Thompson, E. L., & Raftos, D. A. (2017). Contrasting impacts of ocean acidification and warming on the molecular responses of CO_2-resilient oysters. *BMC Genomics*, *18*(1), 431. Available from https://doi.org/10.1186/s12864-017-3818-z.

Götze, S., Matoo, O. B., Beniash, E., Saborowski, R., & Sokolova, I. M. (2014). Interactive effects of CO_2 and trace metals on the proteasome activity and cellular stress response of marine bivalves *Crassostrea virginica* and *Mercenaria mercenaria*. *Aquatic Toxicology*, *149*, 65−82. Available from https://doi.org/10.1016/j.aquatox.2014.01.027.

Graham, H., Rastrick, S. P. S., Findlay, H. S., Bentley, M. G., Widdicombe, S., Clare, A. S., & Caldwell, G. S. (2016). Sperm motility and fertilisation success in an acidified and hypoxic environment. *ICES Journal of Marine Science*, *73*(3), 783−790. Available from https://doi.org/10.1093/icesjms/fsv171.

Gray, J. S., Wu, R. S. S., & Ying, Y. O. (2002). Effects of hypoxia and organic enrichment on the coastal marine environment. *Marine Ecology Progress Series*, *238*, 249−279. Available from https://doi.org/10.3354/meps238249.

Grenier, C., Román, R., Duarte, C., Navarro, J. M., Rodriguez-Navarro, A. B., & Ramajo, L. (2020). The combined effects of salinity and pH on shell biomineralization of the edible mussel *Mytilus chilensis*. *Environmental Pollution*, *263*, 114555. Available from https://doi.org/10.1016/j.envpol.2020.114555.

Grilo, T. F., Lopes, A. R., Sampaio, E., Rosa, R., & Cardoso, P. G. (2018). Sex differences in oxidative stress responses of tropical topshells (*Trochus histrio*) to increased temperature and high pCO_2. *Marine Pollution Bulletin*, *131*, 252−259. Available from https://doi.org/10.1016/j.marpolbul.2018.04.031.

Gu, H., Shang, Y., Clements, J., Dupont, S., Wang, T., Wei, S., Wang, X., Chen, J., Huang, W., Hu, M., & Wang, Y. (2019). Hypoxia aggravates the effects of ocean acidification on the physiological energetics of the blue mussel *Mytilus edulis*. *Marine Pollution Bulletin*, *149*, 110538. Available from https://doi.org/10.1016/j.marpolbul.2019.110538.

Guo, X., Huang, M., Shi, B., You, W., & Ke, C. (2020). Effects of ocean acidification on toxicity of two trace metals in two marine molluscs in their early life stages. *Aquaculture Environment Interactions*, *12*, 281−296. Available from https://doi.org/10.3354/aei00362.

Han, Z. X., Wu, D. D., Wu, J., Lv, C. X., & Liu, Y. R. (2014). Effects of ocean acidification on toxicity of heavy metals in the bivalve *Mytilus edulis* L. *Synthesis and Reactivity in Inorganic, Metal-Organic and Nano-Metal Chemistry*, *44*(1), 133−139. Available from https://doi.org/10.1080/15533174.2013.770753.

Hänsel, M. C., Schmidt, J. O., Stiasny, M. H., Stöven, M. T., Voss, R., & Quaas, M. F. (2020). Ocean warming and acidification may drag down the commercial Arctic cod fishery by 2100. *PLoS One*, *15*(4), e0231589. Available from https://doi.org/10.1371/journal.pone.0231589.

Harney, E., Artigaud, S., Le Souchu, P., Miner, P., Corporeau, C., Essid, H., Pichereau, V., & Nunes, F. L. D. (2016). Non-additive effects of ocean acidification in combination with warming on the larval proteome of the Pacific oyster, *Crassostrea gigas*. *Journal of Proteomics*, *135*, 151−161. Available from https://doi.org/10.1016/j.jprot.2015.12.001.

Harrington, A. M., & Hamlin, H. J. (2019). Ocean acidification alters thermal cardiac performance, hemocyte abundance, and hemolymph chemistry in subadult American lobsters *Homarus americanus* H. Milne Edwards, 1837 (Decapoda: Malcostraca: Nephropidae). *Journal of Crustacean Biology*, *39*(4), 468−476. Available from https://doi.org/10.1093/jcbiol/ruz015.

Harvey, B. P., Gwynn-Jones, D., & Moore, P. J. (2013). Meta-analysis reveals complex marine biological responses to the interactive effects of ocean acidification and warming. *Ecology and Evolution*, *3*(8), 2782. Available from https://doi.org/10.1002/ece3.728.

Henry, R. P., Lucu, C., Onken, H., & Weihrauch, D. (2012). Multiple functions of the crustacean gill: Osmotic/ionic regulation, acid-base balance, ammonia excretion, and bioaccumulation of toxic metals. *Frontiers in Physiology*, *3*, 431. Available from https://doi.org/10.3389/fphys.2012.00431.

Hernandez, J. C., Clemente, S., Garcia, E., & McAlister, J. S. (2020). Planktonic stages of the ecologically important sea urchin, *Diadema africanum*: Larval performance under near future ocean conditions. *Journal of Plankton Research*, *42*(3), 286—304. Available from https://doi.org/10.1093/plankt/fbaa016.

Hofmann, G. E., & Todgham, A. E. (2009). Living in the now: Physiological mechanisms to tolerate a rapidly changing environment. *Annual Review of Physiology*, *72*, 127—145. Available from https://doi.org/10.1146/annurev-physiol-021909-135900.

Hopkins, F. E., Suntharalingam, P., Gehlen, M., Andrews, O., Archer, S. D., Bopp, L., Buitenhuis, E., Dadou, I., Duce, R., Goris, N., Jickells, T., Johnson, M., Keng, F., Law, C. S., Lee, K., Liss, P. S., Lizotte, M., Malin, G., Murrell, J. C., & Williamson, P. (2020). The impacts of ocean acidification on marine trace gases and the implications for atmospheric chemistry and climate. *Proceedings of the Royal Society A: Mathematical, Physical and Engineering Sciences*, *476*(2237), 20190769. Available from https://doi.org/10.1098/rspa.2019.0769.

Howald, S., Cominassi, L., LeBayon, N., Claireaux, G., & Mark, F. C. (2019). Future ocean warming may prove beneficial for the northern population of European seabass, but ocean acidification will not. *Journal of Experimental Biology*, *222*(21), jeb213017. Available from https://doi.org/10.1242/jeb.213017.

Huang, X., Jiang, X., Sun, M., Dupont, S., Huang, W., Hu, M., Li, Q., & Wang, Y. (2018). Effects of copper on hemocyte parameters in the estuarine oyster *Crassostrea rivularis* under low pH conditions. *Aquatic Toxicology*, *203*, 61—68. Available from https://doi.org/10.1016/j.aquatox.2018.08.003.

Huang, X., Liu, Y., Liu, Z., Zhao, Z., Dupont, S., Wu, F., Huang, W., Chen, J., Hu, M., Lu, W., & Wang, Y. (2018). Impact of zinc oxide nanoparticles and ocean acidification on antioxidant responses of *Mytilus coruscus*. *Chemosphere*, *196*, 182—195. Available from https://doi.org/10.1016/j.chemosphere.2017.12.183.

IPCC (2014). *Climate Change 2014: Synthesis report. Contribution of working group I, II and III to the fifth assessment report of the intergovernmental panel on climate change* [Core Writing Team, R.K. Pachauri and L.A. Meyer (eds.)]. IPCC, Geneva, Switzerland, p. 151.

Ivanina, A. V., & Sokolova, I. M. (2013). Interactive effects of pH and metals on mitochondrial functions of intertidal bivalves *Crassostrea virginica* and *Mercenaria mercenaria*. *Aquatic Toxicology*, *144—145*, 303—309. Available from https://doi.org/10.1016/j.aquatox.2013.10.019.

Ivanina, A. V., Hawkins, C., & Sokolova, I. M. (2014). Immunomodulation by the interactive effects of cadmium and hypercapnia in marine bivalves *Crassostrea virginica* and *Mercenaria mercenaria*. *Fish & Shellfish Immunology*, *37*(2), 299—312. Available from https://doi.org/10.1016/j.fsi.2014.02.016.

Ivanina, A. V., Hawkins, C., & Sokolova, I. M. (2016). Interactive effects of copper exposure and environmental hypercapnia on immune functions of marine bivalves *Crassostrea virginica* and *Mercenaria mercenaria*. *Fish & Shellfish Immunology*, *49*, 54—65. Available from https://doi.org/10.1016/j.fsi.2015.12.011.

Ivanina, A. V., Hawkins, C., Beniash, E., & Sokolova, I. M. (2015). Effects of environmental hypercapnia and metal (Cd and Cu) exposure on acid-base and metal homeostasis of marine bivalves. *Comparative Biochemistry and Physiology Part C: Toxicology and Pharmacology*, *174—175*(1), 1—12. Available from https://doi.org/10.1016/j.cbpc.2015.05.001.

Jansson, A., Norkko, J., Dupont, S., & Norkko, A. (2015). Growth and survival in a changing environment: Combined effects of moderate hypoxia and low pH on

juvenile bivalve *Macoma balthica*. *Journal of Sea Research*, *102*, 41−47. Available from
https://doi.org/10.1016/j.seares.2015.04.006.

Jesus, T. F., Rosa, I. C., Repolho, T., Lopes, A. R., Pimentel, M. S., Almeida-Val,
V. M. F., Coelho, M. M., & Rosa, R. (2018). Different ecophysiological responses of
freshwater fish to warming and acidification. *Comparative Biochemistry and Physiology
-Part A: Molecular and Integrative Physiology*, *216*, 34−41. Available from https://doi.
org/10.1016/j.cbpa.2017.11.007.

Jezierska, B., èugowska, K., & Witeska, M. (2009). The effects of heavy metals on embryonic development of fish (a review). *Fish Physiology and Biochemistry*, *35*(4), 625−640.
Available from https://doi.org/10.1007/s10695-008-9284-4.

Jia, X. Y., He, L. H., Jing, R. L., & Li, R. Z. (2009). Calreticulin: Conserved protein and
diverse functions in plants. *Physiologia Plantarum*, *136*(2), 127−138. Available from
https://doi.org/10.1111/j.1399-3054.2009.01223.x.

Jiang, L., Guo, M. L., Zhang, F., Zhang, Y. Y., Zhou, G. W., Lei, X. M., Yuan, X. C.,
Sun, Y. F., Yuan, T., Cai, L., Lian, J. S., Liu, S., Qian, P. Y., & Huang, H. (2020).
Impacts of elevated temperature and pCO$_2$ on the brooded larvae of *Pocillopora damicornis* from Luhuitou Reef, China: Evidence for local acclimatization. *Coral Reefs*, *39*
(2), 331−344. Available from https://doi.org/10.1007/s00338-020-01894-x.

Jin, X., Liu, F., Wang, Y., Zhang, L., Li, Z., Wang, Z., Giesy, J. P., & Wang, Z. (2015).
Probabilistic ecological risk assessment of copper in Chinese offshore marine environments from 2005 to 2012. *Marine Pollution Bulletin*, *94*(1−2), 96−102. Available from
https://doi.org/10.1016/j.marpolbul.2015.03.005.

Johnson, M. D., & Carpenter, R. C. (2012). Ocean acidification and warming decrease
calcification in the crustose coralline alga *Hydrolithon onkodes* and increase susceptibility
to grazing. *Journal of Experimental Marine Biology and Ecology*, *434−435*, 94−101.
Available from https://doi.org/10.1016/j.jembe.2012.08.005.

Keppel, A. G., Breitburg, D. L., Wikfors, G. H., Burrell, R. B., & Clark, V. M. (2015).
Effects of co-varying diel-cycling hypoxia and pH on disease susceptibility in the eastern oyster *Crassostrea virginica*. *Marine Ecology Progress Series*, *538*, 169−183. Available
from https://doi.org/10.3354/meps11479.

Khan, F. U., Hu, M., Kong, H., Shang, Y., Wang, T., Wang, X., Xu, R., Lu, W., &
Wang, Y. (2020). Ocean acidification, hypoxia and warming impair digestive parameters of marine mussels. *Chemosphere*, *256*, 127096. Available from https://doi.org/
10.1016/j.chemosphere.2020.127096.

Kidwell, D. M., Lewitus, A. J., Brandt, S., Jewett, E. B., & Mason, D. M. (2009).
Ecological impacts of hypoxia on living resources. *Journal of Experimental Marine
Biology and Ecology*, *381*, S1−S3. Available from https://doi.org/10.1016/j.
jembe.2009.07.009.

Kim, J. H., Kim, N., Moon, H., Lee, S., Jeong, S. Y., Diaz-Pulido, G., Edwards, M. S.,
Kang, J. H., Kang, E. J., Oh, H. J., Hwang, J. D., & Kim, I. N. (2020). Global warming offsets the ecophysiological stress of ocean acidification on temperate crustose coralline algae. *Marine Pollution Bulletin*, *157*, 111324. Available from https://doi.org/
10.1016/j.marpolbul.2020.111324.

Kim, T. W., Barry, J. P., & Micheli, F. (2013). The effects of intermittent exposure to
low-pH and low-oxygen conditions on survival and growth of juvenile red abalone.
Biogeosciences, *10*(11), 7255−7262. Available from https://doi.org/10.5194/bg-10-
7255-2013.

Klionsky, D. J., & Emr, S. D. (2000). Autophagy as a regulated pathway of cellular degradation. *Science (New York, N.Y.)*, *290*(5497), 1717−1721. Available from https://doi.
org/10.1126/science.290.5497.1717.

Knapp, J. L., Auerswald, L., Hoffman, L. C., & Macey, B. M. (2019). Effects of chronic
hypercapnia and elevated temperature on the immune response of the spiny lobster,

Jasus lalandii. Fish & Shellfish Immunology, *93*, 752−762. Available from https://doi. org/10.1016/j.fsi.2019.05.063.

Ko, G. W. K., Dineshram, R., Campanati, C., Chan, V. B. S., Havenhand, J., & Thiyagarajan, V. (2014). Interactive effects of ocean acidification, elevated temperature, and reduced salinity on early-life stages of the pacific oyster. *Environmental Science & Technology*, *48*(17), 10079−10088. Available from https://doi.org/10.1021/es501611u.

Kong, H., Clements, J. C., Dupont, S., Wang, T., Huang, X., Shang, Y., Huang, W., Chen, J., Hu, M., & Wang, Y. (2019). Seawater acidification and temperature modulate anti-predator defenses in two co-existing *Mytilus* species. *Marine Pollution Bulletin*, *145*, 118−125. Available from https://doi.org/10.1016/j.marpolbul.2019. 05.040.

Kong, H., Jiang, X., Clements, J. C., Wang, T., Huang, X., Shang, Y., Chen, J., Hu, M., & Wang, Y. (2019). Transgenerational effects of short-term exposure to acidification and hypoxia on early developmental traits of the mussel *Mytilus edulis*. *Marine Environmental Research*, *145*, 73−80. Available from https://doi.org/10.1016/j. marenvres.2019.02.011.

Kunz, K. L., Claireaux, G., Pörtner, H. O., Knust, R., & Mark, F. C. (2018). Aerobic capacities and swimming performance of polar cod (*Boreogadus saida*) under ocean acidification and warming conditions. *Journal of Experimental Biology*, *221*(21). Available from https://doi.org/10.1242/jeb.184473.

Küpper, H., Šetlík, I., Spiller, M., Küpper, F. C., & Prášil, O. (2002). Heavy metal-induced inhibition of photosynthesis: Targets of in vivo heavy metal chlorophyll formation. *Journal of Phycology*, *38*(3), 429−441. Available from https://doi.org/10.1046/j.1529-8817.2002.t01-1-01148.x.

Lagos, N. A., Benítez, S., Duarte, C., Lardies, M. A., Broitman, B. R., Tapia, C., Tapia, P., Widdicombe, S., & Vargas, C. A. (2016). Effects of temperature and ocean acidification on shell characteristics of *Argopecten purpuratus*: Implications for scallop aquaculture in an upwelling-influenced area. *Aquaculture Environment Interactions*, *8*, 357−370. Available from https://doi.org/10.3354/AEI00183.

Langdon, C., & Atkinson, M. J. (2005). Effect of elevated $p\mathrm{CO}_2$ on photosynthesis and calcification of corals and interactions with seasonal change in temperature/irradiance and nutrient enrichment. *Journal of Geophysical Research C: Oceans*, *110*(9), 1−16. Available from https://doi.org/10.1029/2004JC002576.

Larsen, B. K., Pörtner, H.-O., & Jensen, F. B. (1997). Extra- and intracellular acid-base balance and ionic regulation in cod (*Gadus morhua*) during combined and isolated exposures to hypercapnia and copper. *Marine Biology*, *128*, 337−346. Available from https://doi.org/10.1007/s002270050099.

Laubenstein, T. D., Rummer, J. L., Nicol, S., Parsons, D. M., Pether, S. M. J., Pope, S., Smith, N., & Munday, P. L. (2018). Correlated effects of ocean acidification and warming on behavioral and metabolic traits of a large pelagic fish. *Diversity*, *10*(2), 35. Available from https://doi.org/10.3390/D10020035.

Lawlor, J. A., & Arellano, S. M. (2020). Temperature and salinity, not acidification, predict near-future larval growth and larval habitat suitability of Olympia oysters in the Salish Sea. *Scientific Reports*, *10*(1), 13787. Available from https://doi.org/10.1038/s41598-020-69568-w.

Lenz, B., Fogarty, N. D., & Figueiredo, J. (2019). Effects of ocean warming and acidification on fertilization success and early larval development in the green sea urchin *Lytechinus variegatus*. *Marine Pollution Bulletin*, *141*, 70−78. Available from https://doi. org/10.1016/j.marpolbul.2019.02.018.

Leo, E., Dahlke, F. T., Storch, D., Pörtner, H. O., & Mark, F. C. (2018). Impact of ocean acidification and warming on the bioenergetics of developing eggs of Atlantic herring

Clupea harengus. Conservation Physiology, 6(1), coy050. Available from https://doi.org/10.1093/conphys/coy050.

Leung, J. Y. S., Russell, B. D., & Connell, S. D. (2020). Linking energy budget to physiological adaptation: How a calcifying gastropod adjusts or succumbs to ocean acidification and warming. *Science of the Total Environment*, 715, 136939. Available from https://doi.org/10.1016/j.scitotenv.2020.136939.

Lewis, C., Clemow, K., & Holt, W. V. (2013). Metal contamination increases the sensitivity of larvae but not gametes to ocean acidification in the polychaete *Pomatoceros lamarckii* (Quatrefages). *Marine Biology*, 160(8), 2089−2101. Available from https://doi.org/10.1007/s00227-012-2081-8.

Lewis, C., Ellis, R. P., Vernon, E., Elliot, K., Newbatt, S., & Wilson, R. W. (2016). Ocean acidification increases copper toxicity differentially in two key marine invertebrates with distinct acid-base responses. *Scientific Reports*, 6, 21554. Available from https://doi.org/10.1038/srep21554.

Li, S., Liu, C., Huang, J., Liu, Y., Zheng, G., Xie, L., & Zhang, R. (2015). Interactive effects of seawater acidification and elevated temperature on biomineralization and amino acid metabolism in the mussel *Mytilus edulis. Journal of Experimental Biology*, 218(22), 3623−3631. Available from https://doi.org/10.1242/jeb.126748.

Li, Y., Wang, W. X., & Wang, M. (2017). Alleviation of mercury toxicity to a marine copepod under multigenerational exposure by ocean acidification. *Scientific Reports*, 7, 324. Available from https://doi.org/10.1038/s41598-017-00423-1.

Liao, H., Yang, Z., Dou, Z., Sun, F., Kou, S., Zhang, Z., Huang, X., & Bao, Z. (2019). Impact of ocean acidification on the energy metabolism and antioxidant responses of the Yesso scallop (*Patinopecten yessoensis*). *Frontiers in Physiology*, 9, 1967. Available from https://doi.org/10.3389/fphys.2018.01967.

Lifavi, D. M., Targett, T. E., & Grecay, P. A. (2017). Effects of diel-cycling hypoxia and acidification on juvenile weakfish *Cynoscion regalis* growth, survival, and activity. *Marine Ecology Progress Series*, 564, 163−171. Available from https://doi.org/10.3354/meps11966.

Lockwood, B. L., & Somero, G. N. (2012). Functional determinants of temperature adaptation in enzymes of cold-versus warm-adapted mussels (genus *Mytilus*). *Molecular Biology and Evolution*, 29(10), 3061−3070. Available from https://doi.org/10.1093/molbev/mss111.

Lockwood, B. L., Sanders, J. G., & Somero, G. N. (2010). Transcriptomic responses to heat stress in invasive and native blue mussels (genus *Mytilus*): Molecular correlates of invasive success. *Journal of Experimental Biology*, 213(20), 3548−3558. Available from https://doi.org/10.1242/jeb.046094.

Lopes, A. R., Faleiro, F., Rosa, I. C., Pimentel, M. S., Trubenbach, K., Repolho, T., Diniz, M., & Rosa, R. (2018). Physiological resilience of a temperate soft coral to ocean warming and acidification. *Cell Stress & Chaperones*, 23(5), 1093−1100. Available from https://doi.org/10.1007/s12192-018-0919-9.

López, I. R., Kalman, J., Vale, C., & Blasco, J. (2010). Influence of sediment acidification on the bioaccumulation of metals in *Ruditapes philippinarum. Environmental Science and Pollution Research*, 17(9), 1519−1528. Available from https://doi.org/10.1007/s11356-010-0338-7.

Lu, Y., Wang, L., Wang, L., Cong, Y., Yang, G., & Zhao, L. (2018). Deciphering carbon sources of mussel shell carbonate under experimental ocean acidification and warming. *Marine Environmental Research*, 142, 141−146. Available from https://doi.org/10.1016/j.marenvres.2018.10.007.

Ma, H., Zou, D., Wen, J., Ji, Z., Gong, J., & Liu, C. (2018). The impact of elevated atmospheric CO_2 on cadmium toxicity in *Pyropia haitanensis* (Rhodophyta).

Environmental Science and Pollution Research, *25*(33), 33361−33369. Available from https://doi.org/10.1007/s11356-018-3289-z.

Ma, J., Wang, W., Liu, X., Wang, Z., Gao, G., Wu, H., Li, X., & Xu, J. (2020). Zinc toxicity alters the photosynthetic response of red alga *Pyropia yezoensis* to ocean acidification. *Environmental Science and Pollution Research*, *27*(3), 3202−3212. Available from https://doi.org/10.1007/s11356-019-06872-7.

Manríquez, P. H., González, C. P., Brokordt, K., Pereira, L., Torres, R., Lattuca, M. E., Fernández, D. A., Peck, M. A., Cucco, A., Antognarelli, F., Marras, S., & Domenici, P. (2019). Ocean warming and acidification pose synergistic limits to the thermal niche of an economically important echinoderm. *Science of the Total Environment*, *693*, 133469. Available from https://doi.org/10.1016/j.scitotenv.2019.07.275.

Marangoni, L. F. B., Pinto, M. Md. A. N., Marques, J. A., & Bianchini, A. (2019). Copper exposure and seawater acidification interaction: Antagonistic effects on biomarkers in the zooxanthellate scleractinian coral *Mussismilia harttii*. *Aquatic Toxicology*, *206*, 123−133. Available from https://doi.org/10.1016/j.aquatox.2018.11.005.

Marques, J. A., de Barros Marangoni, L. F., & Bianchini, A. (2017). Combined effects of sea water acidification and copper exposure on the symbiont-bearing foraminifer *Amphistegina gibbosa*. *Coral Reefs*, *36*(2), 489−501. Available from https://doi.org/10.1007/s00338-017-1547-z.

Marques, J. A., Flores, F., Patel, F., Bianchini, A., Uthicke, S., & Negri, A. P. (2020). Acclimation history modulates effect size of calcareous algae (*Halimeda opuntia*) to herbicide exposure under future climate scenarios. *Science of the Total Environment*, *739*, 140308. Available from https://doi.org/10.1016/j.scitotenv.2020.140308.

Massom, R. A., & Stammerjohn, S. E. (2010). Antarctic sea ice change and variability − Physical and ecological implications. *Polar Science*, *4*(2), 149−186. Available from https://doi.org/10.1016/j.polar.2010.05.001.

McAllen, R. J., Taylor, A. C., & Davenport, J. (1998). Osmotic and body density response in the harpacticoid copepod *Tigriopus brevicornis* in supralittoral rock pools. *Journal of the Marine Biological Association of the United Kingdom*, *78*(4), 1143−1153. Available from https://doi.org/10.1017/s0025315400044386.

Melzner, F., Thomsen, J., Koeve, W., Oschlies, A., Gutowska, M. A., Bange, H. W., Hansen, H. P., & Körtzinger, A. (2013). Future ocean acidification will be amplified by hypoxia in coastal habitats. *Marine Biology*, *160*(8), 1875−1888. Available from https://doi.org/10.1007/s00227-012-1954-1.

Michaelidis, B., Ouzounis, C., Paleras, A., & Pörtner, H. O. (2005). Effects of long-term moderate hypercapnia on acid-base balance and growth rate in marine mussels *Mytilus galloprovincialis*. *Marine Ecology Progress Series*, *293*, 109−118. Available from https://doi.org/10.3354/meps293109.

Michalak, M., Parker, J. M. R., & Opas, M. (2002). Ca^{2+} signaling and calcium binding chaperones of the endoplasmic reticulum. *Cell Calcium*, *32*(5−6), 269−278. Available from https://doi.org/10.1016/S0143416002001884.

Millero, F. J., Woosley, R., Ditrolio, B., & Waters, J. (2009). Effect of ocean acidification on the speciation of metals in seawater. *Oceanography*, *22*(4), 72−85. Available from https://doi.org/10.5670/oceanog.2009.98.

Montgomery, D. W., Simpson, S. D., Engelhard, G. H., Birchenough, S. N. R., & Wilson, R. W. (2019). Rising CO_2 enhances hypoxia tolerance in a marine fish. *Scientific Reports*, *9*, 15152. Available from https://doi.org/10.1038/s41598-019-51572-4.

Moore, M. N., Icarus Allen, J., & McVeigh, A. (2006). Environmental prognostics: An integrated model supporting lysosomal stress responses as predictive biomarkers of animal health status. *Marine Environmental Research*, *61*(3), 278−304. Available from https://doi.org/10.1016/j.marenvres.2005.10.005.

Moreira, A., Figueira, E., Soares, A. M. V. M., & Freitas, R. (2016). The effects of arsenic and seawater acidification on antioxidant and biomineralization responses in two closely related *Crassostrea* species. *Science of the Total Environment, 545−546*, 569−581. Available from https://doi.org/10.1016/j.scitotenv.2015.12.029.

Morrell, B. K., & Gobler, C. J. (2020). Negative effects of diurnal changes in acidification and hypoxia on early-life stage estuarine fishes. *Diversity, 12*(1), 25. Available from https://doi.org/10.3390/d12010025.

Mukherjee, J., Wong, K. K. W., Chandramouli, K. H., Qian, P. Y., Leung, P. T. Y., Wu, R. S. S., & Thiyagarajan, V. (2013). Proteomic response of marine invertebrate larvae to ocean acidification and hypoxia during metamorphosis and calcification. *Journal of Experimental Biology, 216*(24), 4580−4589. Available from https://doi.org/10.1242/jeb.094516.

Murray, C. S., Wiley, D., & Baumann, H. (2019). High sensitivity of a keystone forage fish to elevated CO_2 and temperature. *Conservation Physiology, 7*(1), coz084. Available from https://doi.org/10.1093/conphys/coz084.

Nan, F., Yu, F., Xue, H., Wang, R., & Si, G. (2015). Ocean salinity changes in the northwest Pacific subtropical gyre: The quasi-decadal oscillation and the freshening trend. *Journal of Geophysical Research C: Oceans, 120*(3), 2179−2192. Available from https://doi.org/10.1002/2014JC010536.

Nardi, A., Benedetti, M., d'Errico, G., Fattorini, D., & Regoli, F. (2018). Effects of ocean warming and acidification on accumulation and cellular responsiveness to cadmium in mussels *Mytilus galloprovincialis*: Importance of the seasonal status. *Aquatic Toxicology, 204*, 171−179. Available from https://doi.org/10.1016/j.aquatox.2018.09.009.

Nardi, A., Benedetti, M., Fattorini, D., & Regoli, F. (2018). Oxidative and interactive challenge of cadmium and ocean acidification on the smooth scallop *Flexopecten glaber*. *Aquatic Toxicology, 196*, 53−60. Available from https://doi.org/10.1016/j.aquatox. 2018.01.008.

Navarro, J. M., Villanueva, P., Rocha, N., Torres, R., Chaparro, O. R., Benitez, S., Andrade-Villagran, P. V., & Alarcon, E. (2020). Plastic response of the oyster *Ostrea chilensis* to temperature and pCO_2 within the present natural range of variability. *PLoS One, 15*(6), e0234994. Available from https://doi.org/10.1371/journal.pone.0234994.

Nielson, C., Hird, C., & Lewis, C. (2019). Ocean acidification buffers the physiological responses of the king ragworm *Alitta virens* to the common pollutant copper. *Aquatic Toxicology, 212*, 120−127. Available from https://doi.org/10.1016/j.aquatox.2019.05.003.

O'Connor, M. I., Piehler, M. F., Leech, D. M., Anton, A., & Bruno, J. F. (2009). Warming and resource availability shift food web structure and metabolism. *PLoS Biology, 7*(8). Available from https://doi.org/10.1371/journal.pbio.1000178.

Ong, E. Z., Briffa, M., Moens, T., & Van Colen, C. (2017). Physiological responses to ocean acidification and warming synergistically reduce condition of the common cockle *Cerastoderma edule*. *Marine Environmental Research, 130*, 38−47. Available from https://doi.org/10.1016/j.marenvres.2017.07.001.

Ordoñez, A., Kennedy, E. V., & Diaz-Pulido, G. (2017). Reduced spore germination explains sensitivity of reef-building algae to climate change stressors. *PLoS One, 12* (12), e0189122. Available from https://doi.org/10.1371/journal.pone.0189122.

Parker, L. M., Scanes, E., O'Connor, W. A., Coleman, R. A., Byrne, M., Pörtner, H. O., & Ross, P. M. (2017). Ocean acidification narrows the acute thermal and salinity tolerance of the Sydney rock oyster *Saccostrea glomerata*. *Marine Pollution Bulletin, 122*(1−2), 263−271. Available from https://doi.org/10.1016/j. marpolbul.2017.06.052.

Pascal, P. Y., Fleeger, J. W., Galvez, F., & Carman, K. R. (2010). The toxicological interaction between ocean acidity and metals in coastal meiobenthic copepods. *Marine*

Pollution Bulletin, *60*(12), 2201−2208. Available from https://doi.org/10.1016/j. marpolbul.2010.08.018.

Paula, J. R., Repolho, T., Pegado, M. R., Thörnqvist, P. O., Bispo, R., Winberg, S., Munday, P. L., & Rosa, R. (2019). Neurobiological and behavioural responses of cleaning mutualisms to ocean warming and acidification. *Scientific Reports*, *9*(1), 12728. Available from https://doi.org/10.1038/s41598-019-49086-0.

Pequeux, A. (1995). Osmotic regulation in crustaceans. *Journal of Crustacean Biology*, *15*(1), 1−60. Available from https://doi.org/10.2307/1549010.

Piggott, J. J., Townsend, C. R., & Matthaei, C. D. (2015). Reconceptualizing synergism and antagonism among multiple stressors. *Ecology and Evolution*, *5*(7), 1538−1547. Available from https://doi.org/10.1002/ece3.1465.

Pitts, K. A., Campbell, J. E., Figueiredo, J., & Fogarty, N. D. (2020). Ocean acidification partially mitigates the negative effects of warming on the recruitment of the coral, Orbicella faveolata. *Coral Reefs*, *39*(2), 281−292. Available from https://doi.org/10.1007/s00338-019-01888-4.

Pörtner, H. O., Langenbuch, M., & Michaelidis, B. (2005). Synergistic effects of temperature extremes, hypoxia, and increases in CO_2 on marine animals: From Earth history to global change. *Journal of Geophysical Research C: Oceans*, *110*(9), 1−15. Available from https://doi.org/10.1029/2004JC002561.

Prada, F., Caroselli, E., Mengoli, S., Brizi, L., Fantazzini, P., Capaccioni, B., Pasquini, L., Fabricius, K. E., Dubinsky, Z., Falini, G., & Goffredo, S. (2017). Ocean warming and acidification synergistically increase coral mortality. *Scientific Reports*, *7*, 40842. Available from https://doi.org/10.1038/srep40842.

Rabouille, C., Conley, D. J., Dai, M. H., Cai, W. J., Chen, C. T. A., Lansard, B., Green, R., Yin, K., Harrison, P. J., Dagg, M., & McKee, B. (2008). Comparison of hypoxia among four river-dominated ocean margins: The Changjiang (Yangtze), Mississippi, Pearl, and Rhône rivers. *Continental Shelf Research*, *28*(12), 1527−1537. Available from https://doi.org/10.1016/j.csr.2008.01.020.

Rastrick, S. P. S., Collier, V., Graham, H., Strohmeier, T., Whiteley, N. M., Strand, Ø. O., & Woodson, C. B. (2018). Feeding plasticity more than metabolic rate drives the productivity of economically important filter feeders in response to elevated CO_2 and reduced salinity. *ICES Journal of Marine Science*, *75*(6), 2117−2128. Available from https://doi.org/10.1093/icesjms/fsy079.

Raven, J. A., & Falkowski, P. G. (1999). Oceanic sinks for atmospheric CO_2. *Plant, Cell & Environment*, *22*(6), 741−755. Available from https://doi.org/10.1046/j.1365-3040.1999.00419.x.

Ricardo, B. (2018). *Chapter 1 − Basic concepts. Marine pollution: Sources, fate and effects of pollutants in coastal ecosystems* (pp. 3−20). Elsevier BV, https://doi.org/10.1016/b978-0-12-813736-9.00001-5.

Ries, J. B., Cohen, A. L., & McCorkle, D. C. (2009). Marine calcifiers exhibit mixed responses to CO_2-induced ocean acidification. *Geology*, *37*(12), 1131−1134. Available from https://doi.org/10.1130/G30210A.1.

Rivera-Ingraham, G. A., & Lignot, J. H. (2017). Osmoregulation, bioenergetics and oxidative stress in coastal marine invertebrates: Raising the questions for future research. *Journal of Experimental Biology*, *220*(10), 1749−1760. Available from https://doi.org/10.1242/jeb.135624.

Rodríguez-Romero, A., Basallote, M. D., De Orte, M. R., DelValls, T. Á., Riba, I., & Blasco, J. (2014). Simulation of CO_2 leakages during injection and storage in subseabed geological formations: metal mobilization and biota effects. *Environ Int*, *68*, 105−117. Available from https://doi.org/10.1016/j.envint.2014.03.008.

Russell, M. P. (2013). Echinoderm responses to variation in salinity. *Advances in Marine Biology*, *66*, 171−212. Available from https://doi.org/10.1016/B978-0-12-408096-6.00003-1.

Sabine, C. L., Feely, R. A., Gruber, N., Key, R. M., Lee, K., Bullister, J. L., Wanninkhof, R., Wong, C. S., Wallace, D. W. R., Tilbrook, B., Millero, F. J., Peng, T. H., Kozyr, A., Ono, T., & Rios, A. F. (2004). The oceanic sink for anthropogenic CO_2. *Science (New York, N.Y.)*, *305*(5682), 367−371. Available from https://doi.org/10.1126/science.1097403.

Sage, R. F. (2020). Global change biology: A primer. *Global Change Biology*, *26*(1), 3−30. Available from https://doi.org/10.1111/gcb.14893.

Sampaio, E., Lopes, A. R., Francisco, S., Paula, J. R., Pimentel, M., Maulvault, A. L., Repolho, T., Grilo, T. F., Pousão-Ferreira, P., Marques, A., & Rosa, R. (2018). Ocean acidification dampens physiological stress response to warming and contamination in a commercially-important fish (*Argyrosomus regius*). *Science of the Total Environment*, *618*, 388−398. Available from https://doi.org/10.1016/j.scitotenv.2017.11.059.

Sanford, E. (1999). Regulation of keystone predation by small changes in ocean temperature. *Science (New York, N.Y.)*, *283*(5410), 2095−2097. Available from https://doi.org/10.1126/science.283.5410.2095.

Scanes, E., Parker, L. M., O'Connor, W. A., Gibbs, M. C., & Ross, P. M. (2018). Copper and ocean acidification interact to lower maternal investment, but have little effect on adult physiology of the Sydney rock oyster *Saccostrea glomerata*. *Aquatic Toxicology*, *203*, 51−60. Available from https://doi.org/10.1016/j.aquatox.2018.07.020.

Schneider, S. H. (1989). The greenhouse effect: Science and policy. *Science (New York, N.Y.)*, *243*(4892), 771−781. Available from https://doi.org/10.1126/science.243.4892.771.

Schwemmer, T. G., Baumann, H., Murray, C. S., Molina, A. I., & Nye, J. A. (2020). Acidification and hypoxia interactively affect metabolism in embryos, but not larvae, of the coastal forage fish *Menidia menidia*. *The Journal of Experimental Biology*, *223*(22), jeb228015. Available from https://doi.org/10.1242/jeb.228015.

Sezer, N., Kılıç, Ö., Sıkdokur, E., Çayır, A., & Belivermiş, M. (2020). Impacts of elevated pCO_2 on Mediterranean mussel (*Mytilus galloprovincialis*): Metal bioaccumulation, physiological and cellular parameters. *Marine Environmental Research*, *160*, 104987. Available from https://doi.org/10.1016/j.marenvres.2020.104987.

Shang, Y., Lan, Y., Liu, Z., Kong, H., Huang, X., Wu, F., Liu, L., Hu, M., Huang, W., & Wang, Y. (2018). Synergistic effects of nano-ZnO and low pH of sea water on the physiological energetics of the thick shell mussel *Mytilus coruscus*. *Frontiers in Physiology*, *9*, 757. Available from https://doi.org/10.3389/fphys.2018.00757.

Shang, Y., Wang, X., Kong, H., Huang, W., Hu, M., & Wang, Y. (2019). Nano-ZnO impairs anti-predation capacity of marine mussels under seawater acidification. *Journal of Hazardous Materials*, *371*, 521−528. Available from https://doi.org/10.1016/j.jhazmat.2019.02.072.

Shi, W., Zhao, X., Han, Y., Che, Z., Chai, X., & Liu, G. (2016). Ocean acidification increases cadmium accumulation in marine bivalves: A potential threat to seafood safety. *Scientific Reports*, *6*, 20197. Available from https://doi.org/10.1038/srep20197.

Shrivastava, J., Ndugwa, M., Caneos, W., & De Boeck, G. (2019). Physiological trade-offs, acid-base balance and ion-osmoregulatory plasticity in European sea bass (*Dicentrarchus labrax*) juveniles under complex scenarios of salinity variation, ocean acidification and high ammonia challenge. *Aquatic Toxicology*, *212*, 54−69. Available from https://doi.org/10.1016/j.aquatox.2019.04.024.

Shumway, S. E. (1977). Effect of salinity fluctuation on the osmotic pressure and Na^+, Ca^{2+} and Mg^{2+} ion concentrations in the hemolymph of bivalve molluscs. *Marine Biology*, *41*(2), 153−177. Available from https://doi.org/10.1007/BF00394023.

Siddiqui, S., & Bielmyer-Fraser, G. K. (2015). Responses of the sea anemone, *Exaiptasia pallida*, to ocean acidification conditions and copper exposure. *Aquatic Toxicology*, *167*, 228−239. Available from https://doi.org/10.1016/j.aquatox.2015.08.012.

Sokolova, I. M. (2013). Energy-limited tolerance to stress as a conceptual framework to integrate the effects of multiple stressors. *Integrative and Comparative Biology, 53*(4), 597−608. Available from https://doi.org/10.1093/icb/ict028.

Somero, G. N. (2012). The physiology of global change: Linking patterns to mechanisms. *Annual Review of Marine Science, 4,* 39−61. Available from https://doi.org/10.1146/annurev-marine-120710-100935.

Sordo, L., Santos, R., Barrote, I., & Silva, J. (2019). Temperature amplifies the effect of high CO_2 on the photosynthesis, respiration, and calcification of the coralline algae *Phymatolithon lusitanicum. Ecology and Evolution, 9*(19), 11000−11009. Available from https://doi.org/10.1002/ece3.5560.

Sswat, M., Stiasny, M. H., Jutfelt, F., Riebesell, U., & Clemmesen, C. (2018). Growth performance and survival of larval Atlantic herring, under the combined effects of elevated temperatures and CO_2. *PLoS One, 13*(1), e0191947. Available from https://doi.org/10.1371/journal.pone.0191947.

Stevens, A. M., & Gobler, C. J. (2018). Interactive effects of acidification, hypoxia, and thermal stress on growth, respiration, and survival of four North Atlantic bivalves. *Marine Ecology Progress Series, 604,* 143−161. Available from https://doi.org/10.3354/meps12725.

Stevenson, A., Archer, S. K., Schultz, J. A., Dunham, A., Marliave, J. B., Martone, P., & Harley, C. D. G. (2020). Warming and acidification threaten glass sponge *Aphrocallistes vastus* pumping and reef formation. *Scientific Reports, 10*(1), 8176. Available from https://doi.org/10.1038/s41598-020-65220-9.

Stewart, R. I. A., Dossena, M., Bohan, D. A., Jeppesen, E., Kordas, R. L., Ledger, M. E., Meerhoff, M., Moss, B., Mulder, C., Shurin, J. B., Suttle, B., Thompson, R., Trimmer, M., & Woodward, G. (2013). Mesocosm experiments as a tool for ecological climate-change research. *Advances in Ecological Research, 48,* 71−181. Available from https://doi.org/10.1016/B978-0-12-417199-2.00002-1.

Storey, K. B. (1988). Suspended animation: The molecular basis of metabolic depression. *Canadian Journal of Zoology, 66*(1), 124−132. Available from https://doi.org/10.1139/z88-016.

Stramma, L., Schmidtko, S., Levin, L. A., & Johnson, G. C. (2010). Ocean oxygen minima expansions and their biological impacts. *Deep-Sea Research Part I: Oceanographic Research Papers, 57*(4), 587−595. Available from https://doi.org/10.1016/j.dsr.2010.01.005.

Strzałka, K., Kostecka-Gugała, A., & Latowski, D. (2003). Carotenoids and environmental stress in plants: Significance of carotenoid-mediated modulation of membrane physical properties. *Russian Journal of Plant Physiology, 50*(2), 168−173. Available from https://doi.org/10.1023/A:1022960828050.

Stubler, A. D., & Peterson, B. J. (2016). Ocean acidification accelerates net calcium carbonate loss in a coral rubble community. *Coral Reefs, 35*(3), 795−803. Available from https://doi.org/10.1007/s003380016-1436-x.

Sui, Y., Hu, M., Huang, X., Wang, Y., & Lu, W. (2015). Anti-predatory responses of the thick shell mussel *Mytilus coruscus* exposed to seawater acidification and hypoxia. *Marine Environmental Research, 109,* 159−167. Available from https://doi.org/10.1016/j.marenvres.2015.07.008.

Sui, Y., Hu, M., Shang, Y., Wu, F., Huang, X., Dupont, S., Storch, D., Pörtner, H. O., Li, J., Lu, W., & Wang, Y. (2017). Antioxidant response of the hard shelled mussel *Mytilus coruscus* exposed to reduced pH and oxygen concentration. *Ecotoxicology and Environmental Safety, 137,* 94−102. Available from https://doi.org/10.1016/j.ecoenv.2016.11.023.

Sui, Y., Kong, H., Huang, X., Dupont, S., Hu, M., Storch, D., Pörtner, H. O., Lu, W., & Wang, Y. (2016). Combined effects of short-term exposure to elevated CO_2 and decreased O_2 on the physiology and energy budget of the thick shell mussel *Mytilus coruscus. Chemosphere,* *155,* 207—216. Available from https://doi.org/10.1016/j. chemosphere.2016.04.054.

Sui, Y., Kong, H., Shang, Y., Huang, X., Wu, F. L., Hu, M., Lin, D., Lu, W., & Wang, Y. (2016). Effects of short-term hypoxia and seawater acidification on hemocyte responses of the mussel *Mytilus coruscus. Marine Pollution Bulletin, 108*(1—2), 46—52. Available from https://doi.org/10.1016/j.marpolbul.2016.05.001.

Svetlichny, L., Hubareva, E., & Khanaychenko, A. (2012). *Calanipeda aquaedulcis* and *Arctodiaptomus salinus* are exceptionally euryhaline osmoconformers: Evidence from mortality, oxygen consumption, and mass density patterns. *Marine Ecology Progress Series, 470,* 15—29. Available from https://doi.org/10.3354/meps09907.

Swiney, K. M., Long, W. C., & Foy, R. J. (2017). Decreased pH and increased temperatures affect young-of-theyear red king crab (*Paralithodes camtschaticus*). *ICES Journal of Marine Science,* 74(4), 1191—1200. Available from https://doi.org/10.1093/icesjms/fsw251.

Tatara, C. P., Newman, M. C., McCloskey, J. T., & Williams, P. L. (1997). Predicting relative metal toxicity with ion characteristics: *Caenorhabditis elegans* LC50. *Aquatic Toxicology, 39*(3—4), 279—290. Available from https://doi.org/10.1016/S0166-445X (97)00030-1.

Taulbee, W. K., Nietch, C. T., Brown, D., Ramakrishnan, B., & Tompkins, M. J. (2009). Ecosystem consequences of contrasting flow regimes in an urban effects stream mesocosm study. *Journal of the American Water Resources Association, 45*(4), 907—927. Available from https://doi.org/10.1111/j.1752-1688.2009.00336.x.

Tesdal, J. E., Abernathey, R. P., Goes, J. I., Gordon, A. L., & Haine, T. W. N. (2018). Salinity trends within the upper layers of the subpolar North Atlantic. *Journal of Climate, 31*(7), 2675—2698. Available from https://doi.org/10.1175/JCLI-D-17-0532.1.

Thangaraj, S., & Sun, J. (2020). Transcriptomic reprogramming of the oceanic diatom *Skeletonema dohrnii* under warming ocean and acidification. *Environmental Microbiology, 23*(2), 980—995. Available from https://doi.org/10.1111/1462-2920.15248.

Tomanek, L. (2010). Variation in the heat shock response and its implication for predicting the effect of global climate change on species' biogeographical distribution ranges and metabolic costs. *Journal of Experimental Biology, 213*(6), 971—979. Available from https://doi.org/10.1242/jeb.038034.

Tomasetti, S. J., Morrell, B. K., Merlo, L. R., & Gobler, C. J. (2018). Individual and combined effects of low dissolved oxygen and low pH on survival of early stage larval blue crabs, *Callinectes sapidus. PLoS One, 13*(12), e0208629. Available from https://doi.org/ 10.1371/journal.pone.0208629.

Torstensson, A., Hedblom, M., Andersson, J., Andersson, M. X., & Wulff, A. (2013). Synergism between elevated pCO_2 and temperature on the antarctic sea ice diatom *Nitzschia lecointei. Biogeosciences, 10*(10), 6391—6401. Available from https://doi.org/ 10.5194/bg-10-6391-2013.

Trenberth, K. E., & Fasullo, J. T. (2013). An apparent hiatus in global warming? *Earth's Future, 1*(1), 19—32. Available from https://doi.org/10.1002/2013ef000165.

Turner, A. (2010). Marine pollution from antifouling paint particles. *Marine Pollution Bulletin, 60*(2), 159—171. Available from https://doi.org/10.1016/j.marpolbul.2009. 12.004.

Uthicke, S., Liddy, M., Nguyen, H. D., & Byrne, M. (2014). Interactive effects of near-future temperature increase and ocean acidification on physiology and gonad development in adult Pacific sea urchin, *Echinometra* sp. A. *Coral Reefs, 33*(3), 831—845. Available from https://doi.org/10.1007/s00338-014-1165-y.

Van Colen, C., Jansson, A., Saunier, A., Lacoue-Labathe, T., & Vincx, M. (2018). Biogeographic vulnerability to ocean acidification and warming in a marine bivalve. *Marine Pollution Bulletin, 126,* 308—311. Available from https://doi.org/10.1016/j. marpolbul.2017.10.092.

Velez, C., Figueira, E., Soares, A. M. V. M., & Freitas, R. (2016). Combined effects of seawater acidification and salinity changes in *Ruditapes philippinarum*. *Aquatic Toxicology*, *176*, 141—150. Available from https://doi.org/10.1016/j.aquatox.2016.04.016.

Visconti, G., Gianguzza, F., Butera, E., Costa, V., Vizzini, S., Byrne, M., & Gianguzza, P. (2017). Morphological response of the larvae of *Arbacia lixula* to near-future ocean warming and acidification. *ICES Journal of Marine Science*, *74*(4), 1180—1190. Available from https://doi.org/10.1093/icesjms/fsx037.

Wallace, R. B., Baumann, H., Grear, J. S., Aller, R. C., & Gobler, C. J. (2014). Coastal ocean acidification: The other eutrophication problem. *Estuarine, Coastal and Shelf Science*, *148*, 1—13. Available from https://doi.org/10.1016/j.ecss.2014.05.027.

Wang, M., Lee, J. S., & Li, Y. (2017). Global proteome profiling of a marine copepod and the mitigating effect of ocean acidification on mercury toxicity after multigenerational exposure. *Environmental Science & Technology*, *51*(10), 5820—5831. Available from https://doi.org/10.1021/acs.est.7b01832.

Wang, X., Shang, Y., Kong, H., Hu, M., Yang, J., Deng, Y., & Wang, Y. (2020). Combined effects of ocean acidification and hypoxia on the early development of the thick shell mussel *Mytilus coruscus*. *Helgoland Marine Research*, *74*(1). Available from https://doi.org/10.1186/s10152-020-0535-9.

Wenger, R. H. (2000). Mammalian oxygen sensing, signalling and gene regulation. *Journal of Experimental Biology*, *203*(8), 1253—1263.

Wilhelmsson, D., Thompson, R. C., Holmström, K., Lindén, O., & Eriksson-Hägg, H. (2013). *Marine pollution. Managing ocean environments in a changing climate: Sustainability and economic perspectives* (pp. 127—169). Elsevier Inc., https://doi.org/10.1016/B978-0-12-407668-6.00006-9.

Willmer, P. (2002). Physiology: How molecules make organisms work. *Science (New York, N. Y.)*, *296*(5567), 473. Available from https://doi.org/10.1126/science.1070910.

Wong, J. M., & Hofmann, G. E. (2020). The effects of temperature and pCO_2 on the size, thermal tolerance and metabolic rate of the red sea urchin (*Mesocentrotus franciscanus*) during early development. *Marine Biology*, *167*(3). Available from https://doi.org/10.1007/s00227-019-3633-y.

Wu, F., Cui, S., Sun, M., Xie, Z., Huang, W., Huang, X., Liu, L., Hu, M., Lu, W., & Wang, Y. (2018). Combined effects of ZnO NPs and seawater acidification on the haemocyte parameters of thick shell mussel *Mytilus coruscus*. *Science of the Total Environment*, *624*, 820—830. Available from https://doi.org/10.1016/j.scitotenv.2017.12.168.

Wu, F., Xie, Z., Lan, Y., Dupont, S., Sun, M., Cui, S., Huang, X., Huang, W., Liu, L., Hu, M., Lu, W., & Wang, Y. (2018). Short-term exposure of *Mytilus coruscus* to decreased pH and salinity change impacts immune parameters of their haemocytes. *Frontiers in Physiology*, *9*, 166. Available from https://doi.org/10.3389/fphys.2018.00166.

Wu, R. S. S. (2002). Hypoxia: From molecular responses to ecosystem responses. *Marine Pollution Bulletin*, *45*(1—12), 35—45. Available from https://doi.org/10.1016/s0025-326x(02)00061-9.

Xu, J., Sun, J., Beardall, J., & Gao, K. (2020). Lower salinity leads to improved physiological performance in the coccolithophorid *Emiliania huxleyi*, which partly ameliorates the effects of ocean acidification. *Frontiers in Marine Science*, *7*, 704. Available from https://doi.org/10.3389/fmars.2020.00704.

Young, C. S., & Gobler, C. J. (2020). Hypoxia and acidification, individually and in combination, disrupt herbivory and reduce survivorship of the gastropod, *Lacuna vincta*. *Frontiers in Marine Science*, *7*, 547276. Available from https://doi.org/10.3389/fmars.2020.547276.

Yuan, W., Gao, G., Shi, Q., Xu, Z., & Wu, H. (2018). Combined effects of ocean acidification and warming on physiological response of the diatom *Thalassiosira pseudonana* to light challenges. *Marine Environmental Research*, *135*, 63−69. Available from https://doi.org/10.1016/j.marenvres.2018.01.016.

Zakroff, C. J., & Mooney, T. A. (2020). Antagonistic interactions and clutch-dependent sensitivity induce variable responses to ocean acidification and warming in squid (*Doryteuthis pealeii*) embryos and paralarvae. *Frontiers in Physiology*, *11*, 501. Available from https://doi.org/10.3389/fphys.2020.00501.

Zhang, H., Cheung, S. G., & Shin, P. K. S. (2014). The larvae of congeneric gastropods showed differential responses to the combined effects of ocean acidification, temperature and salinity. *Marine Pollution Bulletin*, *79*(1−2), 39−46. Available from https://doi.org/10.1016/j.marpolbul.2014.01.008.

Zhang, X., Xu, D., Huang, S., Wang, S., Han, W., Liang, C., Zhang, Y., Fan, X., Zhang, X., Wang, Y., Wang, W., Egan, S., Saha, M., Li, F., & Ye, N. (2020). The effect of elevated pCO$_2$ on cadmium resistance of a globally important diatom. *Journal of Hazardous Materials*, *396*, 122749. Available from https://doi.org/10.1016/j.jhazmat.2020.122749.

Zhao, X., Han, Y., Chen, B., Xia, B., Qu, K., & Liu, G. (2020). CO$_2$-driven ocean acidification weakens mussel shell defense capacity and induces global molecular compensatory responses. *Chemosphere*, *243*, 125415. Available from https://doi.org/10.1016/j.chemosphere.2019.125415.

Zhao, X., Shi, W., Han, Y., Liu, S., Guo, C., Fu, W., Chai, X., & Liu, G. (2017). Ocean acidification adversely influences metabolism, extracellular pH and calcification of an economically important marine bivalve, *Tegillarca granosa*. *Marine Environmental Research*, *125*, 82−89. Available from https://doi.org/10.1016/j.marenvres.2017.01.007.

Zhu, Z., Qu, P., Gale, J., Fu, F., & Hutchins, D. A. (2017). Individual and interactive effects of warming and CO$_2$ on *Pseudo-nitzschia subcurvata* and *Phaeocystis antarctica*, two dominant phytoplankton from the Ross Sea, Antarctica. *Biogeosciences*, *14*(23), 5281−5295. Available from https://doi.org/10.5194/bg-14-5281-2017.

CHAPTER SIX

A brief summary of what we know and what we do not know about the impacts of ocean acidification on marine animals

Guangxu Liu and Wei Shi
College of Animal Sciences, Zhejiang University, Hangzhou, P.R. China

Introduction

Since the Industrial Revolution, atmospheric carbon dioxide (CO_2) concentration has rapidly risen from around 280 ppm during preindustrial times to more than 380 ppm nowadays, primarily as a result of anthropogenic activities such as land use changes, cement production, and the combustion of fossil fuels (Orr et al., 2005). According to the estimation by the Intergovernmental Panel on Climate Change (IPCC), about a third of anthropogenic atmospheric CO_2 emitted over the last 260 years has been absorbed by the world's ocean (Doney et al., 2009; Orr et al., 2005). The CO_2 taken up by the ocean surface water would form carbonic acid (H_2CO_3) and rapidly dissociate to produce hydrogen (H^+) and bicarbonate (HCO_3^-) ions, which would eventually lower the surface seawater pH, calcium carbonate ($CaCO_3$) saturation state, and the carbonate (CO_3^{2-}) concentration (Caldeira & Wickett, 2003; Doney et al., 2009; Orr et al., 2005). By the 1980s the ocean surface pH was falling at a rate of 0.015 pH units per decade (Caldeira & Wickett, 2003). It is shown that the average global ocean surface water pH had already fallen by 0.1 pH units, to 8.10 currently, as compared to the preindustrial value of 8.21, which is equivalent to a 30% increase in H^+ ion concentration (Doney et al., 2009). Furthermore, further decreases of 0.3−0.4 and 0.7−0.8 units in average surface seawater pH are predicted to occur according to the IPCC RCP8.5 "business as usual" scenario by the end of the 21st and 23rd centuries, respectively (Meinshausen et al., 2011). This

Ocean Acidification and Marine Wildlife.
DOI: https://doi.org/10.1016/B978-0-12-822330-7.00004-6

phenomenon of CO_2 has caused alternations in seawater pH and carbonate chemistry and is now widely recognized and termed ocean acidification (OA) (Caldeira & Wickett, 2003).

Many marine organisms, especially the species having calcified structures, are sensitive to changes in seawater carbonate chemistry, and their responses to OA may result in profound ecological shifts in marine ecosystems (Jokiel et al., 2008). As such, various studies have been conducted to explore the potential effects of OA on marine organisms and ecosystems (Maier et al., 2016; Ohki et al., 2013). A large number of studies have demonstrated that many calcifying organisms such as coccolithophores, mollusks, echinoderms, and corals exhibited decreased calcification and growth rates upon exposure to OA conditions in both laboratory and field environments (Lemasson et al., 2017). In addition, the results obtained from previous analyses also suggested that future OA could exert a negative effect on many other physiological aspects of marine organisms including fertilization, larval development, larval settlement, immune responses, and survival (Havenhand et al., 2008a, 2008b; Hernroth et al., 2011; Maier et al., 2016). However, there are still some limitations and knowledge gaps that currently exist over OA's impacts on marine organisms and the whole marine ecosystem. Herein, this chapter summarizes our current understanding of the impacts of OA on marine animals and aims to highlight knowledge gaps and potential future research directions.

 ## A general summary of what we know about the impacts of ocean acidification on marine animals

Since the concept of OA was formally put forward, it has gained continuously increasing attention and interest from scientists, the public, and policymakers (Billé et al., 2013). In the past decades, numerous studies have been conducted to investigate OA's impacts on marine animals (Kroeker et al., 2010, 2013a, 2013,b; Lemasson et al., 2017). To date, our knowledge about the effects of OA on marine organisms has been mainly focused on the following aspects.

1. Fertilization and early development

    OA has been confirmed to impair the fertilization of a wide range of marine species (Ross et al., 2011; Shi et al., 2017). On one hand, OA would increase the difficulties in the fertilization of marine organisms and thus result in reductions in fertilization success (Havenhand et al., 2008a; Hernroth et al., 2011; Maier et al., 2016). It is reported that OA can reduce the gamete quality of various marine animals in terms of sperm motility and gametic longevity, which may decrease the probabilities for sperm—egg collision and fusion (Shi et al., 2017). On the other hand, future OA conditions would also increase the risk of polyspermy in marine organisms, especially marine invertebrates, by disrupting both the fast block (membrane depolarization) and the permanent block (cortical reaction) processes involved in polyspermy blocking (Han et al., 2021). Interestingly, the gametes and fertilization of some species were shown to be robust to OA, suggesting that the sensitiveness of the fertilization process to OA may be species-specific (Havenhand et al., 2008a; Hernroth et al., 2011; Maier et al., 2016). As they lack sufficient capacity to regulate intracellular pH, the early development stage of the embryos of marine animals is generally considered to be one of the most sensitive stages to future OA scenarios (Ross et al., 2011). Firstly, OA would significantly increase the embryonic and/or larval developmental time of marine animals, which may eventually lead to reduced larval survival rates (Dupont et al., 2008). Secondly, OA exposure would result in morphological abnormalities in embryos, especially in those that have calcite skeleton (Visconti et al., 2017). In addition, the settlement success of many marine organisms is also susceptible to OA, as both the organism's ability to recognize settlement cues and the composition of the inducer were shown to be altered by elevated pCO_2 levels (Albright et al., 2010).

2. Biomineralization, metabolism, and growth

Given that increased CO_2 levels in the ocean would decrease the CO_3^{2-} concentration, the saturation state of calcium carbonate (Ω) in seawater, which is crucial for the biomineralization process of marine calcifying organisms, would be altered under OA conditions (Caldeira & Wickett, 2003; Doney et al., 2009). In addition, OA can lead to extracellular acidosis in many marine animals, which would reduce the saturation state of $CaCO_3$ at the calcification sites in the body of these organisms and thus result in the observed reduction in net calcification rates and shell formation (Zhao et al., 2017). As such, a series of investigations have confirmed that various species, including gastropods, bivalves, sea urchins, and corals, exposed to elevated pCO_2 levels would experience significant decreases in net calcification and shell mechanical properties (Chan & Connolly, 2013; Zhao et al., 2017).

Nowadays, OA-induced metabolic depression has been reported in a variety of marine animals including coral, bivalves, crabs, and fish (Chan & Connolly, 2013; Pimentel et al., 2014; Zhao et al., 2017). According to previous studies, OA can suppress the metabolism directly by inhibiting food intake and digestive activities of marine animals (Zhao et al., 2017). Furthermore, OA also affects the food chain of the marine ecosystem by altering marine autotrophs, which then indirectly leads to decreased metabolism (Kroeker et al., 2014). Therefore OA-induced suppression on energy acquisition may result in energy limitations to fueling essential processes for metabolism. In contrast, marine animals may allocate more energy to protect themselves against hypercapnia-induced disturbances in response to environmental stressors like elevated pCO_2 levels (Lannig et al., 2010). Under these circumstances, these organisms would enhance their metabolism to compensate for the elevated energy demand. However, since this short-term upregulation of metabolism may occur at the cost of reduced skeletal integrity and muscle mass, it is unlikely to be sustainable in the long term (Lannig et al., 2010; Wood et al., 2008). As a result, OA-induced metabolic depression would disturb the cellular protein turnover and thus undoubtedly lead to inhibited growth in marine animals (Wood et al., 2008).

3. Immunity and behaviors

Since they live in a complex environment, marine animals, especially invertebrate species, are often challenged by various pathogenic microorganisms (Asplund et al., 2014). However, accumulating evidence suggests that near-future OA scenarios may hinder the immune responses of marine animals such as bivalves, echinoderms, and crustaceans (Calder-Potts et al., 2008; Hernroth et al., 2011). On the basis of previous studies, the immune responses of marine organisms under OA conditions were hampered due to the following reasons. Firstly, exposure to near-future OA scenarios may reduce the total counts and alter the cell-type compositions of hemocytes and coelomocytes, which are regarded as the main immune cells in marine animals (Liu et al., 2016). Secondly, phagocytosis executed by hemocytes, which is the primary line of cellular defense against invasive pathogens, can be inhibited by OA in various marine animals (Brothers et al., 2016; Liu et al., 2016). Thirdly, OA may have significant impacts on the microbial community in the marine environment, which may increase the risk of pathogenic infection in marine organisms (Zha et al., 2017).

As reported in a large number of studies, OA is likely to exert a wide range of effects on various functionally important behaviors of marine animals, which includes predation, escape behavior, swimming behavior, collective behavior, behavioral lateralization, habitat detection, and seeking behavior (Rodriguez-Dominguez et al., 2018; Rong et al., 2020; Wang et al., 2017). According to previous studies, these OA-induced behavioral abnormalities can result from the disturbed neurotransmitter system due to the two mechanisms, given as follows. On one hand, previous studies have indicated that future OA conditions alter the gamma-aminobutyric acid A ($GABA_A$) receptor function of many marine animals and then cause behavioral abnormalities (Tresguerres & Hamilton, 2017). On the other hand, in addition to the overexcitation of GABA receptors, the direct impacts of OA on the in vivo contents of neurotransmitters and key molecules from the signal transduction cascade pathway are also shown to induce behavioral impairments in marine organisms (Rong et al., 2020).

A summary of limitations and knowledge gaps that currently exist

With the growing quantity of investigations and studies concerning the ecological and biogeochemical impacts of OA that have been published in the last decade, it becomes increasingly more important to recognize and learn from previous experiences. However, there are still limitations and debates over several issues. This section summarizes current knowledge gaps and limitations in OA research.

1. In order to stress the importance of implementing more rigorous and sophisticated approaches in OA research, many recent workshops and publications such as the European Project on Ocean Acidification have developed a set of best practices for OA research, which, advocating for treatments in experiments, reflect preindustrial, current, and future global atmospheric CO_2 conditions (Riebesell et al., 2010). Therefore our current understanding of OA's effects on the performance of marine organisms and the ecosystem is mainly based on experiments that are performed with "constant acidification," which involves using stable pH levels consistent with open ocean projections throughout the experiment (Challener et al., 2015). Although the progressive reduction in the pH of the open ocean is confirmed in the long term, due to many biological and physical processes, CO_2 concentrations in seawater are not always constant and can be frequently significantly higher or lower than expected from equilibrium with the current atmospheric CO_2 concentration over short time scales (Hales et al., 2005; Hofmann et al., 2011; Wootton et al., 2008). The CO_2 fluctuations over a 24 -hour period are conspicuous in highly variable coastal environments including coral reefs, intertidal, sandy or rocky shores, upwelling zones, estuaries or fjords, and salt marshes (Hales et al., 2005). In coastal areas the photosynthesis of macroalgae or seagrasses would increase the pH of the surrounding seawater and decrease dissolved inorganic carbon (DIC) through the uptake of CO_2 and HCO_3^-, while respiration by marine organisms would decrease the pH and increase DIC as a result of CO_2 release (Semesi et al., 2009). It has been reported that the diurnal changes in pCO_2 and pH could reach up to 600 ppm and >1 pH units, respectively (Ohde & Van Woesik, 1999; Unsworth et al., 2012). For instance, temporal variabilities in CO_2, ranging from the preindustrial value of 100 ppm (pH 8.6) to the future value of 1300 ppm pCO_2 (pH 7.6), were detected in the surrounding seawater in the Lady Elliot Island reef flat over the course of a day (Shaw et al., 2012). The seawater aragonite saturation state (Ω_{arag}) in this area was shown to vary between 1.1 and 6.5, which significantly exceeded the magnitude of change expected in the open ocean by the end of the century (Shaw et al., 2012). In addition, dramatic seasonal variation in seawater pH (ranging from pH 7.85 to 8.1) and pCO_2 levels (ranging from 325 to 725 ppm) were also detected in Florida Bay (Millero et al., 2001).

2. Currently, most studies have examined the impacts of OA on marine organisms using pCO_2 levels that are predicted to be reached in a hundred years in the future with relatively short exposure times (days, weeks, or a few months) (Kroeker et al., 2010, 2013a; Lemasson et al., 2017). In the real environment, as compared to the experiment period, OA is a relatively long-term process; whether marine organisms have the potential for genetic adaptation or acclimation to the acidified seawater during this process is still unclear (Mazurais et al., 2020; Sundin et al., 2019). To date, very few studies have investigated the long-term or transgenerational acclimation potential of marine organisms to future OA conditions. Some of these studies demonstrated that in contrast to what has been observed during short-term

exposures to acidified water, long-term exposures could have a differ-ent effect on marine organisms (Mazurais et al., 2020; Sundin et al., 2019). For instance, studies conducted on the juvenile spiny chromis damselfish *Acanthochromis polyacanthus* showed that long-term acclima-tion (3 months) to near-future OA did not induce any acid—base dis-ruptions in energetics (Sundin et al., 2019). Similarly, Vargas et al. (2017) found that several marine species including mussels, gastropods, and planktonic copepods collected in environments with naturally contrasting pCO_2 levels (coastal vs. estuarine areas and/or river-influenced areas) showed different tolerance to the same level of high pCO_2 levels, suggesting the potential role of local adaptation and/or adaptive phenotypic plasticity in increasing the resilience of species to environmental change.

In addition, the results regarding the adaptability of marine organ-isms to future OA conditions are contradictory across species, which likely reflects lineage-specific evolution, even in closely related taxa (Appelhans et al., 2014; Vargas et al., 2017). For example, the growth performance (feeding rates and growth rates) of the juvenile sea star *Asterias rubens* decreased substantially with increasing seawater pCO_2 levels (ranging from around 650 to 3500 ppm) after short-term (6 weeks) exposure, and the organism was not able to recover from this negative effect during the long-term experiment (39 weeks). In contrast, Form and Riebesell (2012) reported that although the deep-sea coral Lophelia pertusa underwent an approxi-mately 26%—29% decline in calcification and net dissolution of cal-cium carbonate after short-term (1 week) OA exposure (a pH decrease of 0.1 units), they were capable of acclimating to acidified conditions after long-term (6 months) incubations, leading to even slightly enhanced rates of calcification. Given that marine organisms have evolved in lineage-specific ways and exhibit different pheno-typic responses to OA conditions depending on various factors, stud-ies that expose marine organisms to long-term OA conditions are performed in a limited number of species. Furthermore, the underly-ing adaptive mechanisms behind the reported acclimation in most marine species are also unclear.

3. Although the effects of OA on various physiological functions of marine organisms have been confirmed, the potential affecting mechanisms underpinning these impacts are still largely unclear. For example, OA has been widely reported to impair behavioral responses

such as boldness, behavioral asymmetry (lateralization), and responses to olfactory and auditory cues in a variety of fish species (Domenici et al., 2012; Nilsson et al., 2012). As of now, one accepted mechanistic explanation for the OA-impaired behavioral responses in marine organisms is its effects on the neural system, such as the disturbed $GABA_A$ receptor function, which leads to an influx of Cl^- and HCO^{3-} over the neuronal membrane and induces hyperpolarization of the neuron (Nilsson et al., 2012). Given that the trigger of behavioral responses in an organism consists of a series of steps, namely signal recognition and transduction, signal processing, and responding to signals, OA may have the potential to affect the behavior of marine organisms by interfering with these processes (Dimitrov & Crichton, 1997). However, the responses of many other key effect factors such as the glycine receptor, CO_2-sensing peripheral neuron, and neuronal network dynamics that mediate the behaviors of marine fish under OA conditions still remain unknown. A large number of studies have confirmed that OA would exert negative effects on the fertilization success of many marine invertebrate species (Sewell et al., 2014; Shi et al., 2017). In the previous studies the probability of gamete fusion during fertilization has been confirmed in marine invertebrates by calculation using the fertilization kinetics model (Shi et al., 2017). It has been reported that the structure and function of proteins are sensitive to the pH of the surrounding environment (Dimitrov & Crichton, 1997); thus the recognition proteins on the gametes' membrane surface in marine invertebrates, which are in direct contact with the seawater, are particularly susceptible to OA, which may bring about unexpected consequences. However, whether OA exposure reduces gamete fusion probability by altering gamete recognition proteins and consequently declines fertilization success of marine organisms is unstudied. In conclusion, the knowledge uncovering the underlying causes behind OA-induced phenomena in marine organisms is not yet fully understood so far.

4. To date, studies investigating OA impacts are mainly focused on their direct effects on marine organisms. However, alternations in seawater chemistry would influence the ecological communities through both direct and indirect effects (Kroeker et al., 2010; Nagelkerken et al., 2016). Marine organisms in the ocean may be not only directly affected with regard to their physiology and behavior due to elevated seawater pCO_2 levels but also indirectly affected due to the changes caused in the resources such as habitat, prey, and microbial community that they rely on (Kroeker et al., 2010; Witt et al., 2011). For example, bacterial communities in seawater are closely related to the health status of marine organisms; OA-induced seawater microbiota alternations, especially those in pathogenic bacteria, would affect the immunity of marine organisms (Zha et al., 2017). Microbial community analysis showed that the microbial community in the seawater column was significantly altered by 1 week of OA exposure. The microbes of the microbial community in the control seawater (pH 8.1) were mainly ($>90\%$) composed of Proteobacteria (69.6%), Bacteroidetes (12.4%), and Firmicutes (11.1%), while the abundance of Firmicutes was found to be decreased in the acidified seawater (pH 7.4). Notably, at the generic level, the proportion of classified Vibrio, a major group of pathogens in many marine invertebrates, was significantly increased under acidified seawater, suggesting that OA may increase the risk of pathogenic infection in marine organisms. In addition, Nagelkerken et al. (2016) demonstrated that high pCO_2 facilitated shifts in fish habitat, resulting in marked increases in key resources (macroalgae sand habitats, turf habitat, and higher food abundances) and reduced predators, which together drive population increases of some fish and thus partly mitigate the negative effects of OA. Another example is the coral bleaching under future OA scenarios (Albright et al., 2010). Many corals harbor symbiotic dinoflagellate algae and any impacts on these algae would affect coral–algal symbioses (Brown, 1997). Therefore elevated seawater pCO_2 levels can disrupt coral–algal symbioses and cause bleaching. However, whether coral–algal associations respond to changes in pCO_2 has not been thoroughly explored (Albright, 2018). These research works suggested that the indirect effects of OA may have far-reaching consequences in marine organisms and their population abundances, indicating that the OA-induced ecological changes cannot be predicted by the sum of their direct effects (Nagelkerken et al., 2016). Furthermore, direct and indirect effects exerted by environmental stress including OA may work in either the same or opposite directions, altering the readily detectable outcomes of the direct effects alone (Nagelkerken et al., 2016; Werner & Peacor, 2003). Nevertheless, to the best of our knowledge, the indirect effects of OA and their incorporation with direct effects on marine organisms remain largely unstudied.

Directions for future studies

As discussed above, the risks arising from OA are now widely acknowledged as a major threat to the marine environment; however, the current understanding of the scope and the impacts of OA are still relatively limited. In the following, we summarize some of the key points and propose some potential future research avenues and practices to consider:

1. To date, whether marine species will be able to adapt to the current unprecedented rates of the changing seawater pCO_2 levels is still under debate. For example, fast-generating species such as the pond algae *Chlamydomonas* did not show evidence of adaptation to OA after 1000 generations (Collins & Bell, 2004), while the offspring of the oysters exposed to OA conditions showed more resilience than unexposed ones (Parker et al., 2012). In addition, this potential for adaptation to OA may be species-specific (Thompson et al., 2015). Therefore experiments investigating the adaptation of marine organisms to OA should be conducted in more species from different climate regimes. Meanwhile, both longer experimental periods and studies spanning multiple generations should be applied for OA research in the future to make more accurate predictions about OA's effects on marine organisms.

2. Currently, most experimental OA studies often examine the individuals' level responses in isolation; however, OA may also affect the interspecific interactions between different trophic levels (Kroeker et al., 2013b). For instance, although OA was reported to suppress the antipredation behavior of marine bivalves by reducing their shell hardness and adhesion ability (Kroeker et al., 2010), the foraging behavior of their predators like crabs was also decreased under OA conditions (Sanford et al., 2014). In this regard, experiments exploring individual-based responses of single species alone are unlikely to predict their consequences accurately in the complex ecological environments, where species also interact (Queirós et al., 2015). In the future the impacts of OA on a marine species under elevated pCO_2 conditions should be studied in the context of different trophic levels. In addition, field surveys conducted in naturally acidified conditions, such as volcanic seeps, lagoons, and upwelling areas, may also help to provide insights into the long-term responses of marine organisms to OA.

3. Since OA is occurring on a global scale, the level of vulnerability to OA differs in different areas such as coral reefs, the open sea, coastal areas, and the deep sea. For example, the carbonate chemistry of coastal zones may be much more sensitive with respect to future global change than previously estimated in the open sea (Kerrison et al., 2011; Melzner et al., 2013). However, most researchers have examined OA impacts on marine organisms, including those living in coastal areas, using pCO_2 levels following the predictions of the IPCC in the surface water of the open sea. Therefore region-specific OA projections are needed to make accurate assessments of future CO_2 conditions, and habitat-specific CO_2 signatures should be taken into account when exploring the impacts of OA on marine organisms from different areas (Takeshita et al., 2015). Furthermore, diel, seasonal, and interannual variability of seawater CO_2 levels should also be simulated when studying organisms living in highly variable environments.

4. To date, numerous studies have reported the negative effects of OA on a wide range of marine organisms. However, as they live in a complex environment, OA is not the only stressor in the marine environment that marine organisms could be exposed to during their life cycle (Shi et al., 2018). In the past few decades, marine ecosystems have also been impacted by other emerging global threats such as pharmaceutical compounds, endocrine-disrupting chemicals, nanoparticles, and microplastics (Kumar et al., 2020). It can be expected that in some cases, these stressors may interact with OA and thus modulate the biological impacts exerted by any single variable. Consequently, more studies are required to examine the combined effects of these emerging stressors with OA on marine organisms.

5. In recent years, rapid progress in the development of biological sciences such as sequencing technologies has provided many valuable insights into solving unanswered questions of various aspects (Kim, 2016). In the future, sequencing technologies should be applied more widely to investigate the genetic response of a wide range of cellular processes to identify the early responses of marine organisms to OA that do not cause measurable or expected structural, physiological, or behavioral changes. For example, comparative transcriptomics has proven to be an effective method for examining organism—environment interactions in various species, and metabolomics can be a useful tool for observing the OA-induced endogenous metabolite perturbations in marine organisms (Moya et al., 2015; Zha et al., 2017).

6. As the effects of OA on marine organisms have been extensively studied, various marine species have been demonstrated to be extremely vulnerable to OA. According to the general consensus, the impact of climate change is one of the biggest challenges to world food security (e.g., food availability, access to food, stability of food supplies, and food utilization) (Ramasamy, 2011). Although marine seafood provides an important and economical protein source for human consumption and is a primary protein source for over one billion of the poorest people in the world, the way that the biological impacts of the changing seawater pCO_2 levels translate into effects on the productivity of the entire fish stock and aquaculture still remains unclear. Therefore studies that predict the impacts of OA on future global fishery production and the subsequent socioeconomic effects are highly recommended in the future.

References

Albright, R. (2018). *Ocean acidification and coral bleaching* (pp. 295–323). Springer International Publishing. Available from https://doi.org/10.1007/978-3-319-75393-5_12.

Albright, R., Mason, B., Miller, M., & Langdon, C. (2010). Ocean acidification compromises recruitment success of the threatened Caribbean coral acropora palmata. *Proceedings of the National Academy of Sciences of the United States of America, 107*(47), 20400–20404. Available from https://doi.org/10.1073/pnas.1007273107.

Appelhans, Y. S., Thomsen, J., Opitz, S., Pansch, C., Melzner, F., & Wahl, M. (2014). Juvenile sea stars exposed to acidification decrease feeding and growth with no acclimation potential. *Marine Ecology Progress Series, 509*, 227–239. Available from http://www.int-res.com/abstracts/meps/v509/p227–239/.

Asplund, M. E., Baden, S. P., Russ, S., Ellis, R. P., Gong, N., & Hernroth, B. E. (2014). Ocean acidification and host-pathogen interactions: Blue mussels, *Mytilus edulis*, encountering *Vibrio tubiashii*. *Environmental Microbiology, 16*(4), 1029–1039. Available from http://search.ebscohost.com/login.aspx?direct = true&db = aph&AN = 953228 55&lang = zh-cn&site = ehost-live.

Billé, R., Kelly, R., Biastoch, A., Harrould-Kolieb, E., Herr, D., Joos, F., Kroeker, K., Laffoley, D., Oschlies, A., & Gattuso, J.-P. (2013). Taking action against ocean acidification: A review of management and policy options. *Environmental Management, 52*(4), 761–779. Available from https://doi.org/10.1007/s00267-013-0132-7.

Brothers, C. J., Harianto, J., McClintock, J. B., & Byrne, M. (2016). Sea urchins in a high-CO_2 world: The influence of acclimation on the immune response to ocean warming and acidification. *Proceedings of the Royal Society B: Biological Sciences, 283*(1837), 20161501. Available from https://doi.org/10.1098/rspb.2016.1501.

Brown, B. E. (1997). Coral bleaching: Causes and consequences. *Coral Reefs, 16*(1), S129–S138. Available from https://doi.org/10.1007/s003380050249.

Caldeira, K., & Wickett, M. E. (2003). Anthropogenic carbon and ocean pH. *Nature, 425*(6956), 365. Available from https://doi.org/10.1038/425365a.

Calder-Potts, R., Widdicombe, S., Parry, H., Spicer, J., & Pipe, R. (2008). Effects of ocean acidification on the immune response of the blue mussel *Mytilus edulis*. *Aquatic Biology, 2*, 67–74. Available from https://doi.org/10.3354/ab00037.

Challener, R., Robbins, L., & McClintock, J. (2015). Variability of the carbonate chemistry in a shallow, seagrass-dominated ecosystem: Implications for ocean acidification experiments. *Marine and Freshwater Research*, *67*. Available from https://doi.org/10.1071/MF14219.

Chan, N. C. S., & Connolly, S. R. (2013). Sensitivity of coral calcification to ocean acidification: A *meta*-analysis. *Global Change Biology*, *19*(1), 282−290. Available from https://doi.org/10.1111/gcb.12011.

Collins, S., & Bell, G. (2004). Phenotypic consequences of 1000 generations of selection at elevated CO_2 in a green alga. *Nature*, *431*(7008), 566−569. Available from https://doi.org/10.1038/nature02945.

Dimitrov, R. A., & Crichton, R. R. (1997). Self-consistent field approach to protein structure and stability. I: pH dependenceof electrostatic contribution. *Proteins: Structure, Function, and Bioinformatics*, *27*(4), 576−596, https://doi.org/10.1002/(SICI)1097-0134(199704)27:4 < 576::AID-PROT10 > 3.0.CO;2-H.

Domenici, P., Allan, B., McCormick, M. I., & Munday, P. L. (2012). Elevated carbon dioxide affects behavioural lateralization in a coral reef fish. *Biology Letters*, *8*(1), 78−81. Available from https://doi.org/10.1098/rsbl.2011.0591.

Doney, S. C., Fabry, V. J., Feely, R. A., & Kleypas, J. A. (2009). Ocean acidification: The other CO_2 problem. *Annual Review of Marine Science*, *1*(1), 169−192. Available from https://doi.org/10.1146/annurev.marine.010908.163834.

Dupont, S., Havenhand, J., & Thorndyke, M. (2008). CO_2-driven acidification radically affects larval survival and development in marine organisms. *Abstracts of the Annual Main Meeting of the Society of Experimental Biology, 6th - 10th July 2008, Marseille, France*, *150*(3 Supplement), S170. Available from https://doi.org/10.1016/j.cbpa.2008.04.450.

Form, A. U., & Riebesell, U. (2012). Acclimation to ocean acidification during long-term CO_2 exposure in the cold-water coral *Lophelia pertusa*. *Global Change Biology*, *18*(3), 843−853. Available from https://doi.org/10.1111/j.1365-2486.2011.02583.x.

Hales, B., Takahashi, T., & Bandstra, L. (2005). Atmospheric CO_2 uptake by a coastal upwelling system. *Global Biogeochemical Cycles*, *19*. Available from https://doi.org/10.1029/2004GB002295.

Han, Y., Shi, W., Tang, Y., Zhao, X., Du, X., Sun, S., Zhou, W., & Liu, G. (2021). Ocean acidification increases polyspermy of a broadcast spawning bivalve species by hampering membrane depolarization and cortical granule exocytosis. *Aquatic Toxicology*, *231*, 105740. Available from https://doi.org/10.1016/j.aquatox.2020.105740.

Havenhand, J. N., Buttler, F.-R., Thorndyke, M. C., & Williamson, J. E. (2008a). Near-future levels of ocean acidification reduce fertilization success in a sea urchin. *Current Biology*, *18*(15), R651−R652. Available from https://doi.org/10.1016/j.cub.2008.06.015.

Havenhand, J. N., Buttler, F.-R., Thorndyke, M. C., & Williamson, J. E. (2008b). Near-future levels of ocean acidification reduce fertilization success in a sea urchin. *Current Biology*, *18*(15), R651−R652. Available from https://doi.org/10.1016/j.cub.2008.06.015.

Hernroth, B., Baden, S., Thorndyke, M., & Dupont, S. (2011). Immune suppression of the echinoderm *Asterias rubens* (L.) following long-term ocean acidification. *Aquatic Toxicology*, *103*(3), 222−224. Available from https://doi.org/10.1016/j.aquatox.2011.03.001.

Hofmann, G. E., Smith, J. E., Johnson, K. S., Send, U., Levin, L. A., Micheli, F., Paytan, A., Price, N. N., Peterson, B., Takeshita, Y., Matson, P. G., Crook, E. D., Kroeker, K. J., Gambi, M. C., Rivest, E. B., Frieder, C. A., Yu, P. C., & Martz, T. R. (2011). High-frequency dynamics of ocean pH: A multi-ecosystem comparison. *PloS One*, *6*(12), e28983. Available from https://doi.org/10.1371/journal.pone.0028983.

Jokiel, P. L., Rodgers, K. S., Kuffner, I. B., Andersson, A. J., Cox, E. F., & Mackenzie, F. T. (2008). Ocean acidification and calcifying reef organisms: A mesocosm investigation. *Coral Reefs, 27*(3), 473−483. Available from https://doi.org/10.1007/s00338-008-0380-9.

Kerrison, P., Hall-Spencer, J. M., Suggett, D. J., Hepburn, L. J., & Steinke, M. (2011). Assessment of pH variability at a coastal CO_2 vent for ocean acidification studies. *Estuarine Coastal & Shelf Science, 94*(2), 129−137. Available from http://search.ebscohost.com/login. aspx?direct = true&db = aph&AN = 64872741&lang = zh-cn&site = ehost-live.

Kim, S.K. (2016). Marine OMICS: Principles and applications. in marine OMICS: Principles and applications (pp. 1−724). CRC Press. https://doi.org/10.1201/9781315372303.

Kroeker, K. J., Kordas, R. L., Crim, R., Hendriks, I. E., Ramajo, L., Singh, G. S., Duarte, C. M., & Gattuso, J. (2013a). Impacts of ocean acidification on marine organisms: Quantifying sensitivities and interaction with warming. *Global Change Biology, 19* (6), 1884−1896. Available from http://search.ebscohost.com/login.aspx?direct = true&db = aph&AN = 87391941&lang = zh-cn&site = ehost-live.

Kroeker, K. J., Kordas, R. L., Crim, R., Hendriks, I. E., Ramajo, L., Singh, G. S., Duarte, C. M., & Gattuso, J.-P. (2013b). Impacts of ocean acidification on marine organisms: Quantifying sensitivities and interaction with warming. *Global Change Biology, 19*(6), 1884−1896. Available from https://doi.org/10.1111/gcb.12179.

Kroeker, K. J., Kordas, R. L., Crim, R. N., & Singh, G. G. (2010). Meta-analysis reveals negative yet variable effects of ocean acidification on marine organisms. *Ecology Letters, 13*(11), 1419−1434. Available from https://doi.org/10.1111/j.1461-0248.2010.01518.x.

Kroeker, K. J., Sanford, E., Jellison, B. M., & Gaylord, B. (2014). Predicting the effects of ocean acidification on predator-prey interactions: A conceptual framework based on coastal molluscs. *The Biological Bulletin, 226*(3), 211−222. Available from https://doi.org/10.1086/BBLv226n3p211.

Kumar, M., Chen, H., Sarsaiya, S., Qin, S., Liu, H., Awasthi, M. K., Kumar, S., Singh, L., Zhang, Z., Bolan, N. S., Pandey, A., Varjani, S., & Taherzadeh, M. J. (2020). Current research trends on micro- and nano-plastics as an emerging threat to global environment: A review. *Journal of Hazardous Materials, 409*, 124967. Available from https://doi.org/10.1016/j.jhazmat.2020.124967.

Lannig, G., Eilers, S., Pörtner, H. O., Sokolova, I. M., & Bock, C. (2010). Impact of ocean acidification on energy metabolism of oyster, *Crassostrea gigas*—changes in metabolic pathways and thermal response. *Marine Drugs, 8*(8), 2318−2339. Available from https://doi.org/10.3390/md8082318.

Lemasson, A. J., Fletcher, S., Hall-Spencer, J. M., & Knights, A. M. (2017). Linking the biological impacts of ocean acidification on oysters to changes in ecosystem services: A review. *Ecological Responses to Environmental Change in Marine Systems, 492*, 49−62. Available from https://doi.org/10.1016/j.jembe.2017.01.019.

Liu, S., Shi, W., Guo, C., Zhao, X., Han, Y., Peng, C., Chai, X., & Liu, G. (2016). Ocean acidification weakens the immune response of blood clam through hampering the NF-kappa β and toll-like receptor pathways. *Fish & Shellfish Immunology, 54*, 322−327. Available from https://doi.org/10.1016/j.fsi.2016.04.030.

Maier, C., Popp, P., Sollfrank, N., Weinbauer, M. G., Wild, C., & Gattuso, J.-P. (2016). Effects of elevated CO_2 and feeding on net calcification and energy budget of the mediterranean cold-water coral *Madrepora oculata*. *Journal of Experimental Biology, 219* (20), 3208. Available from https://doi.org/10.1242/jeb.127159.

Mazurais, D., Servili, A., Le Bayon, N., Gislard, S., Madec, L., & Zambonino-Infante, J.-L. (2020). Long-term exposure to near-future ocean acidification does not affect the expression of neurogenesis- and synaptic transmission-related genes in the olfactory bulb of European sea bass (Dicentrarchus labrax). *Journal of Comparative Physiology B, 190*(2), 161−167. Available from https://doi.org/10.1007/s00360-019-01256-2.

Meinshausen, M., Smith, S. J., Calvin, K., Daniel, J. S., Kainuma, M. L. T., Lamarque, J.-F., Matsumoto, K., Montzka, S. A., Raper, S. C. B., Riahi, K., Thomson, A., Velders, G. J. M., & van Vuuren, D. P. P. (2011). The RCP greenhouse gas concentrations and their extensions from 1765 to 2300. *Climatic Change, 109*(1), 213. Available from https://doi.org/10.1007/s10584-011-0156-z.

Melzner, F., Thomsen, J., Koeve, W., Oschlies, A., Gutowska, M. A., Bange, H. W., Hansen, H. P., & Körtzinger, A. (2013). Future ocean acidification will be amplified by hypoxia in coastal habitats. *Marine Biology, 160*(8), 1875−1888. Available from https://doi.org/10.1007/s00227-012-1954-1.

Millero, F., Hiscock, W., Huang, F., Roche, M., & Zhang, J.-Z. (2001). Seasonal variation of the carbonate system in Florida Bay. *Bulletin of Marine Science, 68*, 101−123.

Moya, A., Huisman, L., Forêt, S., Gattuso, J.-P., Hayward, D. C., Ball, E. E., & Miller, D. J. (2015). Rapid acclimation of juvenile corals to CO_2-mediated acidification by upregulation of heat shock protein and Bcl-2 genes. *Molecular Ecology, 24*(2), 438−452. Available from https://doi.org/10.1111/mec.13021.

Nagelkerken, I., Russell, B. D., Gillanders, B. M., & Connell, S. D. (2016). Ocean acidification alters fish populations indirectly through habitat modification. *Nature Climate Change, 6*(1), 89−93. Available from https://doi.org/10.1038/nclimate2757.

Nilsson, G. E., Dixson, D. L., Domenici, P., McCormick, M. I., Sørensen, C., Watson, S.-A., & Munday, P. L. (2012). Near-future carbon dioxide levels alter fish behaviour by interfering with neurotransmitter function. *Nature Climate Change, 2*(3), 201−204. Available from https://doi.org/10.1038/nclimate1352.

Ohde, S., & Van Woesik, R. (1999). Carbon dioxide flux and metabolic processes of a coral reef, Okinawa. *Bulletin of Marine Science, 65*, 559−576.

Ohki, S., Irie, T., Inoue, M., Shinmen, K., Kawahata, H., Nakamura, T., Kato, A., Nojiri, Y., Suzuki, A., Sakai, K., & Van Woesik, R. (2013). Calcification responses of symbiotic and aposymbiotic corals to near-future levels of ocean acidification. *Biogeosciences, 10*, 6807−6814. Available from https://doi.org/10.5194/bg-10-6807-2013.

Orr, J. C., Fabry, V. J., Aumont, O., Bopp, L., Doney, S. C., Feely, R. A., Gnanadesikan, A., Gruber, N., Ishida, A., Joos, F., Key, R. M., Lindsay, K., Maier-Reimer, E., Matear, R., Monfray, P., Mouchet, A., Najjar, R. G., Plattner, G.-K., Rodgers, K. B., & Yool, A. (2005). Anthropogenic ocean acidification over the twenty-first century and its impact on calcifying organisms. *Nature, 437*(7059), 681−686. Available from https://doi.org/10.1038/nature04095.

Parker, L. M., Ross, P. M., O'Connor, W. A., Borysko, L., Raftos, D. A., & Pörtner, H.-O. (2012). Adult exposure influences offspring response to ocean acidification in oysters. *Global Change Biology, 18*(1), 82−92. Available from https://doi.org/10.1111/j.1365-2486.2011.02520.x.

Pimentel, M., Pegado, M. R., Repolho, T., & Rosa, R. (2014). Impact of ocean acidification in the metabolism and swimming behavior of the dolphinfish (Coryphaena hippurus) early larvae. *Marine Biology, 161*. Available from https://doi.org/10.1007/s00227-013-2365-7.

Queirós, A. M., Fernandes, J. A., Faulwetter, S., Nunes, J., Rastrick, S. P. S., Mieszkowska, N., Artioli, Y., Yool, A., Calosi, P., Arvanitidis, C., Findlay, H. S., Barange, M., Cheung, W. W. L., & Widdicombe, S. (2015). Scaling up experimental ocean acidification and warming research: From individuals to the ecosystem. *Global Change Biology, 21*(1), 130−143. Available from https://doi.org/10.1111/gcb.12675.

Ramasamy, S. (2011). *World food security: The challenges of climate change and bioenergy* (pp. 183−213). Springer. Available from https://doi.org/10.1007/978-90-481-9516-9_13.

Riebesell, U., Fabry, V., Hansson, L., & Gattuso, J.-P. (2010). Guide to best practices for ocean acidification research and data reporting. In Oceanography (Vol. 22, p. 260). Washington, DC.

Rodriguez-Dominguez, A., Connell, S. D., Baziret, C., & Nagelkerken, I. (2018). Irreversible behavioural impairment of fish starts early: Embryonic exposure to ocean acidification. *Marine Pollution Bulletin, 133*, 562−567. Available from https://doi.org/ 10.1016/j.marpolbul.2018.06.004.

Rong, J., Tang, Y., Zha, S., Han, Y., Shi, W., & Liu, G. (2020). Ocean acidification impedes gustation-mediated feeding behavior by disrupting gustatory signal transduction in the black sea bream, acanthopagrus schlegelii. *Marine Environmental Research, 162*, 105182. Available from https://doi.org/10.1016/j.marenvres.2020. 105182.

Ross, P., Parker, L., O'Connor, W., & Bailey, E. (2011). The impact of ocean acidification on reproduction, early development and settlement of marine organisms. *Water, 3*, 1005−1030. Available from https://doi.org/10.3390/w3041005.

Sanford, E., Gaylord, B., Hettinger, A., Lenz, E. A., Meyer, K., & Hill, T. M. (2014). Ocean acidification increases the vulnerability of native oysters to predation by invasive snails. *Proceedings. Biological Sciences, 281*(1778), 20132681. Available from https:// doi.org/10.1098/rspb.2013.2681, −20132681.

Semesi, S., Beer, S., & Björk, M. (2009). Seagrass photosynthesis controls rates of calcification and photosynthesis of calcareous macroalgae in a tropical seagrass meadow. *Marine Ecology Progress Series, 382*, 41−47. Available from https://doi.org/10.3354/ meps07973.

Sewell, M. A., Millar, R. B., Yu, P. C., Kapsenberg, L., & Hofmann, G. E. (2014). Ocean acidification and fertilization in the antarctic sea urchin sterechinus neumayeri: The importance of polyspermy. *Environmental Science & Technology, 48*(1), 713−722. Available from https://doi.org/10.1021/es402815s.

Shaw, E., McNeil, B., & Tilbrook, B. (2012). Impacts of ocean acidification in naturally variable coral reef flat ecosystems. *Journal of Geophysical Research C: Oceans, 117*. Available from https://doi.org/10.1029/2011JC007655.

Shi, W., Guan, X., Han, Y., Zha, S., Fang, J., Xiao, G., Yan, M., & Liu, G. (2018). The synergic impacts of TiO_2 nanoparticles and 17β-estradiol (E_2) on the immune responses, E_2 accumulation, and expression of immune-related genes of the blood clam, *Tegillarca granosa*. *Fish & Shellfish Immunology, 81*, 29−36. Available from https:// doi.org/10.1016/j.fsi.2018.07.009.

Shi, W., Han, Y., Guo, C., Zhao, X., Liu, S., Su, W., Wang, Y., Zha, S., Chai, X., & Liu, G. (2017). Ocean acidification hampers sperm-egg collisions, gamete fusion, and generation of Ca^{2+} oscillations of a broadcast spawning bivalve, *Tegillarca granosa*. *Marine Environmental Research, 130*, 106−112. Available from https://doi.org/ 10.1016/j.marenvres.2017.07.016.

Sundin, J., Amcoff, M., Mateos-González, F., Raby, G. D., & Clark, T. D. (2019). Long-term acclimation to near-future ocean acidification has negligible effects on energetic attributes in a juvenile coral reef fish. *Oecologia, 190*(3), 689−702. Available from https://doi.org/10.1007/s00442-019-04430-z.

Takeshita, Y., Frieder, C., Martz, T., Ballard, J., Feely, R., Kram, S., Nam, S., Navarro, M., Price, N., & Smith, J. (2015). Including high-frequency variability in coastal ocean acidification projections. *Biogeosciences, 12*, 5853−5870. Available from https:// doi.org/10.5194/bg-12-5853-2015.

Thompson, E. L., O'Connor, W., Parker, L., Ross, P., & Raftos, D. A. (2015). Differential proteomic responses of selectively bred and wild-type Sydney rock oyster populations exposed to elevated CO_2. *Molecular Ecology, 24*(6), 1248−1262. Available from https://doi.org/10.1111/mec.13111.

Tresguerres, M., & Hamilton, T. J. (2017). Acid−base physiology, neurobiology and behaviour in relation to CO_2-induced ocean acidification. *Journal of Experimental Biology, 220*(12), 2136. Available from https://doi.org/10.1242/jeb.144113.

Unsworth, R., Collier, C., Henderson, G., & McKenzie, L. (2012). Tropical seagrass meadows modify seawater carbon chemistry: Implications for coral reefs impacted by ocean acidification. *Environmental Research Letters*, 7. Available from https://doi.org/10.1088/1748-9326/7/2/024026.

Vargas, C. A., Lagos, N. A., Lardies, M. A., Duarte, C., Manríquez, P. H., Aguilera, V. M., Broitman, B., Widdicombe, S., & Dupont, S. (2017). Species-specific responses to ocean acidification should account for local adaptation and adaptive plasticity. *Nature Ecology & Evolution*, *1*(4), 0084. Available from https://doi.org/10.1038/s41559-017-0084.

Visconti, G., Gianguzza, F., Butera, E., Costa, V., Vizzini, S., Byrne, M., & Gianguzza, P. (2017). Morphological response of the larvae of arbacia lixula to near-future ocean warming and acidification. *ICES Journal of Marine Science*, *74*. Available from https://doi.org/10.1093/icesjms/fsx037.

Wang, X., Song, L., Chen, Y., Ran, H., & Song, J. (2017). Impact of ocean acidification on the early development and escape behavior of marine medaka (*Oryzias melastigma*). *Marine Environmental Research*, *131*, 10−18. Available from https://doi.org/10.1016/j.marenvres.2017.09.001.

Werner, E. E., & Peacor, S. (2003). A review of trait-mediated indirect interactions in ecological communities. *Ecology*, *84*, 1083−1100.

Witt, V., Wild, C., Anthony, K., Diaz-Pulido, G., & Uthicke, S. (2011). Effects of ocean acidification on microbial community composition of, and oxygen fluxes through, biofilms from the Great Barrier Reef. *Environmental Microbiology*, *13*, 2976−2989. Available from https://doi.org/10.1111/j.1462-2920.2011.02571.x.

Wood, H., Spicer, J., & Widdicombe, S. (2008). Ocean acidification may increase calcification rates, but at a cost. *Proceedings Biological Sciences/The Royal Society*, *275*, 1767−1773. Available from https://doi.org/10.1098/rspb.2008.0343.

Wootton, J. T., Pfister, C. A., & Forester, J. D. (2008). Dynamic patterns and ecological impacts of declining ocean pH in a high-resolution multi-year dataset. *Proceedings of the National Academy of Sciences of the United States of America*, *105*(48), 18848−18853. Available from https://doi.org/10.1073/pnas.0810079105.

Zha, S., Liu, S., Su, W., Shi, W., Xiao, G., Yan, M., & Liu, G. (2017). Laboratory simulation reveals significant impacts of ocean acidification on microbial community composition and host-pathogen interactions between the blood clam and *Vibrio harveyi*. *Fish & Shellfish Immunology*, *71*, 393−398. Available from https://doi.org/10.1016/j.fsi.2017.10.034.

Zhao, X., Shi, W., Han, Y., Liu, S., Guo, C., Fu, W., Chai, X., & Liu, G. (2017). Ocean acidification adversely influences metabolism, extracellular pH and calcification of an economically important marine bivalve, *Tegillarca granosa*. *Marine Environmental Research*, *125*, 82−89. Available from https://doi.org/10.1016/j.marenvres.2017.01.007.

Index

Printed in the United States
by Baker & Taylor Publisher Services